METHODS IN AQUATIC MICROBIOLOGY

METHODS IN AQUATIC MICROBIOLOGY

A. G. Rodina

Translated, edited, and revised by

Rita R. Colwell and Michael S. Zambruski

University Park Press · BALTIMORE

Butterworths · LONDON

UNIVERSITY PARK PRESS
International Publishers in Science and Medicine
Chamber of Commerce Building
Baltimore, Maryland 21202

Copyright © 1972 by University Park Press
Printed in U.S.A. by Wickersham Printing Co., Inc.,
Lancaster, Penna.

Library of Congress Cataloging in Publication Data
Rodina, Antonina Gavrilovna.
   Methods in aquatic microbiology.
   "Originally published in 1965 . . . as Metody vodnoĭ
mikrobiologii."
   Bibliography: p.
   1. Aquatic microbiology—Technique.   I. Title.
QR105.R5713   576.1'9'2   78–171255
ISBN 0–8391–0071–X

Published jointly by
UNIVERSITY PARK PRESS, BALTIMORE
and
BUTTERWORTH & CO (PUBLISHERS) LTD, LONDON

Library of Congress Catalog Card No. 78–171255
ISBN 0–8391–0071–X (University Park Press)
ISBN 0–408–70357–1 (Butterworths)

Dedicated to

L. R., J. H. C., and G. B. C.

and to

A. and J. E. Z.

Contents

Preface

The translation of this book from Russian to English was begun in the early spring of 1970. The Russian text was sent to us on loan by Dr. S. P. Meyers and our initial intent was merely to translate a few chapters on methodology, in particular those on sampling and culturing of aquatic microorganisms. Our interest in the text increased as the work of translation proceeded and this, coupled with the fact that there is a dearth of adequate texts on aquatic microbiology, led us to see the task through.

We have endeavored to produce the most accurate translation possible of the facts and data presented by the late Dr. Rodina. The text is well known in Russia and deserves a wide distribution in English-speaking countries.

After the translation was completed, the text was edited to update information where necessary (the last edition appeared in Russia in 1965) and to smooth the flow of the writing for easier reading. No doubt we have not been entirely successful in these tasks. However, we believe that the job was worth doing and the result acceptable, considering the obstacles of language and different microbiological methods and equipment used in Russia.

It is reasonable to expect that this text will serve a variety of needs. Upper-level undergraduates and graduate students in aquatic microbiology (both freshwater and marine) will find the text extremely helpful in their studies. Workers in the field no doubt will welcome the opportunity to view the "state of the art" of aquatic microbiology in the U.S.S.R. Perhaps a better understanding of the research being done in aquatic microbiology by Russian scientists will result from the availability of this text to Western scientists in the field.

Finally, the work involved in preparation and translation has been educational. The completion of this job—at times difficult and seemingly never ending—provided a pleasant sense of accomplishment for both of us.

R. R. Colwell and M. S. Zambruski

Acknowledgments

We are especially indebted to several of our friends and colleagues who contributed significantly to the successful completion of this book through their encouragement and assistance. The late Professor Michael Krupensky of the School of Languages and Linguistics at Georgetown University provided invaluable aid in translating some rather obscure Russian passages. We acknowledge the kind assistance of Dr. Robert Lager, also of the School of Languages and Linguistics. Mr. Henry Krussel assisted in the arduous task of proofing parts of the translation. We also acknowledge the generous cooperation of Mr. Leo Larouche, Science Librarian, Georgetown University.

We are particularly grateful to Drs. Aleksandra Grigorievna Kuchaeva and I. Mitskevich who spent an afternoon, during a Moscow visit of one of us (M. S. Z.), graciously answering questions on methods and equipment unfamiliar to microbiologists outside Russia.

Finally, we acknowledge the secretarial assistance of Mrs. Frances Ingram during the preparation of the manuscript.

Parts of the manuscript were read by several of our colleagues: microscopy, Dr. George B. Chapman; radioisotopes and tracer techniques, Dr. Irving Gray; measurement of pH and Eh, Dr. Jack London and Dr. Jack H. Colwell; and nomenclature, Dr. Erwin F. Lessel. Dr. Richard Y. Morita also provided valuable comment during the final stages of the preparation of the manuscript. To these, our friends, we extend our thanks.

One of us (R. R. C.) was supported in part during the preparation of the manuscript by grant GB–18274 from the National Science Foundation.

Translators' Note

Two items required special concern in translating this book: author index and bibliography. There is no accepted standard international system of transliteration used in all foreign scientific publications; therefore we chose merely to use accepted spellings of authors' names—as they had already appeared in print—rather than employ an arbitrary transliteration system when dealing with names. Sources used were the *Science Citation Index, Biological Abstracts, Mikrobiologiia*, and numerous texts. However, wherever it was necessary, or wherever names could not be verified in print elsewhere (especially Slavic names), the Library of Congress system of transliteration without the diacritical marks was used. This system was also employed throughout the text for references to original Russian terms and for bibliographical material.

The following procedure was generally employed in translating the bibliography. The author(s) is given first, followed by the transliterated title of the work and then our translation of the title, enclosed in brackets. Next the Russian publication concerned, issue number, and/or publisher are transliterated. This information is then translated, as far as is practical, and enclosed in brackets. Providing both the transliteration and translation should prove useful for readers interested in corresponding with scientists in the U.S.S.R. It must be noted that page references for various issues of *Doklady, Mikrobiologiia*, and all other Russian works correspond to pages in the original Russian edition, not to the English translations (such as *Microbiology*) which may be available.

Introduction

Use of various microbiological methods, whether in field or laboratory studies, is determined by the program of investigation to be carried out. For the study of contamination and self-purification of bodies of water, there are methods which are distinguished from those utilized for the analysis of productivity of water masses; still other methods are used to explain the role of microorganisms in corrosion of metals, concrete edifices, and so forth. The most important trends in the work being done in aquatic microbiology are outlined briefly below.

Every body of water is characterized by its own peculiar bacterial population, which is determined by the specific conditions in the water mass. The development of microorganisms is closely related to a variety of factors involving the medium: oxygen content; mineral and organic nutrients; pH; temperature; and development of phytoplankton and zooplankton. Communities of microorganisms in various unpolluted water masses have their own peculiar characteristics, which are related to the conditions of the water mass itself. These communities constitute an inherent part of the biocenoses of a given body of water. However, although the cenoses of animal and plant organisms may be known and rather fully characterized, bacterial cenoses require analysis because very little information is available. A variety of methods (including direct microscopy and inoculations) must be used for detection of these cenoses and determination of typical populations for the various types of water masses (rivers, lakes, reservoirs) and the changes produced in them by pollution.

The character and intensity of bacterial metabolism are the basis of nutrient cycles which occur in bodies of water. The biochemical activity of microorganisms is associated with transformation and regeneration of nutrients, including nitrogen and phosphorous, which are essential for development of the primary producers in water masses. Formation of benthic sediments and liberation of gases which change the oxidation-reduction conditions of the bottom-most layers of the water are directly concerned with microbial activity and cannot be understood without knowledge of the processes which cause these conditions. Very little information is available concerning the number of autotrophic microorganisms in water masses and the actual amounts of bacterial protein they produce.

Each water mass has its own characteristic vertical and horizontal distribution of microorganisms. Factors such as the rate of reproduction of microorganisms, their utilization of nutrients, and, basic to this, the production of new organic matter in the form of microbial cells; and the predation of microbial cells by zooplankton, are involved in the overall balance of organic matter in a body of water. Great quantities of living organic matter in the form of microbial cells are carried by rivers into lakes, reservoirs, and seas; recent studies have indicated that the bacterial flow into the sea is in terms of hundreds of tons per year. Thus, this flow has great significance, especially for productivity of river estuaries.

Microorganisms play a significant role in the corrosion of metal and concrete structures and in the fouling of water pipes and of ships. Knowledge of the principles of microbial growth, as well as the qualitative composition of bacterial cenoses, is essential in developing effective measures for combatting these phenomena.

The interrelationships between microorganisms and phytoplankton, aquatic plants and microorganisms, and microorganisms and the animal population of water masses have fundamental significance in the cycling of elements in a body of water. In fact, the clarification of the role of bacteria in the nutrition of marine animals has permitted development of practical measures for the production of these animals as a source of food.

Medicinal ("therapeutic") muds are a product of bacterial activity, as is regeneration of depleted nutrients. Knowledge of the biology of the induction of these processes is required for creation of conditions advantageous to their development.

One of the fundamental concerns of aquatic microbiology is productivity. Knowledge of the basic principles of productivity of water masses (including fish productivity) is based on an understanding of the reproduction of all forms of life, beginning with microorganisms as well as their ecology and interdependencies. Because the principles of hydrofauna development can be understood and exploited only after their interaction with the environment and with bacteria and plankton is understood, microbiology is an integral part of the complex of hydrobiological studies. Microbiological studies should provide data on concentration of bacteria necessary for invertebrate nutrition since invertebrates commonly are used by fish as food. Microbiological processes themselves are now often studied with the aim of increasing productivity. Such microbiological studies are of primary importance in dealing with eutrophic and fertilized water masses. At the present time the geographical spread of eutrophication has broadened concomitantly with the decline in yield of fish in many water masses. Eutrophic conditions have spread not only in ponds, but also in the seas and lakes, including entire lakes and sections of rivers.

In highly detailed investigations, microorganisms emerge as the first link joining the living world with the abiotic milieu; and, in a majority of cases, microbiological data, more so than chemical data, serve as an index for all changes which arise in a given environment. This is because microorganisms are extremely sensitive "reagents" for searching out alterations of the environment.

Microbiological studies of fertilized ponds used for fish culture are valuable in determining appropriate times of fertilizer application and the optimal quantities to be applied. The rates of reproduction of bacteria and their metabolic activity determine the time for replacing one plant fertilizer with another. The aquatic plant population in a pond at a given time is determined by the densities of the developing bacterial populations, since the consumption of oxygen by the microorganisms can lead to oxygen deficiency. Fish kills may occur as a result of sulfate reduction at the time of application of organic fertilizers to ponds containing water rich in sulfates.

Mineral fertilizers or nutrients can be applied effectively

to ponds if data are available both on the concentrations required for those elements involved in phytoplankton growth, and on the rate of microbiological processes responsible for transforming nutrients into food for fish. The rates of reproduction and the energy required by the processes effected by various groups of bacteria determine, to a great extent, the utilization of various mineral nutrients. Metabolism will not proceed in an identical fashion in different ponds of various climatic zones of a given geographical area. Only an accumulation of a vast amount of data from samples collected using proper microbiological methods will provide a scientific basis for selecting nutrients essential for specific types of ponds or a particular class of water.

The most significant among the physiological groups of microorganisms which cause transformation of nutrients are discussed in this book. Bacteria which mineralize organic matter are distinguished by their great numbers and high metabolic activity at the time of application of plant fertilizers and during decomposition of algae. In periods of maximum development, heterotrophic bacteria comprise up to 50 per cent of the entire mass of microorganisms in a pond. All of the bacteria serve as nutrient for aquatic plants.

The nitrogen balance of water masses is closely related to the activity of nitrogen-fixing organisms. Moreover, the cells of nitrogen-fixing bacteria are preferred by grazers as a form of bacterial forage. Also significant is the excretion by these bacterial cells of substances which stimulate development of algae.

The process of nitrification is highly useful for fertilized ponds. It can be enhanced by the presence of one kind of fertilizer and weakened or inhibited by another.

Microbiological studies are useful in demonstrating the value and utility of applying nitrogen fertilizers. Investigation of the rate of denitrification under certain conditions in any given pond can show whether or not the applied nitrogen fertilizers are being utilized as intended.

Another line of study, distinct from those described above, is sanitary or public health bacteriology. Bacteriological investigations are an integral part of any work involving water supplies of a population, for example, in preservation of water from contamination, and purification of drinking and waste waters. The program of operation is usually limited to the accumulation of data necessary for determining the sanitary quality of the water mass. The methods developed for this particular aspect of microbiology limit the range of study to microorganisms of sanitary significance, such as those which characterize the degree of contamination or pollution of a given body of water. One of the important factors of an investigation to detect pollution is the speed with which data on contamination can be obtained. There are two widely employed fundamental indicators: the number of heterotrophic bacteria growing on a standard nutrient agar at 37 C, and the number of coliforms, the indicators of fecal contamination of water.

There is great need for more and better bacteriological indicators. However, this aspect of any given study is controlled by the composition of the discharges entering a given water mass. Specifically, the application of direct microscopic methods reveals special features of bacterial communities in unpolluted water. It can also be used to determine less commonly studied cenoses of portions of water masses contaminated by city sewage and water from industries discharging sewage containing various organic substances or minerals or both. The methods usually include: (1) direct count of microorganisms; (2) study of the microbial communities of water and sediments, using submerged slides; and (3) study of the composition of the latter by fluorescent microscopy. Direct counts of bacteria in water and sediments give a true representation of the quantity of viable (and culturable) bacteria, and of the general composition of the bacterial mass, namely, number of cocci, rods, yeast, and sulfur bacteria.

Use of submerged slides enables observation of bacterial communities—in particular, those types that do not grow on ordinary nutrient media but nevertheless often achieve extensive populations and are, therefore, characteristic types associated with certain conditions of a water mass or an individual section thereof. Submerged slides allow detection of interrelationships between microorganisms in bacterial cenoses. The nature of the pollution determines the nature of the bacterial communities. At the present time it is still impossible to determine either cenoses which are typical for a specific type of contamination or pollution, or the dominant types of bacteria related to the degree to which saprophytes are present in water masses.

Communities of microorganisms developing in sediments are vital indicators of contamination of water masses and individual water mass zones. Special conditions which result in characteristic changes in the composition of the bacterial cenoses of the sediments are created by the precipitation, or settling to the bottom in water masses, of various contaminating substances (cellulose, lignin, bark, proteins, wool fiber, heavy metals) and the adsorption by sediments of various organic and inorganic compounds. Even after long periods, the precipitated compounds can be, and often are, sources of secondary contamination. Sediments represent those biotopes in which streptomycetes, causing earthy smells and a peculiar, but characteristic, taste in water, achieve massive development.

Fluorescence and fluorescent microscopy are very promising tools in aquatic microbiology, in general, and in public health investigations in particular. To date, it is only by means of fluorescent microscopy that the distribution of bacteria in soils can be demonstrated. Selected bacterial strains in a general population of free, active, non-adsorbed cells, i.e. "agents of mineralization," and the number of "low-activity" cells adsorbed by sediments can be directly enumerated by fluorescent microscopy. The study of natural processes of mineralization in various water masses has great significance with respect to the character of the body of water.

The qualitative distribution of microorganisms and their physiology must be known to control the biochemical processes in sewage digestion. The composition of microorganisms in an active sludge determines the quality of biological degradation. Selection of optimal conditions for sludge digestion is made possible by determination of the kinds of bacteria that are active agents of purification and knowledge of their physiological characteristics. In order to effect control measures it is essential to be able to recognize nuisance bacteria—for example, bacteria that cause an activated sludge to float up.

Efficient and correct application of any sewage treatment is possible only through knowledge and measurement of microbial activity. At the present time, the studies which are of special significance are those concerned with elucidating the role of bacteria both in food chains and in the destruction or concentration of radioactive and other dangerous contamination in water masses, for example, heavy metals such as mercury, and pesticides and detergents.

Mention could be made of many additional problems, both practical and theoretical in nature, for which methods of aquatic microbiology offer solutions. However, the short inventory given here attests to the fact that the topics involved are many and varied.

One must note that many problems created by science and technology cannot be solved wholly and precisely by the methods of aquatic and marine microbiology at the present time. For example, in some quarters it is postulated that oil spills in the ocean can be cleaned up by the action of microorganisms. Further research, revisions, and refinements of the methods of aquatic microbiology are necessary before such tasks are accomplished, if indeed they can be. The increasing body of literature in the field of aquatic microbiology indicates that such work is going on continuously. Throughout the text sources are indicated, and the more important work on methods of microbiological investigations is listed at the end of the book. Aquatic microbiology, of necessity, makes use of methodological achievements developed in other areas of microbiology.

Some comment is required concerning methods for investigating brackish and salt water. A listing of methods which conform to those devised for fresh water conditions is presented in the text, as is customary in manuals on marine microbiology. However, even in studies of continental water masses it is necessary to deal with saline or brine solutions or both, mineral sources, pharmaceutically important sediments, and the like. The same methods are used in both marine microbiology and fresh water microbiology. The chief difference is that the salt content is taken into account in preparing media. The following are used most often in the preparation of media: (1) undiluted water from the water mass under study; (2) dilution with distilled water (usually 1:1); or (3) distilled water to which are added the necessary salts present in the water mass under study.

Chapter 1

Sampling for Microbiological Investigations

SELECTING SAMPLING SITES

The selection of sampling locations in a water mass is an important task which must be performed with all possible care. A sound decision in this case can be made only after collecting data on the morphometric features of the water mass: dimensions, depth pattern, geological structure of shores, sediment distribution, and presence of inflow, i.e. runoff, or, in the case of lakes, drainages. In order to locate sites for microbiological sampling, consideration must be made of such basic physico-chemical factors as temperature pattern, oxygen distribution, and pH. Collection of data on vertical and horizontal composition of the water mass should also precede a microbiological investigation.

The presence of large communities or industrial plants on the shores of open water masses and location of steamship docks and livestock pastures and watering places are factors which must be taken into consideration in microbiological investigations. Such information is required even for studies that are not primarily concerned with sanitary conditions or public health because all of the factors cited exert influence both on the numbers and kinds of bacteria in a water mass and on the course of microbiological processes occurring therein.

Each type of water mass has its own special features and these must be considered in designing experiments. In a lake, the best plan is to distribute stations by limnological profiles. In the accepted limnology terminology, a station is any execution of work in a given place at a given time. A limnological profile begins at the coastal area, includes the region extending from the coast to the deep area lying beyond the shore, and continues through this deep region to the other shore. In the water mass itself samplings should be taken from the distinct water layers—epilimnion, metalimnion, and hypolimnion (these regions in a water mass are distinguished by vertical temperature distribution). (See Fig. 1.) The number of stations in a lake depends on its size and depth pattern. The number of stations in each region depends on the conditions in that region. Because of the close interrelationships which exist between microorganisms and the conditions of the environment, the number and types of bacteria are dependent on the conditions existing therein.

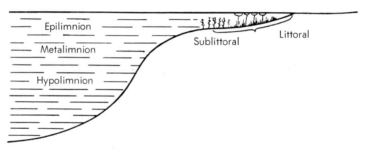

Fig. 1. Various zones in a lake.

The greatest diversity of environmental conditions is observed along the coast. This region is, to a great extent, subject to the influence of shore, atmosphere, and surf. A special type of temperature behavior may be observed in the shallow coastal area. Daily fluctuations in temperature are clearly expressed in this area. Even more prominent are the seasonal fluctuations; for example, in winter a shallow layer of water directly under the ice will have a temperature close to 0 C. In the spring the water in coastal zones is warmed earlier than the water mass of the open part of a lake. In summer the daily temperature in the coastal section is higher than that in areas of significant depth. A much greater oxygen concentration is observed in the shallower depths and is enhanced by the action of surf. In water layers of shallow depth, light penetrates to the bottom, thus facilitating growth of aquatic plants. Thickets of macrophytes in the coastal region create special conditions for the development of microorganisms. The autumnal die-off of macrophytes results in enrichment of the environment with organic substances, a source of nutrition for many groups of bacteria. Great accumulations of phytoplankton (many of which are undergoing decay), another source of organic material for bacteria, are often observed in this region. Because of the diversity of conditions in the coastal zone it is necessary that sampling stations be distributed at short distances from one another.

The region located between the coast and deeper water is characterized by intermediate conditions. Here, possibly, is the beginning of oxygen concentration decrease and also the intermediate range between coastal and deep water temperatures.

The deep portion of a body of water is characterized by more or less stable conditions in comparison with the coast. Relatively unchanging or constant low temperatures, reduced content of organic matter, absence of light, and absence of turbulence and vegetation create special conditions for growth of certain groups of microorganisms and the course of processes effected by them. The floor of lake or ocean deeps is usually covered with a layer of sediment.

Each of the three vertical zones in a lake presents special conditions for growth of microorganisms. The epilimnion is the most thoroughly warmed zone, the richest in oxygen, the one which receives the most light, and the site of production of organic substances by phytoplankton. The metalimnion is often characterized by a sharp drop in temperature. The hypolimnion is the zone in which temperature decreases gradually but steadily to the bottom of the lake.

In the lowermost layers of deep bodies of water the temperature is constant throughout the year. In many lakes, for example, the water of deep layers is mixed with water of upper layers only during periods of circulation of the water mass in spring and autumn. In very deep lakes, however, circulation may not affect the deep layers at all.

Because of these factors, vertical samplings are taken in the epilimnion (in the top layer and at levels of 5 and 10 m), in the metalimnion, and in the hypolimnion (at levels of 25, 50, 100 m and deeper, where depths exceed 100 m, and always at the bottom). The first vertical sampling is usually taken at a depth of 10–15 cm from the surface; the last sampling is usually taken at a distance of 20–30 cm from the bottom. Thus, the number of vertical samplings taken at each station varies according to the depth pattern.

Lake coves or inlets of estuaries often are distinguished, sometimes very sharply, from the lakes or main body of water. They must, therefore, be included in the experimental design.

When taking samples, the boat or ship must be operated in a manner in which a minimum of turbulence will be created. Also, when sampling in lakes, the direction of currents, which is determined by the wind, must be taken into account.

Ponds are distinguished by a shallow depth. This basic factor accounts for the negligible change in the vertical layers of factors such as temperature, light penetration, and specific hydrochemical characteristics. Thus the water in ponds can be considered to be relatively homogeneous. Consequently, the number of samples taken in ponds can usually be limited (provided that the ponds are not very large in area and are not separated into distinct regions). In ponds used for fish breeding, samples are usually taken near the location where water enters, in the center, and where it drains, and, in each case, from two levels, the top-most and bottom-most levels. In very shallow ponds only one vertical sample need be taken, about 20–30 cm from the bottom.

Depending on whether they have been created in rivers or in lakes, reservoirs have the same general characteristics as their source. Usually the basin is asymmetric and the greatest depths are found in front of the embankment of the dam. Reservoirs do not have a clearly expressed sublittoral area and fluctuations in the level are significant.

The vertical breakdown of a water mass is characterized by special features which are related to the character of water drainage. When water flows over a dam, spill currents can be observed in the upper layers; in the rest of the body of water a vertical stratification of temperature may occur. The effect is different when water flow is through a floodgate at the bottom of the water mass. In shallow reservoirs with large surface areas the winds produce a mixing of the water. Because of these conditions, a hydrographical study of a reservoir is carried out by means of longitudinal and transverse profiles.

A river is sharply distinguished from lakes, ponds, and reservoirs by its own peculiar conditions. Every river has a source, an intermediate section (central course), lower reaches, and a mouth. The rate of flow, determined by the slope of the river bed, is different in each of these parts for one and the same river. Current speeds, which to a large degree determine the diversity of conditions, are not identical even over a cross section of the river. Near the shores and bottom the current is usually slower than in the central portion of the river. In a river only the littoral (coastal) zone, where the aquatic flora grows, can be easily distinguished; boundaries of the zone extending to the deep areas are difficult to determine. A river current usually carries water of almost uniform temperature and oxygen saturation; however, differences can be found in individual streams of water, even over a short distance in a given river.

A river is characterized by a bed, which remains filled with water even at the time when the water level is at its lowest, and by a flood plain, which floods with water only during times of high water and flooding. Owing to changes in the bed caused by the river itself, sections of former river beds can be found in some places and these present special conditions for microbial growth.

Conditions differ in the source, central and lower courses, and mouth of a river. Rivers often carry suspended particles in the water which are rich in microorganisms. Because of the mixing action of currents, suspended particles are transported for long distances without precipitation. At the mouth, where a decrease in the rate of the current occurs, the greater part of the material precipitates.

Tributaries, which should always be considered in a river study, exert a great influence on the number of microorganisms in the river, on the content of suspended particles, and on the chemical composition of the water. Rivers, especially powerful ones, are seldom sampled along their entire extent; they are more often sampled at distinct sections which are located near cities. In these cases samples must be taken at locations above the city, across from it, below it in several places over a significant distance (Fig. 2), and above and below any points of sewage discharge. If tributaries are present in the area under study, samples should then be taken above the inflow of the tributary, at its mouth (at a sufficient distance into the tributary so that the receiving river has no influence on the conditions within the tributary), and below the inflow into the river. The samplings should be carried out at preselected track lines (Fig. 3).

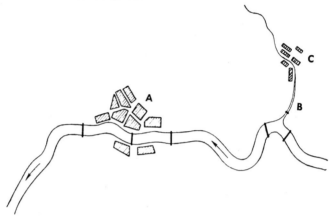

Fig. 2. Distribution of track lines on a river. *A,* Populated area on river. *B,* Tributary. *C,* Populated area on tributary.

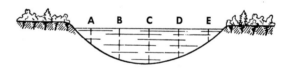

Fig. 3. Distribution of sampling stations on a river track line. (Letters above the vertical lines indicate stations on the track line.)

Waste water from urban industrial plants and runoff of soil particles from agricultural areas along the shore have a significant influence on the content and kinds of microorganisms in river water.

The number of microorganisms in a river does not stay at a constant level throughout the year. It depends greatly on climatic and meterological conditions. The population of microorganisms in a river rises sharply at times of high water and flood. Those factors which cause a rise in the number of microorganisms include erosion associated with heavy rainfall, melting of snow, which also carries away surface impurities, and

overflow of the river itself, which often floods significant portions of dry land, thus washing away particles of soil saturated
with microorganisms and also various organic residues, including pesticides and fertilizers. At times of low water levels, the
numbers of microorganisms may decrease. However, a rise in
water temperature creates conditions more advantageous for
growth of microorganisms. In winter, conditions under the ice
are again different with respect to microbial growth. Because
of these factors, studies of rivers or estuaries should be conducted at different seasons of the year.

Springs are often subjected to microbiological study because they are used either as a water supply or for hydropathy
(health spas). Spring water is distinguished by stability of
composition. However, when spring water surfaces, the number
and types of microorganisms present alter sharply as conditions
change. Samples of the water must be taken where the spring
issues from the earth. At the point where the water exits from
the ground, oxygen content is low. However, the dissolved oxygen level increases as the water flows on. Both water temperature and organic content change and, as a result of movement
along the bottom, water becomes enriched with soil bacteria.

When artesian wells are sampled, the samples are taken
as water is pumped out. The pumping apparatus should first
be disinfected and the duration of pumping should not be less
than 2 to 3 days before sampling.

For well pumps or water piped from springs, the water
should be pumped 10 to 15 minutes before the actual sample
is taken. The sample is then taken by holding a sterilized container to catch water flowing from the tap.

There is a whole series of general environmental data
which must be collected when a given water mass is sampled.
Microbiological sampling should be performed along with
measurement of basic environmental factors such as temperature; transparency; organic content; dissolved oxygen; pH; salinity; calcium, nitrogen, and phosphorus content; and, in
rivers, current flow rate.

Meteorological conditions can influence the data which
are collected; for example, rainfall usually adds eroded soil
particles which are rich in microorganisms, and often other
contamination as well. Therefore, it is absolutely necessary to

take samples at preset times and, in particular, at times when weather conditions are stable.

Samples under ice are taken from holes of large diameter cut into the ice. An apparatus with evacuated vessels should be used to take samples from an ice hole of less than 1 m in diameter with surrounding ice in excess of 0.5 m thickness. The first vertical sample is taken 10 to 15 cm below the ice.

Samples are always taken at the same time of day (usually in the morning hours), unless, of course, the investigation is intended to show diurnal changes in bacterial content.

In small bodies of water, microbiological sampling can be run successfully from a boat. The necessity of placing a winch with a metal cable on board depends on the instruments used. Many instruments can simply be lowered by hand, using a cable marked out in meters and half meters. See Plate 12. When measuring the cable, one must take into account dimensions of the instrument with which the work is undertaken, i.e. the length of the sampler itself.

For work in large, deep bodies of water the ship from which the sampling is done should be equipped with a hoist with a metal cable and a mechanical meter wheel.

The method used for sampling is very important in microbiological investigations of water as well as sediments. Only the most careful observance of all the rules of standard bacteriological procedure will prevent contamination of the samples and provide accurate analyses. Samples must be taken quickly and previously sterilized containers must be carefully wrapped and protected in the paper in which they are sterilized up to the very moment at which the sample is taken. Before a sample is taken, all metallic parts of the instruments not previously sterilized should be flame-sterilized with a small, portable blowtorch or alcohol lamp.

Sample analyses must be made as soon as possible after collection, since the shorter the time between collection and inoculations, the more accurate the results. When water samples are allowed to stand, rapid changes take place both in total numbers of bacteria and in the proportion of groups and species of bacteria present in the sample. These changes begin during the first hour after the samples are taken. The number of bacteria in isolated samples increases greatly even during the first hours after collection. Some species reproduce rapidly and suppress the development of other bacteria; this accounts for sharp changes in the group and species composition of microorganisms in samples.

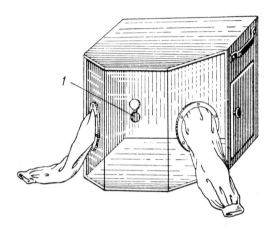

Fig. 4. Portable box for inoculations under field trip conditions. *1,* Opening for the end of a pipette. Several U.S. firms (including Wildlife Supply Company, Saginaw, Michigan) supply a similar apparatus for microbiological work.

To avoid contamination problems and to obtain accurate data, platings for the determination of total numbers of microplankton and species composition are done aboard ship either in a specially designed microbiological laboratory or, in the absence of such facilities, by using a suitable compartment or portable box. These boxes (Fig. 4) may be of various dimensions. Usually they are made from plywood and plastic and are sturdy enough to withstand breakage during transit. An inoculation box is packed in a specially constructed, tightly closed case for protection from dust during transportation. Before inoculations are made, the inside of the box must be sterilized with a portable ultraviolet lamp; when this is not available, the walls and floor of the box are washed down with bactericidal solutions.

To transport samples, the following precautions must be taken: (1) the samples must arrive at the laboratory no later than 1 to 3 hours after collection; (2) in hot weather, samples must be transported either in a double-walled container packed with ice or cooling mixtures—in order that the temperature of the samples in the box does not exceed 4 C—or in a container in which rubber water bottles or plastic bags filled with ice are placed among the samples (in winter, to prevent freezing, sacs filled with warm water are placed in the box); and (3) samples must be contained in sealed bottles or in flasks with ground glass stoppers.

INSTRUMENTS FOR TAKING SAMPLES
OF WATER AND SEDIMENT

Instruments for Sampling Water

A basic requirement for any aquatic microbiological investigation is aseptic collection of samples. There are various instruments which may be used for sampling water. The following rules, however, apply in all cases: Samples must be collected in a sterile container; all glass and rubber parts of instruments must be sterilized; and flasks, bottles, and other sterilized parts of the apparatus must be protected up to the exact moment of sampling by being kept wrapped in the paper in which they were sterilized.

Before taking samples, metallic parts of instruments must be thoroughly flame-sterilized (the instrument should not be wet from previous sampling). After assembling all the parts, the instrument should again be flame-sterilized.

A number of instruments for aquatic sampling have been constructed and described. Some have mechanical devices for removing the stopper at the desired depth and replacing it when the vessel has filled with water. Other arrangements make use of capillary tubes, which are broken off at the desired depth by a messenger-triggered device; in this way a water sample is drawn into the sterile collecting container.

Instruments of the first type require two cables; one lowers the entire instrument, the second raises and lowers the stopper. These, however, can only be used successfully in water which is not very deep. They are not suitable for sampling at significant depths since the cables often become twisted, and it is then impossible to open the sample bottle. An instrument of this type is shown in Figure 5. It is composed of a metal frame on which is placed a bottle (A) with a flat ground glass stopper. The edges of the stopper are held by two or three catches (B). The instrument is lowered on a line (C), which is fastened to a small bow on the instrument, and has a cable to open the bottle (D) which is fastened to a tube (E) that contains a spring which ends at the catches.

As the instrument is lowered the cable (D) is left slack. When the apparatus reaches the designated depth, the cable is pulled and the ground glass stopper is drawn out of the mouth of the vessel, thus permitting water to enter. When the bottle is filled, the cable is slackened and the stopper closes the vessel as a result of the pressure exerted by a spring which is located in the tube (E). A very large metal base (F) holds the instrument in a vertical position. The stages of the sampling can be judged by observing the air bubbles which rise to the surface.

If special instruments are not available for work in small, shallow bodies of water such as ponds, samples may be taken with any large, heavy bottle provided with a rubber stopper containing two holes, through which two glass tubes, one short and one long, can be passed (Fig. 6). The glass tubes are joined by a small piece of rubber tubing. Around the neck of the bottle is fashioned a ring to which a loop of metal cable is fastened to act as a weight. A rope line is then fastened to the metal loop, and a lightweight line is connected to the rubber tube. The bottle is sterilized before use. The instrument is lowered by the rope line. When the bottle reaches the desired level, by pulling on the lightweight line, the rubber tubing is removed and the water enters the bottle through one of the glass tubes. The other tube allows the air in the bottle to escape.

It is much easier to work with instruments with a single cable. Water is collected by opening a tap or by breaking a tube or the end of a container by means of a messenger. One of the drawbacks of this type of apparatus is that the water

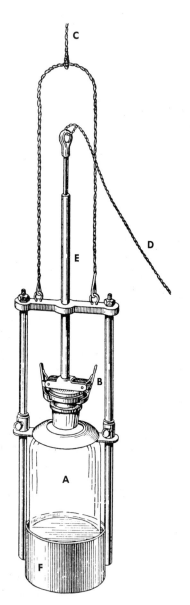

Fig. 6. Bottle for aquatic microbiological sampling.

Fig. 5. Apparatus containing a bottle for sampling water (see text for description of use).

sample is raised in an open container, i.e. the tube (or aperture) through which the water sample was taken remains open. Many instruments of this type are available.

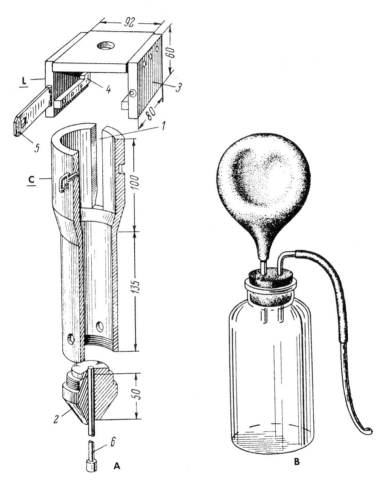

Fig. 7. Stolbunov and Ryabov apparatus for water sampling (see text for explanation).

Stolbunov and Ryabov (1964) have constructed a rather convenient apparatus (Fig. 7, *A*). It consists of a heavy, thick-walled, hollow cylinder (*C*), tapered at the bottom and widened at the top, with a cover (*L*). The cylinder is provided with a piston (*6*), which slides within it. At the upper part of the

cylinder there is a slot (1), through which the rubber tube
of a vessel is led out. Near the bottom of the cylinder are
holes for draining water from the instrument. At the bottom
is a conical plug (2) which screws into the cylinder. Along
the axis of this plug is a passage for the piston, which serves
to remove the vessel containing the sample out of the cylinder.
The cover of the apparatus consists of a baseplate and two
sides (3). In the center of the baseplate is an aperture with
a cross piece, to which a cable is attached. The sides fasten the
lid to the cylinder by means of grooves with flanges (4) and a
locking latch (5).

The container for collecting the water sample with this
instrument is a standard wide-necked bottle approximately 10
cm high and 100 to 150 ml in volume. The bottle is tightly
closed with a fitted rubber stopper through which two thick-
walled glass tubes are inserted, one of which is bent at a right
angle. A rubber bulb is connected to the straight tube, and a
rubber tube 12 to 14 cm long is attached to the bent tube.
Another glass tube, with a sealed end, is inserted into the open
end of the rubber tube. Such vessels should be prepared in
the number needed but should be assembled only just before
taking samples (all parts must be sterilized beforehand). They
are assembled as follows: the bottle is removed from its protective paper; the rubber bulb is squeezed; the cotton pads
which cover the rubber and glass tubes are removed; the glass
tube with the sealed end is quickly attached, at the open end,
to the rubber tube connected to the bent glass tube. The assembled vessel (Fig. 7, B) is then placed in the cylinder in
such a way that the rubber tube, which contains the glass tube
with the sealed end, passes through the groove in the cylinder
wall. The lid is then placed on the bathometer and fastened.
The protruding rubber tube is bent as the sealed glass tube is
fixed on the upper baseplate by means of a special ring. The
top part of the bathometer and the glass tube are swabbed well
with alcohol. The assembled instrument is lowered to the desired depth, at which time a messenger is sent down the cable
(the breaking end of which is sterilized before use). The messenger breaks the glass tube and the released elastic rubber
tube straightens out. At a distance of 12–14 cm from the
bathometer (i.e. at the end of the tube) the water is sucked

in through the tube by the action of the rubber bulb. When the instrument is brought back to the surface, the rubber stopper must be replaced with a sterile stopper without holes or, when bottles with ground glass stoppers are used, by one of the stoppers.

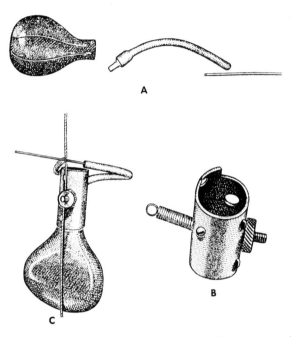

Fig. 8. Sieburth instrument for water sampling (see text for explanation).

Sieburth designed an instrument (Fig. 8) following the same principle. Its simple operation is as follows: there is a carrier mechanism (*B*) consisting of a small brass or stainless steel (preferred) tube with a spring and a grip screw; a compressed rubber bulb (see *A*) connected via rubber tubing to a glass tube, one end of which is fastened at the cable, is lowered into the water; a messenger breaks the glass tube and the bulb draws in water.

When working at great depths, a device suggested by Isachenko (Fig. 9) has proved convenient (see ZoBell, 1946). It consists of a metal (nickel-plated) cylinder (*1*) 30.5 cm in length. At the top of the cylinder is an opening (*2*) for the

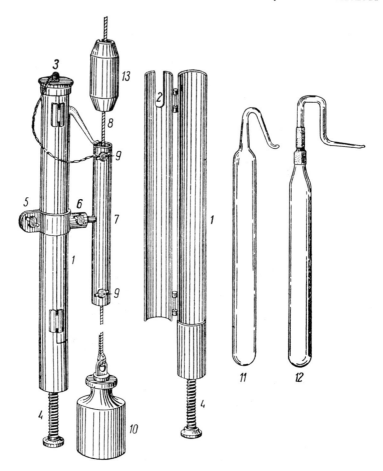

Fig. 9. Apparatus for sampling water using evacuated vials and sampling vessels (Isachenko) (see text for explanation).

neck of a container, and a lid (*3*) which closes it. In the lower part of the cylinder there is a second bottom which can be raised and lowered by means of a screw (*4*). With the aid of a removable ring (*5*), which encircles the cylinder, and a connecting rod (*6*), the entire instrument is joined with another metal tube (*7*) containing a deep channel. Through this channel passes a metal cable (*8*) fastened by two screws (*9*). A 1- to 2-kg weight (*10*) is attached to the end of this cable. A sterile evacuated container (*11*) is placed in the cylinder in such a way that the end of the curved neck of the vessel lies in a special depression in the metal tube (7) containing the fastened cable. The container should rest securely on the moveable bottom; the neck of the container should not touch the edge of the groove in the cylinder wall. The glass container is constructed with cylindrical glass tubes 26 mm in diameter, 22–27 cm long from base to neck, and 100–110 ml in capacity.

After lowering the apparatus to the desired depth, a messenger (13) weighing approximately 300 g is sent down. This breaks off the end of the curved neck of the glass container, which then immediately fills with water. When the apparatus is brought back to the surface, the glass container is sealed up or the opening is stoppered with sterilized cotton. Each container of this type can be used for only one sampling. It cannot be used a second time simply because the broken tube extending from the container makes further sampling impossible. A sampler which allows the same container to be used many times has been devised. The container is first cut at the neck, and then a curved glass tube is attached to it with rubber tubing (Fig. 9, 12). The vessel is evacuated and the end of the tube is sealed. The entire vessel is wrapped in paper and sterilized in an autoclave. After a sample is taken, it is necessary only to replace the glass tube, the end of which was broken by the messenger. This adaptation significantly reduces the cost of sampling.

Samples can also be collected in large test tubes closed by rubber stoppers, through which pass curved tubes sealed at one end after evacuation of the air in the vessel. ZoBell has designed a sampler along these lines. It is distinguished by the fact that ordinary bottles and flasks are used instead of special containers (Fig. 10). The bottle (1) in this apparatus is closed with a rubber stopper (2) fitted with a curved glass tube (3). This tube is connected to a piece of rubber tubing (4) which, in turn, is connected to another glass tube (5) that is sealed at one end. The entire assembled apparatus is sterilized in an autoclave, with the assembled stopper lying loose in the neck of the bottle. When sterilization is completed, the bottle is removed from the autoclave and quickly sealed with the stopper. ZoBell indicates that in this way 50–90% of the air in the bottle is evacuated. The bottle is attached to a brass

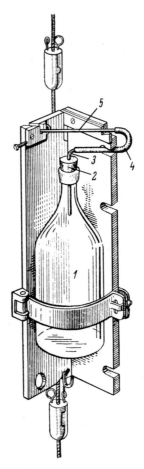

Fig. 10. Apparatus for sampling water (ZoBell) (see text for explanation).

frame when the actual sampling is done. The sampling itself is actuated by means of a messenger. Sorokin (1960b) has constructed a deep-water sampler on this same principle.

It must be noted that devices employing vessels connected by rubber tubing or closed by rubber stoppers are unsuitable for working at great depths. When these vessels are subjected to the great pressures of deep-water masses, the water begins to filter in at connecting points or presses the rubber stoppers down into the evacuated vessels. Consequently, when working at great depths, sealed evacuated glass or poly-ethylene-coated stainless steel should be used. ZoBell suggests using thick-walled rubber containers. See Plate 11.

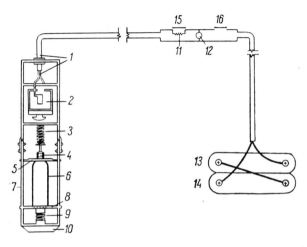

Fig. 11. Apparatus for opening and closing a container de-signed by Lewis, McNail, and Summerfelt (1963). 1, Rubber-insulated electric cable; 2, solenoid; 3, spring; 4, cork stopper; 5, bottle holder; 6, bottle; 7, metal frame; 8, moveable plat-form; 9, spring; 10, weighted base; 11, resistance rheostat; 12, signal light; 13 and 14, batteries (viewed from above) and connecting cables; 15 and 16, switches.

Lewis, McNail, and Summerfelt (1963) offer another method for opening and closing a container that involves using an electric current from two 6-v dry electric batteries. The apparatus (Fig. 11) consists of a framework which contains a vessel and a solenoid assembled above the vessel. The solenoid is connected by means of a rubber-insulated electric cable with a system of electric switches. The vessel is closed with a stopper and held in place by a spring platform from below and by a collar from above. When the apparatus is lowered to the de-sired depth, the electric current is turned on and the stopper rises; when the vessel is filled, the current is turned off and the vessel is closed.

None of the apparatus described can be used for taking samples in springs which flow out into small streams of water. Here sampling is done with a large sterilized bottle that is

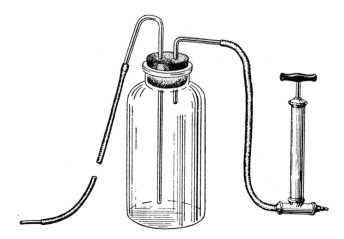

Fig. 12. Bottle prepared for sampling spring water.

closed with a rubber stopper which contains two glass tubes (Fig. 12). One of these tubes is connected by rubber tubing to a pump (Fig. 12, *right*) to evacuate the bottle; the other glass tube is connected via rubber tubing to a small glass tube, which is introduced into the water of the streamlet. The entire apparatus must be sterilized beforehand.

In rivers with rapid currents, ordinary sampling apparatus cannot be used because they will drift. In these cases, special weighted instruments must be used when working at deep levels. When the depths are not great, the best method is to attach a sterile bottle by means of a special net to a pole, which is fixed in the water bed while the samples are taken. The bottles are placed so that the openings face against the direction of the current.

A water sampler devised by Niskin (1962) consists of a large metal hinge fitted with a sterile plastic bag with a tube. When the sampler is tripped by a messenger, the tube opens aseptically and the hinge flips apart, sucking in the water sample. When the plastic bag is full, a clamp closes the intake tube. An advantage of the Niskin sampler is that it goes down sterile and flat between the metal hinge. Unfortunately, the Niskin sampler is known to tear and also to delay in opening. However, it has the distinct advantage of being able to take up to 5 liters of water (see Plate 12).

Instruments for Sampling Sediments

At present aquatic microbiologists have not yet devised special instruments for sampling sediments aseptically. Samples of sediment are taken with various apparatus constructed for other purposes (such as enumeration of benthic fauna, study of the microzonal composition of sediment, detecting animal remains, etc.). Therefore, the discussion of the instruments referred to here may be found in handbooks on sampling sediments for studying benthos, sediment formation, and so forth, or in oceanography manuals, viz., marine geology.

The choice of apparatus for microbiological sampling of sediments is determined by the purposes of the investigation as well as by the depth of the water mass and the character of the sediment. Because the topmost layer of the sediment

Fig. 13. Sounding corer for sampling sediments.

is the site of the most intensive microbiological activity, it is important always to sample sediment without stirring it or causing changes in the sediment layers.

Soft sediments in shallow water masses (ponds and shallow lakes) can be sampled with a sounding corer-type instrument (Fig. 13). This consists of a metal tube 8–10 cm in diameter and 50–70 cm in length, the bottom edge of which is sharpened. A small bow is fastened to the upper end by hinges which permit the bow to be moved to the side when the sample is withdrawn from the corer. A pole of appropriate length fastened to the upper part of the bow permits the instrument to be driven into relatively compact sediment to the desired depth. The entire apparatus is hauled up with the sediment sample. Use of translucent coring tubes enables detection of the distribution of sediment layers in the water bed.

Sediment samples may also be taken by bottom-grab, which is suitable for sampling sediments of various densities at any depth. At present there are many grab systems. For microbiological investigations, however, the only suitable equipment is that in which the cover of the grab can be opened from above. Investigations have shown that the number of bacteria and the diversity of constituent groups decreases rapidly as the lower layers of the sediment are reached. Therefore the topmost, the most active layer (in particular, the sediment-water interface) should be sampled in microbiological analyses.

A grab sampler designed by Zabolotsky consists of a housing with opening jaws below and a springless rod above (Fig. 14, A). At the upper part of the housing are lids capable of being opened and closed easily. The grab is lowered with the jaws raised. When pressure is exerted on the rod the grab digs into the water bed. The pole is rotated and pressure is again applied, this time resulting in the jaws closing. The closed grab is brought to the surface, the lid at the top is opened, and the topmost layer of sediment is placed into a sterile glass jar by means of a spatula, flamed after swabbing with a cotton wad wetted with alcohol.

A cable grab, similar to that of Borutzky (Fig. 14, B) can be used when working in loose sediments. The grab is lowered with the jaws opened (1), and small cables are attached to the free ends of the small levers of the release apparatus.

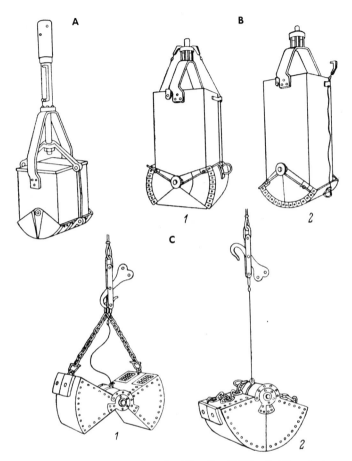

Fig. 14. Various types of grabs (Zhadin, 1956). *A,* Closed view of grab with springless pole, designed by Zabolotsky. *B,* Grabs designed by Borutzky: *1,* opened; *2,* closed. *C,* Grab shovel: *1,* opened; *2,* closed.

When the grab plunges into the sediment, a messenger is sent along the main cable, striking the spring of the release apparatus and thereby freeing the small cables. When the cables are released, the jaws close (*2*). The grab is raised in the closed position onto the boat or ship. The top layer of sediment is removed as described earlier.

The grab-type apparatus shown in Fig. 14, *C* is also suitable for sampling, provided that the grating on one of the

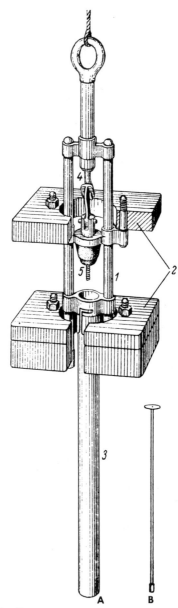

Fig. 15. A, Perfilyev stratometer (see text for explanation).
B, Piston for pushing through the sediment sample.

upper sides is replaced by a removable plate, thus permitting the top level of the sediment sample to be taken.

For more complex studies the corer-type sampler designed by Sukachev (Kordé and Pyavchenko, 1950) may be used. With a corer it is possible to take a sediment sample in the form of a column of desired length. A simple type of sampler, shown in Plate 13, has a removable autoclavable core catcher.

Various types of stratometers may also be used for taking core sediment samples. With one of these instruments the investigator can obtain a core of sediment of considerable length (up to 1 m) in which the layers of benthic sediments can be seen and studied.

The stratometer most widely used in the USSR is that designed by Perfilyev (Fig. 15, A). The instrument consists of a frame (1) which supports removable weights (2) and a tube (3) which does the actual sampling. Weighting of the tube can be increased or decreased depending on the density of the sediment and the desired length of the core sample. The apparatus is provided with a special hammercock (4) connected to a rubber stopper (5), which closes the tube when the hammercock is released. After the sample of the sediment is taken and the instrument is hauled to the surface, it is necessary to close off the bottom while it is still under water. To do this, a stopper of the corresponding diameter is placed under the bottom of the tube and is used to seal it. Only in this way can the apparatus be taken from the water.

For very compacted sediments, Perfilyev has designed an impact core sampler.

When taking samples for microbiological investigations, before the actual sampling is done the tubes of the stratometers must be swabbed with a cotton wad soaked in alcohol and then flamed. A rod of the needed length, the end of which is wrapped with sterile absorbent cotton, is used for this procedure. After the sample has been taken, it must not be left in the metal tubes even for a short time. In fact core liners of an inert plastic (autoclavable) are preferred. In any case, immediately after the instrument is brought to the surface, the sediment sample must be transferred to a glass vessel by any means available.

One way to do this is merely to push the sample out into a sterile glass tube of the same diameter as the original cylinder and immediately close off both ends of the glass tube with sterile rubber stoppers. Thus, the column of sediment in the glass tube can be examined easily. Later the tube may be cut at designated levels and inoculations may be prepared with sediment taken from the centermost portion of one of these layers. However, regardless of the method used to divide the sediment into layers, inoculations must be prepared only from the centermost part of the core sample and never from parts which come into contact with the walls of the sampling tube.

A special metal or wooden piston is used to push the soil sample out of the tube. A metal or wooden plate with the same diameter as the sampling tube is securely fastened to one end of the piston, which should be marked in centimeters and millimeters. These markings help in calculating the depth levels of the sediment layers.

A sediment sample may be separated into portions (Fig. 16) immediately after collection and at the same time transferred to sterile glass vessels. This is accomplished by introducing a sterile glass tube into the central part of the sediment (A) and removing the center of the sediment core. The filled glass tube is closed at both ends with rubber stoppers (B).

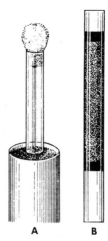

A   B

Fig. 16. Dividing the sediment sample. *A,* A sterile tube is introduced into the sample. *B,* A tube with an aliquot of the sediment.

Instruments for Sampling Water
and Sediment Simultaneously

A variety of instruments have been suggested for use in collecting sediment and water which is in direct contact with benthic sediments. Such an instrument was devised by Baranov and Bure (Baranov, 1956). The basic parts of the apparatus (Fig. 17) include a 1-liter capacity Plexiglas cylinder (*1*) with a case (*2*); a lowering and closing mechanism (*3*); a small cap which closes the cylinder (*4*); a sleeve pipe (*5*) for increas-

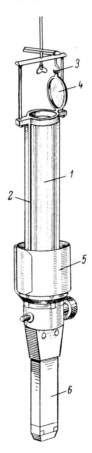

Fig. 17. Baranov apparatus for sampling sediment and benthic water (Baranov, 1956) (see text for explanation).

ing the weight of the instrument; an attachment for sampling the soil (6); and a messenger. The device is lowered with the cap open. This cap is closed when the messenger is sent down and strikes it. The sampler can be used without the lower part if only water samples are to be taken.

In all instruments of the type described, samples are collected in glass or Plexiglas cylinders, which allow the sample itself to be examined and the benthic water to be separated by eye.

Chapter 2

Methods of Microscopy [1]

Access to a microscope of high quality and the knowledge of how to use its potential magnification and resolution are of primary importance to the microbiologist.

Contemporary microscopes are characterized by their complex construction; they are constantly being updated to fulfill the demand for newer, more modern models. At the present time two Russian models of biological microscopes are widely used: MBI-3 and MBI-6.[2] Comparable microscopes used in the United States are manufactured by Zeiss, American Optical Company, Leitz, and Bausch and Lomb, among others.

LIGHT MICROSCOPY

Microscope MBI-3 (Fig. 18) is a complex instrument designed for investigating transparent objects in transmitted light in a bright field using direct and oblique illumination.

Microscope MBI-6 (Fig. 19) is a universal research microscope for examining and photographing objects in transmitted light in bright and dark fields, with phase contrast, and

---

[1] A description of the general construction of a microscope and the path of the light rays within it is not given here, since this information is included in courses in general microbiology and cytology.

[2] MBI is the transliterated Russian catalogue code of these particular microscopes. Similar codes (OI, KF, MFA, ML, DRSh, SVDSh, MBS, BGS, VSE, LP) for appropriate instruments are indicated in the text as they appeared in the original.

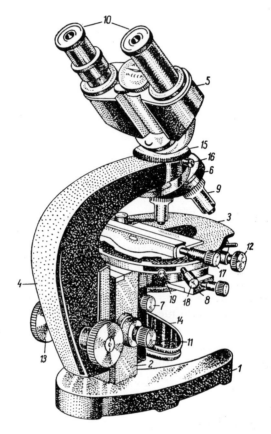

Fig. 18. Microscope MBI-3, general view. *1*, Base; *2*, housing containing the fine focus mechanism; *3*, specimen stage; *4*, tube support; *5*, binocular attachment; *6*, nosepiece for objectives; *7*, condenser focus knob; *8*, condenser; *9*, objective; *10*, eyepieces of the binocular attachment; *11*, knob for fine focus; *12*, knob for shifting the upper part of the stage; *13*, knob for rough focus; *14*, mirror; *15*, cap of the tube support; *16*, screw for securing the binocular attachment; *17*, knob for positioning slides; *18*, centering screws of the stage; *19*, set screw of the stage.

in reflected light in bright and dark fields. A research microscope should have a wide selection of optics. In addition, a microscope used in fieldwork should be constructed so that it rests on a shock-absorbing desk and withstands a fair amount of abuse.

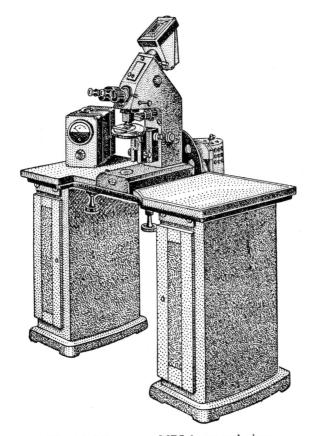

Fig. 19. Microscope MBI-6, general view.

Basic Principles for Working with the Light Microscope

When working with any microscope, the following factors are essential: (1) correct illumination of the specimen; (2) correct selection of the optical system: objective, eyepiece, and condenser; and (3) ability to use thin slides and cover glasses (slides not thicker than 1.4 mm; cover glasses not thicker than 0.15–0.17 mm).

Illumination of the Specimen. It is recommended that only artificial light and a plane mirror be used in the microscope

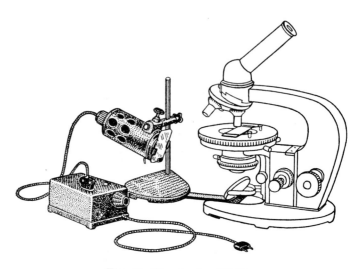

Fig. 20. Illuminator OI-19.

for illuminating the specimen. For this purpose a low-voltage lamp with a thick filament is usually employed. The source of light in itself must be a relatively small area, actually constituting, during preliminary alignment of lamp and microscope, the image of the lamp filament projected onto the closed condenser iris diaphragm. The intensity of the light must be great. The best alternative is to use one of the illuminators—OI-7, OI-19 (Fig. 20), OI-20—manufactured by optical instrument companies. These instruments have a collector lens which allows the image of the filament to be focused in the plane of the iris diaphragm of the microscope condenser in order to obtain parallel rays of light. The collector is provided with an iris diaphragm, the diameter of which, when fully opened, must correspond with the diameter of the condenser iris diaphragm. When fully closed, the diameter of the condenser iris diaphragm should be 0.5 mm. In the final alignment of the lamp and microscope, the image of the lamp diaphragm is focused on the plane of the specimen. An attachment which allows the bulb of the lamp to be centered facilitates alignment. Illumination of the specimen should be done by Kohler's method, the only procedure which guarantees ideal illumination (Fig. 21).

Arranging the illumination in a microscope is initiated by aligning the light source in the following manner. A white

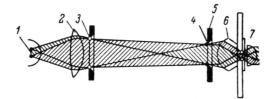

Fig. 21. Schematic representation of the path of the rays when using Kohler's method of illuminating a specimen (Peshkov, 1962). *1*, Bulb filament; *2*, collector; *3*, field diaphragm; *4*, image of the filament in the plane of the aperture diaphragm; *5*, aperture diaphragm; *6*, condenser; *7*, image of the edges of the field diaphragm in the plane of the specimen.

screen (a piece of white paper) is placed 25–30 cm from the lamp. The bulb is turned on fully and the collector is moved back and forth until the enlarged image of the filament is focused on the screen. The image should appear as a wide strip approximately 2 cm in length, surrounded by a uniform bluish halo (as a result of spherical and chromatic aberration). The uniformity of the width of the halo around the image indicates the accuracy of the lamp alignment.

The next step is the reciprocal alignment of the light source and the Abbe illuminator. This is done by first placing the illuminator 30–40 cm in front of the microscope. A disc of white paper is then placed on the flat mirror of the microscope and the ray of light is directed from the bulb onto this disc. The enlarged image of the filament spiral is focused on the center of the paper disc. It is essential that a sharp image of the filament be obtained by shifting the lamp socket and by moving the collector of the lamp in or out.

It is next necessary to produce the filament image on the surface of the fully closed iris aperture diaphragm of the microscope. This is very important since this diaphragm is located precisely in the plane of the forward main focus of the condenser (i.e. at the entrance aperture of the microscope). After removing the paper disc from the mirror, the light from the illuminator is directed onto the closed aperture diaphragm of the microscope with the help of the mirror. The filament image is then obtained by moving the collector lens of the illuminator in or out. To accomplish this last step of focusing the image and for convenience in observing the diaphragm, a small mirror is placed on the right leg of the microscope at an angle such that, when the little mirror is viewed from above, it reflects the closed aperture diaphragm for the observer.

It is essential to check the accuracy of the illumination alignment. To do this, the aperture diaphragm (iris diaphragm of the microscope) is opened all the way. Then the intensity of the light is decreased until it is reddish yellow in color. A specimen is placed on the stage of the microscope and brought into focus with a low-power objective (10×) combined with a medium-power eyepiece (10×). The mirror is rotated slightly and the image of the collector lens and the field diaphragm occurring with it are sought in the field of vision. Because the objective is focused on the specimen, in order to focus the image of the collector in the plane of the specimen, it is necessary either to lower the condenser or to raise it slightly. If the illumination is adjusted properly, then both the specimen and the lens of the collector will be visible simultaneously, the latter appearing as a uniformly illuminated circle. If the collector lens is only partially illuminated or appears dark, covered with illuminated spots, then the lamp should be tilted slightly around its horizontal axis until the proper position is reached. The alignment is accurate only when the specimen and the image of the lens surface are visible simultaneously to the eye of the investigator. When the field diaphragm is constricted, its interlocking edges should appear sharply defined. When the final focus of the condenser is made on the specimen, the image of the constricted aperture of the field of vision diaphragm should be surrounded by a red or azure border and should be quite sharp. The small circle of light (the size of which depends on the degree of the opening of the field diaphragm) is then fixed in the center of the field of vision by moving the mirror. Only then is it possible to begin observations.

When working with low magnifications, Peshkov (1962) recommends replacement of the high-aperture part of the condenser of microscope MBI-3 with the low-aperture condenser of the eyeglass type. In this way the condenser has a decreased aperture and an extended focal point. The image of the object is precisely superposed optically with the image of the aperture of the field diaphragm while the aperture diaphragm is opened fully. While looking into the microscope, the field diaphragm should be opened until the edges disappear beyond the edge of the field of the ocular. This type of condenser is especially useful in microphotography.

Arbitrarily raising or lowering the condenser of the illuminator for the purpose of regulating the strength of the light should always be avoided, since this leads to distortion of the images observed in the microscope.

Selection of Optical System.   The correct use of objectives and eyepieces determines to a large extent the quality of the image in the microscope. Different combinations of objectives and oculars yield different magnifications. When observing bacteria, high-power objectives (90×) and lower power eyepieces (7× or 10×) are usually used. Such combinations enable a sharp increase in the resolving power of the microscope and observation of objects up to 0.4 μ in size.

Use of Thin Slides and Cover Glasses.   The thickness of slides used is an extremely important factor to be considered. When the numerical aperture of the system is in full use, a thick slide interferes with the correct focusing of the condenser and sharply decreases the clarity of the microscope image. It is best to use slides no greater than 1.4 mm thick. Cover glasses should not be thicker than 0.15–0.17 mm; otherwise it is impossible to employ a powerful short-focus immersion objective, the working distance of which is smaller than the thickness of an inappropriate cover glass.

Special Methods of Microscopy

Dark-Field Microscopy.   Observations in a dark field are made with the aid of a special condenser (Fig. 22). This darkfield condenser allows only high-aperture (very oblique) marginal rays to pass. These rays, falling on the surface of the condenser, are reflected and concentrated at its focal point, which must be located in the plane of the object under observation. Peshkov (1962) suggests that any high-aperture Abbe condenser can be changed into a dark-field condenser by unscrewing it and covering its central portion (not less than two-thirds)

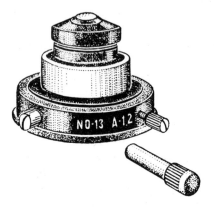

Fig. 22. Dark-field condenser OI-13, external view. At lower right is the screw.

Fig. 23. Slip of paper for dark-field microscopy (Arkhangelskii, from Timakov and Goldfarb, 1958).

with a small disc of black paper. The diameter of the paper circle is determined empirically. Four projections (Fig. 23) are left on the circle to insure that it occupies the central position between the lenses.

When observations are made in a dark field, oblique rays from the condenser, which pass through the specimen and meet dense particles such as bacterial cells, deflect from these particles; as a result the particles become visible, i.e. are illuminated, in the dark field. The clarity of the image depends on the light source and the accuracy of the setting.

Here the best light sources are the point lamps of the OI-9 and OI-19 type illuminators.[3] The light for dark-field observations is established according to the bright field. The Abbe condenser is then replaced by a dark-field condenser, and the aperture diaphragm of the lamp is opened completely.

Observations in a dark field demand the use of low-aperture objectives: high dry systems with homogeneous immersion of 1/7, numerical aperture 0.75 (Peshkov, 1962).

Dark-field observations are made with a substage condenser: a drop of immersion oil is placed on the upper lens of

[3] Equivalent equipment is readily available in the United States.

the condenser, which is raised slightly until the oil spreads over the slide. The microscope is then focused on the specimen.

The thickness of the specimen is an extremely important factor. The thicker the specimen, the greater the refracting material within, and the less the contrast regarding the image.

Phase-Contrast Microscopy. The method of phase-contrast microscopy is one means of increasing the contrast of the image in the microscope of microscopic objects, such as bacteria, which are transparent in visible light. The majority of micro-organisms are transparent in visible light. The light passing through them is either not affected at all, or is altered very little in intensity. Alterations in the light path when it passes through such objects cause only a slight phase change in the light waves, which is not appreciated by the conventional microscope; as a result the objects appear with a lack of contrast.

With phase-contrast microscopy, it is possible to transform minor phase shifts so as to provide contrast (Fig. 24).

The phase-contrast apparatus (Fig. 25) is available from various manufacturers of optical instruments. It consists of phase objectives (1), each of which contains a ring-shaped phase plate; a condenser (2) with a revolving disc, ring diaphragms corresponding to those of the objective, and centering screws which superpose the image of the condenser diaphragm with the phase plate of the objective; and an auxiliary lens (3) with which the coincidence of the ring diaphragm and the phase plate is checked.

When working with the phase-contrast microscope, it is extremely important that the specimen be well illuminated. Therefore, illumination should be made according to Kohler's principles.

The steps to follow with a phase-contrast apparatus are:

1. In the microscope the condenser and objectives are replaced with the phase equipment or with built-in phase equipment; the phase objective is then swung into position.

2. The specimen is placed on the stage of the microscope and the Kohler illumination is set under low magnification (10×).

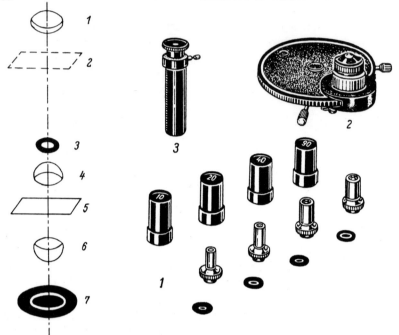

Fig. 24. Diagram of
phase-contrast mi-
croscope. 1, Ocular;
2, image plane; 3,
phase ring; 4, ob-
jective; 5, object; 6,
condenser; 7, dia-
phragm.

Fig. 25. KF-4 apparatus for observations made
with phase contrast (see text for explanation).

3. The diaphragm of the condenser is set to correspond
with the diaphragm of the low-power (10×) objective. The
objective is then focused on the specimen.

4. The ocular is replaced by the auxiliary lens unless a
phase telescope is built into the microscope. Raising the tube
of the auxiliary lens or "phase telescope" provides a sharp focus
in the output aperture, which appears as a luminous circle inside of which the darker ring-like phase plate is concentrically
located.

5. The condenser disc is turned in such a way that the
number corresponding to the objective which will be used for
the investigation is in position. The objective is then set. With
the help of the centering screws, the ring-like diaphragm is

superposed with the phase plate so that the latter is accurately concentric with the luminous ring of the diaphragm.

6. The auxiliary lens or phase telescope is removed; the ocular is replaced; and the investigation is undertaken.

Whenever the objective is changed, it is essential to set the corresponding ring diaphragm and, with the aid of the auxiliary lens, to check the accuracy of the coincidence of the condenser ring diaphragm with the phase plate.

Phase contrast simplifies taking a direct count of microorganisms, since it provides a sharper, more clearly defined image.

Anoptral (Phase/Dark-Field) Microscopy.   The anoptral instrument, developed by Wilska (1954), yields a more sharply contrasted image of very fine objects than does phase contrast. When making observations with the anoptral system, the outline of objects is sharp; the object acquires dimension, and details show up in all shades—from light brown to near white. According to the degree to which the details of the object are distinguished in optical density from their surroundings, the image will be more clearly defined.

The anoptral (phase/dark-field) arrangement is essentially a type of negative phase contrast. An apparatus of this type is produced in Russia under the label MFA-2.

The basic characteristic of the anoptral system is the use of special achromatic phase/dark-field objectives and a special condenser with ring diaphragms fixed in a nosepiece. The difference in phase/dark-field objectives lies in the fact that they have a diffraction ring of specific dimensions on one of the surfaces of the lenses. Ordinary Huygenian oculars are used with phase/dark-field objectives.

This type of microscopy requires the use of an auxiliary lens in the tube instead of an ocular.

When dry objectives of 20× and 45× are used, phase/dark-field contrast is conducted in the same manner as regular phase contrast. Kohler illumination is required.

When 1.23 (70×) and 1.25 (90×) objectives are used, the system is arranged as follows:

1. The condenser is set.

2. The objective is selected and focused on the specimen while the aperture diaphragm is closed completely.
3. The iris diaphragm of the condenser is opened fully.
4. The ocular is replaced with an auxiliary lens; by adjusting the ocular of the auxiliary eyepiece, the latter is focused on the diffraction ring of the objective (during this time the focusing of the microscope should not be racked down).
5. By turning the nosepiece of the condenser, the light diaphragm corresponding to the objectives is put into use; besides the diffraction ring of the objective, the bright ring of the diaphragm (Fig. 26, A) should also be visible in the auxiliary eyepiece.
6. The bright ring is superposed onto the dark one (Fig. 26, B) by means of the condenser centering screws.
7. The auxiliary eyepiece is replaced by the ocular.

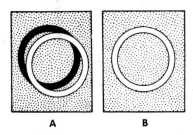

A   B

Fig. 26. Superpositioning the diffraction circle of the objective with the ring of the diaphragm. *A,* Rings not superposed; *B,* rings superposed.

When working with the 1.23 (70×) (water immersion) objective, a drop of water should be placed on the front lens of the condenser, so that the drop will cover the entire lens. Also, after raising the condenser until it comes in contact with the specimen, a drop of water should be placed on the front lens of the objective and the microscope focused to obtain sharp image.

When working with a 1.25 (90×) (oil immersion) objective, a drop of immersion oil is placed on the condenser and specimen.

Fluorescent Microscopy.   Fluorescent microscopy is one of the most promising contemporary methods of microscopy. In fluorescent microscopy many microstructures, indistinguishable in ordinary microscopy, become visible and clearly defined either from their natural fluorescence, or as a result of the action of fluorescing compounds which combine with specific cell components and transmit to them the power to fluoresce. With the intense, contrasting brightness fluorescent microscopy makes it possible to detect very small numbers of microorganisms in an environment. Determination may also be made of distribution and, in many cases, physiological condition of the microorganisms.

Fluorescence is the luminescence of a substance by various forms of energy. Fluorescence occurs when energy—absorbed by atoms, molecules, or ions of various substances which are capable of fluorescing—is transformed into light. Having absorbed energy, molecules of a fluorescent substance convert to an excited state which continues for a given, usually very brief, time and ultimately results in reversion to a lower energy level. This return to the original normal state is accompanied by the emission of excess energy in the form of light, i.e. fluorescence. There are various ways to transfer the energy necessary for exciting fluorescence to molecules of a substance capable of fluorescing: beams of light rays, electron bombardments, x-rays, and radioactive emission.

The simplest means of inducing fluorescence is illumination of a fluorescent substance by ultraviolet rays of a specific wavelength, or by short-wave rays (the blue-violet of the visible spectrum). The spectral composition of light is outlined below.

Invisible Light, Ultraviolet (120–380 mμ)
    Far ultraviolet (120–200 mμ)
    Intermediate ultraviolet (200–300 mμ)
    Near ultraviolet (300–380 mμ)

Visible Light (380–750 mμ)
    Violet (380–430mμ)
    Blue (430–460 mμ)
    Azure (460–500 mμ)
    Green (500–550 mμ)

Yellow (550–600 mμ)
Orange (600–660 mμ)
Red (660–750 mμ)

Invisible Light, Infrared (750–2000 mμ)

In a contemporary fluorescent microscope, fluorescence of the specimen is excited either by near ultraviolet light, or by rays of the blue-violet part of the spectrum (with a wavelength of 380–430 mμ).

Depending on its nature, the illuminated substance begins emitting visible light of a distinct color. Usually, under illumination, the substance—having absorbed light—transforms the received emission of one wavelength into an emission of another wavelength, usually longer.

Fluorescence is distinguished from phosphorescence on the basis of persistence of afterglow. If the afterglow persists in the range 10^{-9} to 10^{-7} sec, the luminescence is called fluorescence; if it is longer, it is called phosphorescence.

For fluorescent microscopy fluorescence and one of its forms—photoluminescence—are employed.

When the afterglow is as short as described above, the image of the fluorescent specimen in the fluorescent microscope is observed only at the time of irradiation with the exciting light rays. When irradiation ceases, the specimen becomes invisible. The appearance of fluorescence is thus fundamental to fluorescent microscopy.

On the basis of the nature of the rays causing fluorescence, the following are distinguished: (1) fluorescent microscopy in the near ultraviolet; (2) fluorescent microscopy in visible light. The first method of fluorescent microscopy is usually employed to study natural, so-called primary fluorescence of biological objects. The second method—fluorescent microscopy in visible light—is most widely used in microbiology. The specimen, when irradiated with blue and violet rays of the visible spectrum, emits light. The rays which excite fluorescence are absorbed by a yellow filter on the ocular. A dark background, against which objects under observation clearly luminesce through the yellow filter, is thus obtained. As a result, a highly contrasted image is provided.

Fluorescent microscopy in visible light is used for observing secondary (externally introduced) fluorescence of specimens which are treated with fluorochromes (fluorescent dyes).

Plant and algal cells possess primary fluorescence because of the presence of chlorophyll.

Living algae fluoresce with a bright, blood-red light. A few minerals fluoresce. Many particles of sediments possess fluorescence which, though weak, is nevertheless sufficiently distinctly expressed. Only a few microorganisms possess their own distinctly expressed fluorescence. The majority of them fluoresce weakly. The method of Repetigny, Sonea, and Frappier (1961) has been suggested for determining the intensity of fluorescence of microorganisms. These authors propose that the difference in fluorescence—color, character, and intensity—can be employed in classifying bacteria and identifying distinct species.

Secondary (or induced) fluorescence has great significance in investigation of microorganisms. Living structures which do not possess their own fluorescence are transformed into fluorescent structures when treated with fluorochromes, which are bound with proteins and transmit characteristic fluorescence.

When studying fluorescence, it is necessary to remember that not all absorbed energy is emitted in the form of fluorescent energy. The output of fluorescence (the proportion of energy of emission to the energy of the absorbed light) depends both on the fluorescent substance itself and on various external conditions: temperature, pH of the solvent, concentration of the substance, and impurities. When working with fluorescent microscopy, it is therefore necessary to consider all conditions present.

Apparatus. Fluorescent microscopy is carried out either with the aid of a special microscope, or with an apparatus which includes a microscope, light source, and a system of light filters which intercept rays with a wavelength greater than ultraviolet, i.e. short-wave rays of visible light. Because red light and infrared rays pass through the majority of yellow filters, an additional filter—which absorbs long-wave rays—is placed in the path of the exciting rays. Blue-glass filters containing copper sulfate are usually used. Filters containing copper sulfate in solution may also be used.

A container filled with distilled water should be placed in the path of the strong beam of light emitted by the light source to the filters in order to prevent the blue filter from heating up.

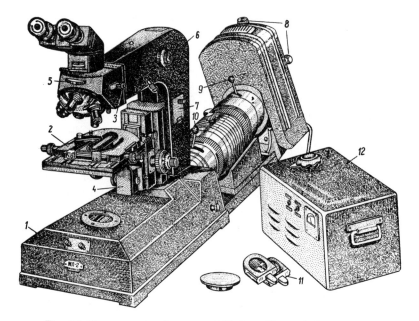

Fig. 27. Fluorescent microscope ML-2. *1,* Base of the microscope; *2,* Specimen stage; *3,* knob for shifting the light-separating plate and the circular mirror; *4,* supporting stand for the stage and attached condenser; *5,* disc for variable magnification; *6,* field diaphragm knob; *7,* light filter housing; *8,* screws for centering the lamp; *9,* knob for shifting the collector; *10,* field diaphragm iris knob; *11,* light filter; *12,* electrical input PRL-5.

The Russian fluorescent microscope ML-2 (Fig. 27) is especially convenient for microbiological investigations. Equivalent fluorescent microscope apparatus are widely available in the United States. The ML-2 is designed for studying microbiological and other specimens in fluorescent light excited by the blue-violet part of the spectrum and by ultraviolet rays with a wavelength up to 360 mμ. This microscope permits observation of specimens illuminated from above through the objective in incident light. Thus the presence of microorganisms on non-transparent objects can be determined. This method of illumination has a number of advantages. The light filters are used for protection from the relatively small portion of exciting emission. The microscope can also be used to examine specimens in transmitted light, when the rays which excite fluorescence are directed onto this specimen from below through the condenser of the microscope. A phase-contrast apparatus may then also be used. The microscope is provided with a camera which makes possible microphotography.

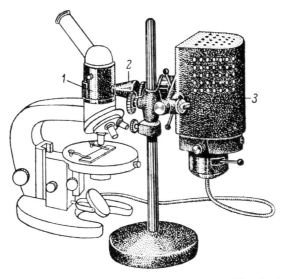

Fig. 28. Opaque illuminator OI-17. *1,* Opaque illuminator; *2,* conical tube; *3,* light source.

Fluorescent microscopy can be conducted in an ordinary microscope when the opaque illuminator OI-17 (Fig. 28)—used with incident light—and a light source are available. The fluorescent opaque illuminator is inserted in the microscope between the tube and the nosepiece. When using the opaque illuminator, it is absolutely necessary to have a powerful source of invisible ultraviolet and visible blue-violet radiation, for example, the DRSh-250 lamps (illuminator OI-18, produced with the opaque illuminator, does not provide sufficient light output for microbiological investigations when used with the SVDSh-120 lamp). The mercury-quartz lamps DRSh-250 used at the present time are a gas-discharge source of light with a discharge of mercury vapor. To ignite the vapor, it is necessary to administer higher voltage than for steady combustion. Normal operation of the lamp is established a short time after it is turned on.

In certain cases it is possible to work with a microscope when illuminating the object through the microscope mirror. In this case, an ordinary illuminator of the OI-7 type or the OI-9 type can be used. When more powerful light sources are needed, the OI-18 illuminator with the mercury-quartz lamp or equivalent is the best choice. To isolate the blue-violet portion of the spectrum with wavelengths from 380 to 430 mμ, glass light filters, placed on the oculars (blue-light filters) are used. When such filters are not available, it is possible to use a liquid filter, prepared according to the Meisel method, as follows: a solution of concentrated ammonia is added gradually to a saturated solution of chemically pure copper sulfate in distilled water. A di-ammonium sulfate precipitates. The ammonia solution is added until the precipitate dissolves. The working solution is obtained in this way. The solution is diluted with distilled water according to the thickness of the container to be used. The solution must be perfectly transparent. When the prepared blue liquid filter is superposed with the yellow-glass filter, which is already on the ocular, total or almost total darkening results when a beam of bright light is passed through both filters. The concentration of the liquid filter is adjusted either by dilution with distilled water or by addition of some of the basic solution.

Regardless of the method of illumination used in fluorescent microscopy, it is always most expedient to use objectives with coated optics. The use of objectives with a large aperture creates more optimal conditions for illuminating the specimen for observation in incident light. Huygenian oculars have an advantage over compensating ones. The greatest brightness of the image is obtained when stronger magnification oculars are used in combination with weaker objectives. Use of a monocular microscope with a vertical tube helps to achieve this goal. Occasionally, insufficient fluorescence of microscopic specimens makes the use of binocular attachments impossible.

When working with immersion objectives in fluorescent microscopy, non-fluorescent immersion oil [4] must be used. Ordinary oil and its substitutes are unsuitable because of their own fluorescence. A series of non-fluorescent oil substitutes is available—anisole (methyl phenyl ether); dimethylphthalate (dimethyl ether of O-phthalic acid); mineral oil (Mikhailov and Dyakov, 1961). Several types of purified glycerin, as well as mineral and liquid paraffin oil, may be used. Fluorescence-quenchers should be added to ordinary immersion oil. Nitrobenzene (from 2 to 10 drops per 1 g of oil) is useful as a quenching agent.

Fluorochromes. A variety of organic compounds and dyes for reacting biological specimens with fluorochromes is available. Those most widely used are the acridine group (acridine orange, acridine yellow, coriphosphine, acriflavine hydrochloride) and the thiozole group (primuline). Fluorochromes are applied in water solutions (distilled water). They are subdivided into three basic groups according to their ability to dissociate in water solutions:

1. alkaline fluorochromes: acridine orange, auramine 00, acridine yellow, coriphosphine, aurophosphine, berberine sulfate $[(C_{20}H_{18}O_4N)_2 \, SO_4 \cdot 3H_2O]$; neutral red; acriflavine hydrochloride
2. neutral fluorochromes: rhodamine B
3. acidic fluorochromes: primuline (yellow thiozole dye); fluorescin; sodium fluorescein; eosin; and erythrosine.

The concentration of the fluorochrome has great significance: depending on the concentration, the color of fluorescence can change. Fluorochromes are usually used in low concentrations (1:500–1:100,000).

Many fluorochromes change the tint of fluorescence de-

---

[4] Non-fluorescent oil, sold in East Germany, has indices of refraction of 1.520 (for examining specimens without a cover glass) and 1.515 (for examination with a cover glass).

pending on the pH of the object. Various components of microbial cells combine and adsorb fluorochromes differently; as a result a difference in the color and brightness of the fluorescence occurs. A specimen may be treated simultaneously or sequentially with several fluorochromes.

When attempting to obtain a highly contrasted image of the distinct parts of the cell, it is necessary to resort to special extinguishers. These include methylene blue, aniline blue, fuchsin, Congo red, and weak solutions of iodine.

Solutions of fluorochromes can be kept for many months without change if they are stored in dark vessels in a cool place. Phenol may be added to the fluorochrome solutions to a final concentration of 3%.

Treating specimens with fluorochromes is extremely simple. A small quantity of liquid or sediment containing microorganisms is placed on a slide in a drop of a previously introduced fluorochrome. The specimen is covered with a cover glass and in 0.5–1 minute is examined with the microscope.

Fixed specimens can also be treated with fluorochromes. Fixation is carried out by various means: heating, alcohol, 10% aqueous solution of formalin, Nikiforov mixture (ethyl alcohol and ether in a proportion of 1:1), and others. Solutions containing substances which extinguish fluorescence (such as mercuric chloride and osmic acid) are unsuitable for fixation.

After fixation the specimen is dried, a drop of the fluorochrome is added, and the mixture is allowed to stand for 1–10 minutes. The specimen is washed with distilled water to remove excess fluorochrome and then air-dried. Fixed specimens must be examined in a non-fluorescent medium such as water, a physiological solution of NaCl, or a special medium with the following composition: distilled water, 10 ml; gum arabic, 10 g; glycerin, 5 ml; chloral hydrate, 1 g.

Fluorescent Analysis. The fact that fluorescence is a characteristic of a radiant substance makes it possible to identify the substance according to the character of radiation. Fluorescent analysis is the study of various compounds and the identification of the composition of chemical compounds or biological specimens by measuring differences in fluorescence.

There are several forms of fluorescent analysis, the value of which is an extremely high sensitivity. Fluorescent analysis can be used to determine the presence and character of the primary fluorescence of cultures. In fluorescent analysis small quantities of the substance under study are subjected to irradiation by exciting rays.

The excitation of fluorescence in fluorescent analysis can be conducted by various methods. Most often ultraviolet rays are used. The rays pass through a light filter such as Wood's glass or black glass. The light filter should intercept all visible rays and efficiently allow the passage of ultraviolet rays with a wavelength of 300–380 mμ. To prevent the light filter from heating up, a heat filter is placed between it and the lamp.

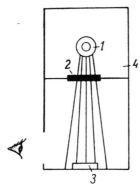

Fig. 29. Diagram of fluorescent analysis apparatus. (See text.)

Figure 29 shows a diagram of an arrangement for fluorescent analysis. A mercury vapor lamp (1) is set in a case (4) and a light filter is placed below the lamp. The light filter (2) operates in such a way that only ultraviolet rays reach the lower part of the instrument. The culture or specimen (3) is located below the stream of ultraviolet rays. The fluorescence of the material under investigation can be observed either directly by eye or through a microscope.

A fluorescent microscope, specially designed and constructed for fluorescent microscopy also can be used and, of course, entails a much simpler procedure.

In aquatic microbiology fluorescent microscopy can be used to measure the quantity of living and dead cells in examination of primary fluorescence of microorganisms (p. 50); to follow sporogenesis, for rapid diagnosis of *Escherichia coli* (p. 402; and to examine the nature of the distribution of microorganisms in sediments. Distribution of microorganisms in sediments is determined with fluorochromes such as acridine orange and coriphosphine. The fluorescent method can be used to identify species of nitrogen-fixing bacteria and to examine for these directly in sediments.

The method proposed by Repetigny, Sonea, and Frappier (1961) can be used to determine the presence of primary fluorescence in microorganisms. Aqueous suspensions of microorganisms, washed free of traces of the medium, are centrifuged and diluted three times with water. About 0.05 ml of the suspension is collected in a capillary 65 mm long with an internal diameter of 1 mm. The middle portion of the capillary is observed under a microscope, using a low-power objective. To show the influence of pH on the intensity of the fluorescence, 1 drop of a buffer solution of M/15 potassium monophosphate is introduced into the suspension. The intensity of the fluorescence is determined by comparison with the fluorescent intensity of fluorochromes of a specific concentration. The characteristic luminescence varies in different bacterial species and the color of the fluorescence of cultures is determined by microscopic observation.

ELECTRON MICROSCOPY

With electron microscopy it is possible to obtain an image of objects magnified up to 100,000 times; objects indistinguishable in ordinary microscopes can be viewed by electron microscopy. This type of microscopy is achieved by using, in place of a light source, a stream of moving electrons emitted by a special source (electron gun). The stream of electrons (electron beam), while moving in a vacuum, propagates like a light beam, in a straight line.

There are several types of electron microscopes and these are differentiated according to the method of observing the object under study (for details see Spivak, 1954 or any one of the large number of excellent and more recent books on the subject).

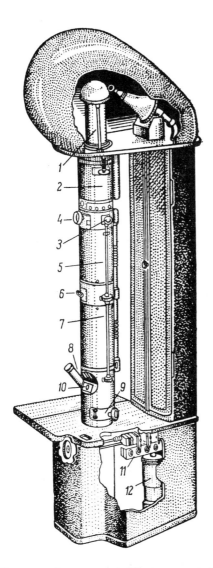

Fig. 30. Electron microscope. *1,* Electron gun; *2,* condenser
lens; *3,* object chamber; *4,* hatch for changing specimens; *5,*
objective lens; *6,* observation of the intermediate image; *7,*
projection lens; *8,* light microscope; *9,* hatch for removing
photo-plate; *10,* observation of the final picture; *11,* vacuum
regulator; *12,* pump.

In biological research the electron microscopes used most often are of the transmission type, in which the object is observed in a transmitted stream of electrons.

A microscope of this type operates so that the beam of electrons permeates the object. The first electron image of the object arises in the objective electromagnetic lens, on which the beam of electrons, focused by the condenser lens, falls after penetrating the object (Fig. 30). The image appears on a fluorescent screen, as a result of the screen fluorescing under the impact of the electrons. To place the specimen in precise focus, an intermediate image can be observed through an inspection lens. From the intermediate screen the electrons pass through a projection lens which magnifies the image and presents it in another plane. The final image, which is observed through an inspection window, arises in this plane, on a second fluorescent screen. Additional lenses permit still greater magnification.

It is possible to record photographically the image obtained by placing a camera, provided with an apparatus for changing the plates, under the screen. The electron micrographic negative obtained in the electron microscope can be further magnified 4 to 5 times by ordinary accepted photographic methods. With the electron microscope, accurate measurement of dimensions and shape of objects and the most minute details of cell structure can be obtained.

Electron microscopy requires specially prepared specimens. For a successful study of bacteria, it is necessary to have a clean specimen free of impurities which can distort the microscopic picture when the specimens are dried. Methods have been worked out for preparing electron microscopic specimens (Kushnir, 1958; Pekhov, 1962a, 1962b, and other more recent papers in the literature), for preparing supporting films, and for the introduction of material onto them. For this type of preparation, specimens may range in thickness from about 20 A to several microns.

A new world of microorganisms of a unique form and of dimensions smaller than 1 μ (Volarovich and Tropin, 1963) has been found in soils and sediments with the electron microscope. The potential has only begun to be tapped since very few electron microscopic examinations of aquatic samples have been performed.

In the late 1940s, a technique for the preparation of biological material by ultrathin sectioning was developed. This method involved special fixation and embedding, followed by the cutting of sections 100–1000 A thick. In the last 20 years, this method has done more than any other to produce a renaissance of interest in the fine structure of cells of all types. See Plates 3 and 4.

More recent developments in the field of electron microscopy include the freeze-etching technique of specimen preparation and observation and scanning electron microscopy. A three-dimensional view of biological materials is afforded by these methods and greater use of these techniques and instruments in microbial ecology will be seen in the immediate future. See Plate 5.

Chapter 3

Methods of
Culturing Microorganisms

Culture methods in specific nutrient media are used to study microorganisms. Culture media for microorganisms not only are a source of nutrition and energy, but also represent the growth environment of the cells. Transport of nutrients into cells is an extremely complicated process: some substances (proteins, fats, polysaccharides) enter the cell only in a water-soluble state and must, therefore, be rendered soluble by action of exoenzymes of the cells; others (e.g. substances soluble in lipids) enter through the lipid fraction of membranes and need not be previously dissolved. Transport of minerals into a cell depends on such factors as the degree of ionic dissociation, pH of the surrounding medium, and charge on the cytoplasmic membrane. Therefore, the composition of the medium, pH, conditions of aeration—oxidation-reduction potential of the medium, osmotic pressure, and other factors—are important in the growth of microorganisms; indeed, they determine growth.

Autotrophic microorganisms require mineral media of a strictly determined composition in which mineral salts are added in precisely specified quantities. Autotrophic bacteria use carbon dioxide as a source of carbon. Heterotrophic microorganisms require organic compounds; some need simple compounds, others require complex ones. As a source of carbon, heterotrophic bacteria use organic substances of a varied chemical structure, but for many of them, CO_2 is also necessary to complete complex exchange reactions.

The requirements of microorganisms with respect to nitrogen source are varied: some can obtain the nitrogen necessary for synthesizing cellular protein only from complex proteins; others obtain their nitrogen requirement from simple mineral compounds; and a few can assimilate the molecular nitrogen of the atmosphere.

In addition to organic nutrients, bacteria must be provided with inorganic elements such as sulfur, phosphorus, potassium, and magnesium. For the majority of microorganisms, the best sources of sulfur are its oxidized forms, the sulfates. Many microbes use sulfur derived from complex organic compounds, and there are also groups of microorganisms which assimilate reduced compounds of sulfur (sulfobacteria) and elementary sulfur (thiobacteria). The best sources of phosphorus for the majority of microorganisms are the oxidized compounds of phosphorus, the orthophosphates. Some microorganisms utilize organic compounds containing phosphorus. Microorganisms utilize best the phosphates (mono- and diderivatives) as a potassium salt and, therefore, these salts are included in the majority of synthetic media. There is a variety of mineral salts (mainly sulfates) which serve as a source of magnesium.

In addition to the basic inorganic elements, the majority of microorganisms, in order to achieve normal growth, need very small quantities of trace elements (molybdenum, boron, zinc, cobalt, manganese, aluminum, and copper) which are involved in synthesis of enzyme proteins. Moreover, trace elements play an important role in activating enzyme systems. Therefore, it is often absolutely necessary to introduce very small amounts of trace element solutions into a medium.

Many heterotrophic organisms cannot independently synthesize some components of their cellular protoplasm; they must have them provided in complex form. Specifically, vitamins and other supplementary growth factors are necessary for synthesizing different enzymes. Therefore, yeast autolysates (lysed yeast cells rich in vitamins) and extracts or separate vitamins, or a combination of these, are introduced into many culture media.

Winogradsky has introduced into microbiology a method of using enrichment or selective media which satisfy the demands of specific physiological groups of microorganisms, but are unsuitable or barely utilizable for others. A number of selective media, the formulas of which are given below in the

discussion of the corresponding physiological groups, have been
developed and put into use in microbiology.

Several media are widely used and are common in microbiological work. Among the formulas of preparation given
below is that for beef-peptone medium. Among the common
media a whole series of natural substrates are included, especially those deriving from historical microbiological investigations: potato, carrot, milk, egg albumin, corn, and various
plant broths. These complex substrates are no longer used routinely in the modern microbiological laboratory. For the sake of
completeness they are included here. Synthetic media containing either mineral salts alone, or complex organic compounds in
combination with mineral salts, are used more frequently, however. If bacterial cultures grow well in a completely defined
medium, it would be a preferred culture medium.

Culture media are of a specific consistency: liquid, solid,
or semisolid. Solid nutrient media are prepared from the liquid
by adding agar in the amount of 1.5–2%, or gelatin in the
amount of 10–15% (less often, 20%). Semisolid media are
prepared by adding 0.1–0.2% agar. The silica gel plates introduced by Winogradsky are widely used as solid media, particularly for the soil and sediment bacteria.

PREPARATION OF COMMON LABORATORY
CULTURE MEDIA[1]

Beef-Peptone Media

The starting material for preparing beef-peptone media
is beef water, which is prepared in the following manner. Finely
chopped or ground fresh beef, free of bones, tendons, and fat

---

[1] The instructions for preparing a variety of complex media are
given as described by Rodina. However, most of these are available in the
United States in dehydrated form to which only distilled water need be
added in the required amount and the reconstituted media can then be
autoclaved and used. Difco Laboratories, Detroit, Michigan, and the Baltimore Biological Laboratories, Division of Bioquest, Baltimore, Maryland, are two of several commercial firms which supply dehydrated media.

(500 g), is covered in an enamel kettle with 1 liter of tap water and is heated to 50–55 C. The solution is kept warm for 1 hour at a temperature of 50–55 C, or 12 hours at ordinary room temperature.

The extract is filtered through gauze with a layer of cotton, which is then squeezed out. The liquid obtained is boiled and filtered twice (the first time through gauze with cotton; the second time, through filter paper). The filtrate is diluted with water to 1 liter in volume and poured into flasks which are closed with cotton stoppers. The mouths of the flasks with stoppers are covered with paper caps. Flasks with beef broth are sterilized at 120 C for 20 minutes. This type of beef broth can be used at any time for preparing media. Resterilization is unnecessary if the media are prepared immediately. Beef can be replaced by beef extract. Usually 5 g of beef extract, which dissolves in water, are added per liter of the medium.

Beef-peptone broth is prepared as follows. Peptone (10 g) and sodium chloride (3 g) are added to 1 liter of beef water. This is heated, stirred until the peptone completely dissolves, and then removed from the heat. The pH is adjusted to 7.0 by adding drops of a saturated solution of $NaHCO_3$. The liquid is boiled for 5 minutes, after which the pH is measured and the solution is filtered through thick filter paper, without clarifying the broth, or else after an albumin clarification. In the latter case, a clear medium is obtained. The filtered broth is poured into test tubes in amounts of 5 to 6 ml and sterilized at 120 C for 20 minutes.

To clarify media with an albumin solution, the medium is cooled to 50 C, and fresh egg albumin, whipped with double its volume of water, is prepared. The medium is then mixed with this foamy liquid. The mixture is boiled gently and stirred constantly for 10 minutes, and then filtered. After filtration, nutrient broth is prepared according to the above instructions.

Beef-peptone agar is prepared from beef-peptone broth. First 15 g of finely cut or shredded agar are added to 1 liter of broth. The medium is heated until the agar dissolves; the pH is adjusted with a saturated solution of $NaHCO_3$. The medium is filtered through gauze combined with a layer of absorbent cotton and poured into test tubes in specific amounts depending on the purpose: 10 ml if the agar is for pour plates; 5 ml if agar slants are to be prepared; and 7 ml for stab inoculation. The test tubes are closed with cotton plugs or plastic caps and sterilized in an autoclave at 120 C for 10 minutes.

For agar slants, immediately after sterilization the test tubes are arranged on a table in an inclined position so that the surface of the solution is slanted (see Fig. 31). Test tubes designated for stab inoculation are placed in a rack so that the surface remains level.

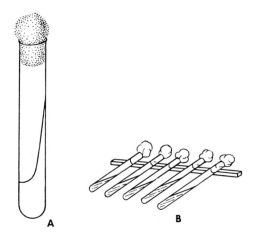

Fig. 31. Agar slants. *A,* Agar slant. *B,* Position of test tubes, with melted agar, for preparing agar slants.

Fish-peptone agar may be used in place of beef-peptone agar. When preparing fish-peptone agar, muscle of any type of fish, except very fatty fish, can be used. However, it is first necessary to prepare fish broth. This is done by cleaning and thoroughly washing a fresh fish. The flesh is separated from the bones, cut, and then weighed out in the required amounts. Five hundred grams of fish are added per liter of water, which is then heated in an autoclave at 120 C for 20 minutes. The liquid is then allowed to precipitate. The liquid is carefully decanted from the precipitate and filtered through filter paper. Water is added to the filtrate to 1 liter total volume. Further steps are carried out as with beef-peptone broth.

Beef-peptone gelatin is prepared from beef-peptone broth by adding gelatin to broth to a final concentration of

10–15%. The concentration of gelatin depends on the temperature at which the study will be conducted. A 10% gelatin liquifies at 24 C, a 15% gelatin at 25 C. In warm summer weather a 15% concentration of beef-peptone gelatin is used for fieldwork. Since incubators are used in most laboratories, the concentration of gelatin need not be considered except as a nutrient. Purified gelatin is added to a cold beef-peptone broth, which is allowed to stand for a certain time—so that the gelatin swells—and then is liquified by careful heating, usually by steaming in an autoclave for several minutes. When the gelatin dissolves completely, the pH is adjusted to 7.0–7.2; the solution is then boiled for 5 minutes and finally allowed to cool to 40–50 C. Egg albumin, whipped with a small quantity of water, is poured into the cooled gelatin, which is then shaken well, and rewarmed. The congealing albumin clears the gelatin. Then the medium is filtered, the pH is checked, and the medium is dispensed into test tubes. Beef-peptone gelatin can be sterilized by flowing steam for 3 days for 30 minutes each day or by autoclaving. Gelatin media should be stored in a cool place or in a refrigerator.

Yeast Autolysates, Yeast Extracts, Yeast Water

Yeast Autolysates. Yeast autolysates are widely used in preparing media. There are many methods of preparation. However, in essence, they are all directed towards creating the most favorable conditions possible for cell autolysis, which is effected by proteolytic enzymes within the cells. In the autolysis of yeast, most of the proteins are introduced into solution in the form of peptones and peptides. Autolysis proceeds most efficiently at a temperature of approximately 45 C and at pH 6.2. The following are a few methods of preparation.

1. One liter of water, boiled and cooled to 60 C, is poured over a 1-kg yeast cake. This is blended into an homogeneous mass, heated in a water bath to 49–50 C, and placed in an incubator for 72 hours at 50 C. After incubation, the mixture is autoclaved at 105 C for one-half hour. The liquid is then filtered several times until transparent. This filtrate con-

tains about 0.9% nitrogen. Water (600 ml) is added to the remaining precipitate. The mixture is shaken well and filtered. Both filtrates are then mixed. The average percentage of nitrogen obtained is 0.6–0.8. The liquid is neutralized to pH 6.8, poured into small flasks, and sterilized for 15 minutes at 115 C. If a yeast autolysate with a lower nitrogen content (0.5%) is needed, the liquid is diluted. In this case nitrogen is determined by the Kjeldahl micromethod.

2. Water (500 ml) is poured over a 500-g yeast cake; the two are thoroughly blended, and placed in an incubator for 2 to 3 days at 45 C. The solution is filtered, and the precipitate on the filter is washed with 1 liter of water. The liquid obtained is diluted with water to a total volume of 2 liters, poured into small flasks, and sterilized for 30 minutes at 120 C.

3. Water (300 ml) is added to a 100-g yeast cake; the two are mixed thoroughly, and placed in an incubator for 48 hours at 53–56 C. From time to time the mixture is stirred. When autolysis is complete, 200 ml of water are added and the mixture is filtered. The filtrate is sterilized for 30 minutes at 120 C.

4. *Formula from the Leningrad Institute of Vaccines and Serums.* Tap water (0.5 liter) heated to 45 C is added to a 1-g yeast cake; these are mixed well, placed in an incubator for 1 day at 50 C, and then filtered. In autolysis by this method, 10 g of granulated sugar may be added to the water.

5. A liquid baker's or brewer's yeast cake is washed and diluted with water to a consistency of thick sour cream. The pH is adjusted to 5.6–6.0 with dilute sulfuric acid. The mixture is then placed in an incubator at 48 C for 20 hours. When autolysis is complete, the mixture is boiled for 5 minutes, allowed to settle, and then filtered.

6. A yeast cake (400 g for 1 liter) is placed in an incubator at 48 C and shaken periodically for 48 hours. It is then filtered.

Yeast Extracts.   Several methods of preparation are given.

1. A 100-g yeast cake is crushed in a mortar with 100 ml of water, placed in a glass beaker in a hot water bath, and

heated for 30–60 minutes. It is then filtered three times through filter paper. The transparent, yellowish liquid obtained is sterilized in small flasks.

2. One liter of water is poured over a 75-g yeast cake, boiled for 30 minutes on an asbestos pad, and filtered through filter paper.

3. Water (200 ml) is poured over a 100-g yeast cake and heated in an autoclave. After sterilization, it is allowed to precipitate. The transparent liquid obtained after settling is complete is then collected with a pipette.

4. Distilled water (400 ml) is poured over 100 g of brewer's yeast; the two are blended thoroughly, stirred while being boiled for 10 minutes, and centrifuged. The slightly yellowish, almost transparent liquid obtained is drawn off with a pipette, transferred to flasks, and sterilized for 10 minutes under flowing steam. From this extract it is possible to prepare a 5–10% yeast broth.

5. Distilled water (400 ml) is added to 100 g of brewer's yeast. The pH of the medium is adjusted to 4.1–4.5. It is then boiled for 10 minutes, filtered, poured into flasks and sterilized at 100 C for 10 minutes on three consecutive days or autoclaved.

Yeast Water. There are various ways of preparing yeast water. Two are cited as follows.

1. A 70-g fresh yeast cake or 10 g of dry yeast are boiled for 30 minutes in 1 liter of distilled water and allowed to precipitate in a tall beaker. The liquid is then decanted and filtered. One liter of water is added to the filtrate, boiled for 30 minutes, filtered into sterile flasks, and sterilized by boiling for 20 minutes on three consecutive days or by autoclaving.

2. Brewer's or baker's yeast cake is placed in an incubator at 50 C for autolysis. After 24 hours the autolysate obtained is diluted at a ratio of 6 liters of water per kilogram of yeast and boiled for 30 minutes. The pH is adjusted to 7.0–7.2. The mixture is sterilized at 120 C for 20 minutes, allowed to stand for 24 hours to permit precipitation, and filtered aseptically. The yeast water is used for preparing liquid media or, when agar is added, for solid media.

METHODS OF STERILIZATION

In any microbiological work (namely, inoculation, isolation, transfer, preservation of pure cultures) sterile media, sterile dishes, and sterile instruments are used to avoid contamination. Transfers are performed in special chambers—boxes treated before use with ultraviolet rays (under aseptic conditions) or laminar flow hoods. In the field, portable boxes (Fig. 4) are used. Among the various procedures for sterilization are the following.

Sterilization of Glassware

Dry heat is used to sterilize glassware. This is done by heating the material at 170 C for 2 hours (from the time the proper temperature is reached) in a Pasteur oven or in drying ovens (set at a temperature of 200 C).

Under field conditions, when drying ovens are not available, glassware may be sterilized in an autoclave or in the hot-air oven of a stove (heating is sufficient when paper inserted in the oven turns brown).

Before sterilization, glassware must be carefully washed, dried, and wrapped in paper. Flasks, bottles, tubes, and pipettes are individually wrapped in paper and are protected by the paper until used. Petri dishes can be wrapped in packages of two or three. Before sterilization pipettes are plugged at the wide end with cotton and wrapped in long, narrow strips of paper, starting from the tip end or slipped into disposable paper pipette bags. At the time the sterile pipettes are used, they are picked up by the wide (mouth tip) end and unwrapped. Sterile disposable glass or plastic pipettes are now available as single units or in multiple packets. These are very useful in fieldwork.

Test tubes must be closed with cotton stoppers or with metal or plastic caps (Fig. 32). A protective covering or cap (i.e. two layers of paper fastened at the neck with twine) must cover the stoppers of bottles and small flasks. This applies to fieldwork and is not necessarily required in the laboratory.

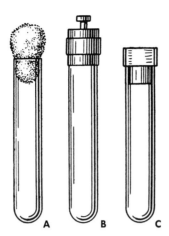

Fig. 32. Stoppers for test tubes. *A,* Cotton; *B,* metal; *C,* plastic.
(A variety of disposable plastic test tube caps are available
from microbiological supply houses in the United States.)

When a drying oven is used for sterilization, it is best to
wrap glassware in thin paper (onionskin or tissue paper); when
an autoclave is used, newspaper or heavy brown wrapping paper
—which will not become sodden—is useful. Bottles closed with
a ground glass stopper should never be sterilized. Instead, the
bottle should be closed with cotton stoppers. The glass stoppers
should be wrapped separately and tied to the vessels, or else a
piece of gauze should be placed between the mouth of the
bottle and the stopper.

Sterilizing Nutrient Media

Nutrient media are sterilized either in an autoclave under pressure or by flowing steam. Sterilization by flowing steam
can be carried out in an autoclave if appropriate adjustments
are made. Flowing steam is used to sterilize media containing
sugar and gelatin. This type of sterilization is conducted over
a 3-day period at the same time each day, so that the period
between each sterilization is always 24 hours. The length of each
sterilization is 30 minutes, timed from the point of vigorous
steaming.

The majority of media are sterilized by saturated steam in pressurized autoclaves. Increasing the pressure in the autoclave increases the temperature of the steam formed inside the autoclave and, consequently, increases the temperature of sterilization:

Manometer readings (atm)	Temperature of steam (C)	Manometer readings (atm)	Temperature of steam (C)
0.0	100.0	1.0	121.0
0.2	105.0	1.2	124.0
0.4	110.0	1.4	126.0
0.5	112.5	1.5	127.0
0.6	114.5	1.8	130.0
0.8	117.0	2.0	132.0

Various glassware, instruments, and apparatus are sterilized in an autoclave. If during the time of sterilization the paper in which the glassware is wrapped becomes moist, then the package is transferred to a desiccator or oven for drying.

Sterilization by flame incineration is employed for needles and loops used in inoculations and for removing material from cultures to prepare smears. This method is also used to sterilize the mouths of test tubes and flasks, as well as stoppers, by subjecting the specified area of the instruments to a short period of flame-sterilization (in an alcohol lamp or gas burner).

Ultrafiltration is a method of "cold sterilization." It is used to free a substratum from bacteria, to study the exchange products of non-filterable bacteria, and to obtain suspensions of viruses. It is necessary to take into consideration the fact that mycoplasma and submicroscopic forms of bacteria pass through filters (Kalina, 1962; Anderson and Heffernan, 1965). A variety of solid, bacteria-retaining filters composed of diverse non-organic materials, as well as semisolid and membrane filters are used for filter sterilization.[2] To obtain liquids free from bacteria, the instruments and the filters must be sterilized before use in an autoclave at 120 C. Filters, such as

---

[2] A brochure on membrane filtration, including field applications, is available from the Millipore Filter Corporation, Bedford, Massachusetts.

those produced by the Millipore Filter Corporation, can be purchased sterile and ready for use. Sterile, disposable filter apparatus are also available and these often are preferred for field use.

Filter candles (Fig. 33) are prepared from kaolin mixed with other substances, for example, diatomaceous earth—siliceous diatoms, consisting mainly of $Si(OH_4)$. Chamberland filters with pores of varying dimensions (L_1, L_2, L_3, and so on up to L_{13}) have been widely used in the past but are no longer popular since membrane filters are more convenient. L_1 filters are used for preliminary filtration. L_5 and L_7 candles are used to free liquids from bacteria. The Leningrad Ceramic Institute produces similar filters. Berkefeld filters contain a more significant percentage of diatomaceous earth and are produced as coarsely porous (V), normally porous (N), and finely porous (W). Only the W type filters retain bacteria. The pore size of the various filters is given in Table 1.

Semisolid asbestos filters resemble compact discs, 3–5 mm thick. They are enclosed in special filtering instruments (Fig. 34).

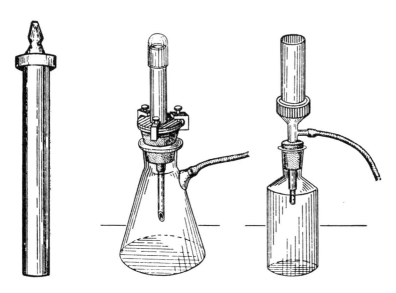

Fig. 33. Filter candle.

Fig. 34. Instrument for filtration.

Table 1. Pore dimensions of various filters
(according to Kalina, 1962)

Filters of the Leningrad Ceramic Institute	Corresponding Make of Chamberland Filters	Pore Dimensions (μ)	Berkefeld Filters	Pore Dimensions (μ)
F_1 (L_1)	L_1	4.5–7.0	V	5
F_2 (L_2)	L_2	2.5–4.5	N	3.5–5
F_3 (L_3)	L_3	1.9–2.5	W	< 3.5
F_5 (L_5)	L_5 (F)	1.3–1.9		
F_7 (L_7)	L_7 (B)	0.9–1.3		
F_{11} (L_{11})	L_{11}	Approx. 0.8		

Three types of Seitz asbestos filters are available: EKS, EKS_1, EKS_2. The EKS_1 and EKS_2 filters retain bacteria but allow viruses and filterable forms to pass through. The Russian-produced F brand asbestos filters retain suspended particles and large bacteria; the SF brand filters block bacterial passage.

At the present time, soft filters are prepared mainly from solutions of nitrocellulose. They are much more widely used than either solid or asbestos filters. Termed membrane filters, they have found numerous applications in the direct counting of aquatic and sediment bacteria and in sanitary-bacteriological analysis of water. In Russia, a factory in Mytishchi produces membranous filters with pores of varying dimensions (see Table 2). In the United States, membrane filters of a variety of pore sizes are available. (See Plate 2.)

Besides the five filter numbers given in Table 2, there is still another, special kind of filter—the "preliminary" filters or pre-filters (no. 6), which can be used to free water of sus-

Table 2. Characteristics of membrane filters

Filter No.	Rate of Filtration for 500 ml of Water		Mean Diameter of Pores (μ)
	Mean Rate (sec)	Range (sec)	
1	540	360–720	0.35
2	270	180–360	0.50
3	130	90–180	0.70
4	70	45–90	0.90
5	25	20–45	1.20

pended particles. These, however, allow most bacteria to pass through. When observed in light, these filters have a very noticeable porous structure (pore diameters range from 3–5 μ).

Special instruments are used for filtration through membrane filters. The filtration is usually performed under positive or negative pressure. Negative pressure is employed when membrane filters are used. When solid filters are being used in the filtration process, either positive or negative pressure can be applied (Fig. 35).

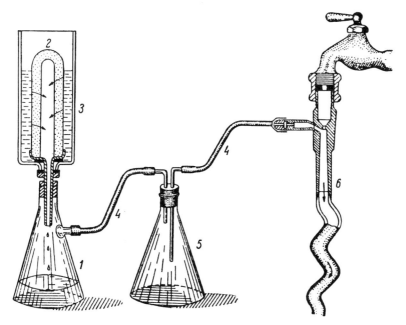

Fig. 35. Diagram of filtration apparatus. *1,* Vessel into which the filtrate enters; *2,* candle; *3,* liquid being filtered; *4,* rubber tube; *5,* intermediate flask; *6,* aspirator.

INOCULATION PROCEDURE

The bacteriological loop is a universal instrument in microbiology, used both for inoculations and for transferring material from a culture for smears. The loop is made from platinum or nichrome wire 0.5–0.6 mm in diameter. Platinum needles are used in stab inoculations; Drigalski spatulas can be used for streak inoculation in petri dishes. When inoculating liquid material, Pasteur pipettes (25–28 cm in length)—which are prepared from soft glass tubing 3–4 mm in diameter—are used in addition to the loop. (Disposable glass Pasteur pipettes are readily available.) In quantitative analyses, when it is necessary to transfer precisely determined volumes of liquid for preparing dilutions, pipettes graduated in 0.1-ml units are used. When preparing inoculations with water and sediment dilutions, a large number of sterile graduated pipettes is needed. Therefore, it is necessary to have in the laboratory a sufficient quantity of pipettes in various sizes: 100, 10, 5, 2, and 1 ml (sterile, disposable glass or plastic pipettes can be purchased ready to use).

Inoculation into liquid media is performed either with pipettes, by immersing the tip 1 mm below the surface of the medium, or with a loop, by touching—with the loop—the wall of the test tube at the edge of the liquid and slightly rubbing the loop against the test tube wall so as to facilitate the transfer of the inoculum into the nutrient medium.

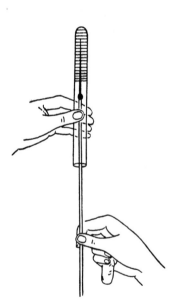

Fig. 36. Stab inoculation.

Inoculations of solid nutrient media are performed on the surface of slanted agar or by stab puncture into a column of agar (Fig. 36).

There should be a small amount of condensed liquid on the bottom of an agar slant. Dried out agar media which are kept for a long period of time are unsuitable for experiments. The inoculation is carried out by tracing, with the loop containing the inoculum, a wide straight line or a zigzag line from the condensed water to the top of the slant (Fig. 37).

There are two methods for inoculating solid media in petri dishes: deep inoculation or surface inoculations of the medium. In the first method, either the water sample being analyzed, a sediment dilution, or a culture dilution is placed in a sterile petri dish and covered with melted agar cooled to 40–45 C. The agar is poured out of the test tube (immediately after the rim of the test tube has been flame-sterilized in an alcohol burner) into the dish directly on the inoculum (Fig. 38). When the inoculum is being placed in the dish and when the agar is added, the lid of the petri dish should be raised as little as possible.

Immediately after the agar is poured in, it is mixed well with the inoculum by gently rotating the dish on the surface of the table. The agar must be distributed over the

Fig. 38. Pouring agar into a petri dish.

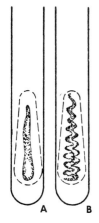

Fig. 37. Inoculation lines on an agar slant. A, Straight; B, zigzag.

bottom of the dish in a thin, uniform layer without air bubbles and must cover the entire bottom. Care must be taken so that no agar falls on the edge of the dish, since this would make it unsuitable for use. When the agar in the dishes hardens, the petri dishes are turned upside down and placed in an incubator (the dishes must be inverted because the agar yields condensing water which will then settle on the lid of the dish, provided that it is thus inverted).

Fig. 39. Inoculation of the surface of an agar medium with a loop.

When inoculating the surface of a solid medium, the nutrient agar is poured beforehand into a petri dish and allowed to solidify. The solid agar is partially dried in a desiccator at approximately 40 C, usually overnight. A specific quantity of inoculum is then deposited on the surface of the partially dried agar with a sterile graduated pipette. This material is smeared over the agar and then spread evenly with a sterile glass loop or bent glass rod. While this is being done, the petri dish should be rotated left and right but only opened as little as is necessary (Fig. 39). Small revolving tables of a diameter of the standard petri dish are available commercially and are useful when a large number of petri plates have to be inoculated. When inoculations are made on a solid medium, sediment is deposited in small clumps, whereas water is applied in drops; both are spread uniformly and firmly over the entire surface, without rubbing.

GROWTH OF MICROORGANISMS

Microorganisms are incubated in an incubator. This apparatus should have a constant temperature throughout. A temperature control device is used to regulate the temperature.

Although there are many systems of thermoregulators, any one used must automatically shut off the electric current when the temperature rises above or below the set mark, and switch it on when appropriate. A temperature variation of ±0.5 C should be the maximum permissible. However, many of the commercially available incubators control only to ±1.0 C. The temperature variation at the top, bottom, and sides of a given incubator should be determined, particularly in studies of fecal bacterial contaminants of water and sediments.

Culture of Aerobes

Those organisms which require atmospheric oxygen for normal development are termed aerobes. Obligately aerobic microorganisms usually grow as films on the surface of liquid and solid media. However, according to recent observations, for many types of aerobes a great deal of aeration is not most conducive to development (Rabotnova, 1957). Current data indicate that aerobes can exist in extremely varied partial oxygen pressures and that optimal conditions for them are determined by the composition of the medium. A polarograph is used in special situations to follow oxygen concentration.

The most favorable conditions for obligate aerobes are generally produced in a medium provided with aeration. Therefore, only a thin film of the media to be used for incubating aerobes is poured into small conical flasks, Winogradsky flasks, or other vessels for maximal aeration. Aerobes grow well on the surface of solid nutrient media; consequently, methods of agar slant inoculation, as well as culturing on the surface of gel plates, agar, and other solid media in petri dishes, are widely used in culturing aerobic bacteria.

When special investigations are being conducted, aerobes are grown in liquid media placed in special shaking devices. These shakers enable the medium to be spread as a thin film over the walls of the flask. The shaking method of incubation is very widely used in microbiology.

In another frequently used method of aeration air bubbles are passed through a liquid medium. The air is either delivered by a compressor or drawn through by an aspirator.

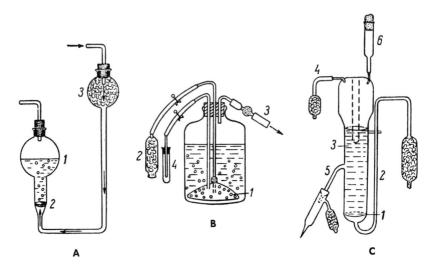

Fig. 40. Various kinds of aerators. *A,* Simple aerator: *1,* culture liquid; *2,* porous plate; *3,* cotton filter. *B,* Aerator permitting sampling: *1,* air diffuser; *2,* cotton filter; *3,* tube joined with a vacuum pump; *4,* tube for sampling. *C,* Liubimov aerator: *1,* porous plate; *2,* air supply tube; *3,* culture medium; *4,* tube for removing used air; *5,* sampling tube; *6,* inoculation tube.

The air is passed through a sterile cotton filter before it enters the medium. The smaller the air bubbles which enter the medium, the better are the conditions of aeration therein. During this time the microorganisms in the entire mass of the medium exist under identical environmental conditions; the cells are constantly being bathed with an aerated nutrient solution. Small air bubbles are produced by passing the air through plates with small pores. This method of culture has been termed subsurface culture. It is widely used in situations in which the intensity of aeration reaches significant proportions. A whole series of laboratory aerators (Fig. 40) and industrial fermentors have been designed.[3] When the spray type of aeration is used, the coefficient of useful action of the air blast is small. Moreover, some media foam or froth, thus hindering air

---

[3] Laboratory instrument manufacturers have available a wide variety of shakers and fermentors. One such firm is New Brunswick Scientific Company, New Brunswick, New Jersey.

diffusion. If the bubble size is increased or antifoaming agents (soybean, castor, and other oils, surfactants, higher alcohols) are added to the media, foaming may be avoided.

An efficient method of aeration is that of vortex aeration: the use of rapidly rotating agitator mixers. The entire liquid in the vessel passes through the vortex zone in the course of a definite period of time and thus becomes aerated. The degree of aeration in these apparatus is regulated by the speed of rotation of the impeller. However, this type of apparatus has a disadvantage: the agitator mixers must be inserted through the neck of the vessel, and this creates difficulty in maintaining sterile culture conditions.

Continuous Culture Methods

At the present time use of the method of continuous culture is constantly increasing. In this method bacterial growth takes place in a medium which is continuously replenished, that is, a medium with conditions as close as possible to the optimum for growth or previously selected conditions for an experiment. In subsurface or standing culture the medium is not renewed or replaced. Hence the chemical composition changes as the microorganisms metabolize the nutrient. Exchange products accumulate, leading to aging of the culture and to death of a portion of the cells. A circulating or continuously replenished substrate presents a series of advantages. The culture receives a continuous flow of fresh nutrient, as the exchange products are removed, so that the culture can be retained a long time without change in physiological properties. The studies of Malek (1956) indicate that, in comparison with a non-continuous culture, a hundredfold increase in the microbial population can be obtained in a continuous culture. Malek and others have devised a variety of continuous culture apparatus (Fig. 41). One special characteristic of these culture apparatus is that a given amount of bacterial cells are carried away with the used medium as a result of the one-way flow of the nutrient. Continuous culture or fermentor units are available commercially in a selection of culture vessel capacities (7-liter, 14-liter, 28-liter).

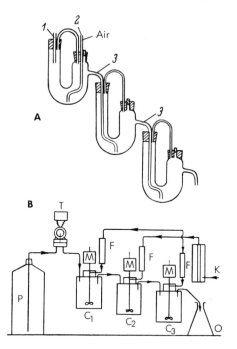

Fig. 41. Continuous culture apparatus (Malek, 1956). *A,* Three-staged apparatus: *1* and *3,* tubes through which the nutrient medium enters; *2,* aeration tube. *B,* Apparatus with vortex aeration: C_1, C_2, C_3, cultivation vessels; *M,* agitators; *P,* vessel containing the nutrient medium; *T,* motor; *K,* compressor forcing air through the filters (*F*); *O,* overflow vessel.

A variety of instruments for culturing microorganisms in cellophane or collodion bags have been designed (Perfilyev and Gabe, 1961). These are bathed on the outside by a stream of nutrient solution. Bacterial cells cannot pass through the semipermeable membranes, but the nutrient substances can and do enter and exchange products are carried out.

Culture of Anaerobes

The enumeration, isolation, and culture of anaerobes is achieved by special methods directed at creating conditions which guarantee the microorganisms both isolation from air and elimination of oxygen from the immediate environment and the nutrient medium.

The nutrient media employed for enumeration and culture of anaerobes are usually the same media as used for aerobic bacteria, with the only difference in the growth conditions. For example, when determining the number of protein-degrading anaerobes in a body of water, the same nutrient media and pour plate method are used as with aerobes. One condition which is absolutely necessary for growth of anaerobes is a low oxidation-reduction potential of the medium. The initial potential (upon inoculation) is especially important, since it determines the initiation of growth. Later the anaerobes, while reproducing, reduce the oxidation-reduction potential owing to the production of reducing compounds (perhaps containing the sulfhydryl group). If the inoculation is performed with a large quantity of cells, reducing substances are included in an amount sufficient to lower the oxidation-reduction potential to the required level. Therefore, the size of the inoculum is very important in anaerobic culture.

Conditions favorable for the growth of anaerobes can be created in various ways. The following are a few of these.

1. The medium may be boiled to remove dissolved oxygen. Anaerobes develop rapidly in a medium which comes in contact with air only when the medium is either freshly sterilized or boiled beforehand. Two essential factors are the rapid cooling of the medium—so that it does not become saturated with air—and prompt inoculation.

2. The medium may be covered with a layer of sterile liquid mineral or paraffin oil. After the inoculation is performed in a preheated medium, a layer of sterile oil is poured aseptically over the medium.

3. A more viscous medium may be created by the addition of agar (0.1–0.2%). In such a medium oxygen diffusion is hampered, and not far below the surface a rather low oxidation-reduction potential is established. After sterilization, when contact is made with the air the oxidation-reduction potential at the surface is approximately 20 and gradually decreases to 7 (Fig. 42). The optimal oxidation-reduction potential, O/R, or Eh, zone for anaerobes lies between 0 and 12 (0 to 14 according to other data). In practice the addition of 0.2% agar permits growth of strict anaerobes at the bottom of a test tube.

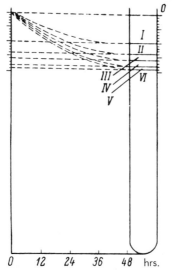

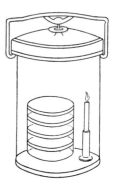

Fig. 42. Oxidation-reduction rH₂ (Eh) distribution in a column of a nutrient medium (Lambin and German, 1961). 0, Surface of the medium; I, level of the medium with $rH_2 = 20$; II, $rH_2 = 14$; III, $rH_2 = 12.5$; IV, $rH_2 = 9.2$; V, $rH_2 = 8.0$; VI, $rH_2 = 7.4$.

Fig. 43. Simple device for anaerobic culture.

4. Reducing substances may be added to the medium. Addition of reducing agents yields very good results. These agents include glucose (1–2%), sodium formate (0.3–0.5%), ascorbic acid (0.1–0.2%), cysteine, and sodium thioglycolate (0.1%). Glucose is a strong reducing agent. Arabinose, lactose, galactose, mannose, and xylose (but not saccharose) possess these same reducing properties. Complex organic media contain reducing agents, for example, peptone media. Addition of ascorbic acid to peptone media permits growth of strict anaerobes in air.

Various instruments (anaerobe jars, containers, and desiccators) are used to prevent penetration of air and to secure conditions of anaerobiosis.

When working in a laboratory, anaerobic culture may be obtained with more sophisticated methods by using instruments which provide fully anaerobic conditions. In field conditions it is often necessary to limit oneself to rather simple procedures. One such method of obtaining conditions which guarantee negligible oxygen content in the area surrounding the inocula is to place the inocula in a container along with a lighted candle (Fig. 43). The container is tightly sealed and the candle burns until combustion uses all the oxygen, and the candle goes out.

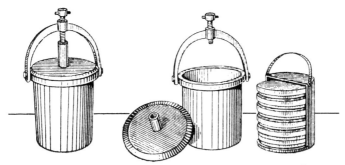

Fig. 44. Aristovskij apparatus for culturing anaerobes.

An apparatus (Fig. 44) designed to provide conditions of anaerobiosis consists of a metal cylinder with an internal diameter corresponding to the diameter of a petri dish. The height of the apparatus varies. A small metal bucket, one-half the vertical wall of which is cut away, is placed snugly inside the cylinder. The inoculated petri dishes are placed one by one in the small bucket, which has a small bow fastened to its upper edge. The cylinder is provided with a metal cover which is screwed to the instrument much like the lid of an autoclave: a rubber pad, situated in a special depression at the upper edge of the cylinder, is placed between the lid and the upper edge of the cylinder. On the bottom surface of the metal lid there are two parallel circular flanges which are forced into the rubber pad when the lid is screwed on.

To establish anaerobiosis when the apparatus is fully loaded, the top and bottom petri dishes are filled with a dry mixture of a chemical oxygen absorber in the form of sodium carbonate and sodium hydrosulfite ($Na_2S_2O_4 \cdot 2H_2O$). The hydrosulfite must be kept in an air-tight package. A layer of the sodium carbonate (about 20 g) is poured on the bottom of the dishes and a uniform layer of the sodium hydrosulfite

(1.5 g) is added. Pyrogallic acid may be substituted for the hydrosulfite. The powder is mixed and wetted with water from an atomizer. The small bucket containing the inocula and the petri dishes with the mixture are placed in the cylinder, which is then hermetically sealed and placed in an incubator. The apparatus is taken out of the incubator at the necessary times (for counting colonies, isolations, and observations).

The hermetic state and degree of oxygen removal of the Aristovskij apparatus can be checked by testing with methylene blue. There are three such test solutions:

1. 6 ml of 0.1 N NaOH diluted with distilled water to 100 ml;
2. 3 ml of a 0.5% water solution of methylene blue diluted with distilled water to 100 ml; and
3. 6 g of glucose dissolved in 100 ml of distilled water, to which a small crystal of phenol is added.

Whenever an indicator solution is needed, the solutions are mixed in a test tube in the proportion 1:1:1. Then, either a strip of gauze is saturated with the mixture, or the test tube containing the mixture is placed in a boiling water bath until the color disappears. The test tube is placed in the anaerobic container before the anaerobic conditions are established. When the container is free of oxygen, the tinted gauze loses its color, but the color of the solution in the test tube does not change. If the gauze remains colored and the solution in the test tube turns blue, this indicates incomplete oxygen removal. It is possible that the color will appear but then disappear when anaerobic conditions develop (the color disappears when pO_2 decreases to 0.05 atm). This test is recommended for all instruments used for anaerobic culture.

An anaerobe apparatus also provides complete anaerobiosis (Fig. 45). It consists of a container (1), frame (2), a netted form (3), and two asbestos discs (4, 5). The container is filled with no more than 2 cm of water. Petri dishes or test tubes are placed in the frame (2); over them are placed a few layers of filter paper (6) and on top an asbestos disc (4). A porcelain dish is placed on the asbestos, and a small piece of yellow phosphorus is placed in the dish. The phosphorus is kept under water and is handled only with forceps. The netted

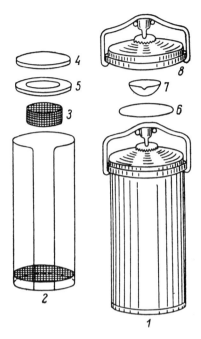

Fig. 45. An anaerobe apparatus (see text for explanation).

form (*3*) is turned over above the dish (*7*), an asbestos disc is laid on the form, and the entire apparatus is quickly closed with the lid (*8*). The phosphorus ignites very rapidly and burns until the oxygen is used up. The water on the bottom of the vessel absorbs the P_2O_5 which is formed. This method of oxygen removal can also be used when no metal container is available, in which case a heat-resistant glass cylinder with a lid can be used instead. This method assures rapid attainment of anaerobiosis; however, every precautionary measure must be taken when working with phosphorus as it is a very dangerous chemical to handle.

In the laboratory it is perhaps most convenient to work with anaerobe jars (Fig. 46). After loading the anaerobe jar the lid is screwed down; one of the stopcocks is connected to a vacuum pump and the other is closed. The extent of evacuation of the air is measured with a manometer. At present there are available special anaerobic incubators in which the temperature is kept constant and a vacuum is maintained.

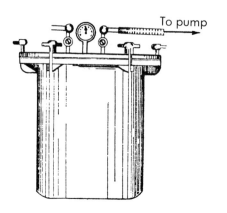

Fig. 46. Anaerobe jar.

Fig. 47. Vacuum-des-
iccator for anaerobes.

A vacuum jar may also be used for growth of anaerobes (Fig. 47).

Anaerobes are more often grown in an atmosphere of inert gas (nitrogen, argon, helium). In this case, both stopcocks of the anaerostat are left open. One of them is connected to a gas cylinder containing an inert gas. When the stopcock connected to the cylinder is opened, the gas enters the anaerobe jar and displaces the air. This usually takes 20–30 minutes, after which the open stopcock is closed; the other stopcock connecting the anaerobe jar with the cylinder is then

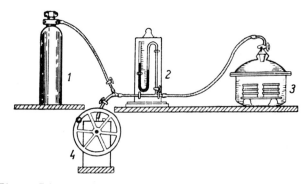

Fig. 48. Diagram of apparatus used to replace air with carbon dioxide. *1*, CO_2 cylinder; *2*, mercury manometer; *3*, vacuum jar with the culture; *4*, vacuum pump.

immediately closed and the gas cylinder is disconnected (Fig. 48).

Removing oxygen can be accomplished with biological methods, most often by culturing aerobes and anaerobes together. *Staphylococcus aureus, Serratia marcescens,* and *Sacchromyces cerevisiae* are usually the aerobes used for this purpose.

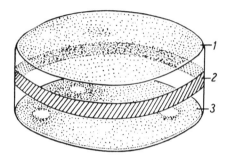

Fig. 49. Combined culture of aerobes and anaerobes. *1,* Aerobic culture; *2,* strip of tape; *3,* anaerobic culture.

Pure cultures of aerobes are first incubated. The procedure for joint culture is as follows. Two petri dishes with the same size bottoms are selected and sterilized. Nutrient agar is poured on the bottom of one of the dishes (Fig. 49, dish 1). After the agar solidifies, it is inoculated with a dense streaking of an anaerobe culture. The medium for the anaerobes is poured in the other dish (dish 3) and inoculated. After it solidifies the lids to both dishes are removed and the bottoms are sealed together at their outer rims with a narrow strip of tape (Fig. 49).

This method is very simple and can be used in almost any circumstance. Its drawback lies in the possibility that anaerobiosis may not be complete enough to avoid killing the anaerobic vegetative cells in the inoculum.

Anaerobes may be enumerated under field conditions by the following technique: The petri dishes are assembled before sterilization, not in the normal way, but rather by placing the bottom in the lid. They are then wrapped in paper and sterilized in the usual manner. Paraffin is sterilized separately in small vessels or test tubes. After adding reducing agents, to

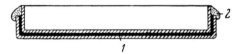

Fig. 50. Culturing anaerobes by Sturm method. 1, Nutrient medium; 2, paraffin.

the same Eh as in the water and sediment, to the medium, the anaerobic and aerobic inoculations are carried out in parallel. The inoculations for enumerating anaerobes are made in the lids of the petri dishes, after which the nutrient agar is poured in. The bottoms of the dishes are then used as lids and are quickly placed in the original lids (Fig. 50). After sterile paraffin is poured around the edges of the dishes, they are placed in an incubator. The agar is poured into the lid of the dish in such a way that the bottom (used as a lid) will lie tightly in the agar; the medium should not, however, be displaced on the edges of the dish. When using a petri dish of a certain diameter, it is possible to determine, before culturing is done, how much agar is needed and to pour this amount into test tubes beforehand. The colonies which develop are counted in the usual manner.

Tubes may also be used for anaerobic culture. Here diluted inoculum is added to an agar medium which has been freshly boiled and rapidly cooled to 40 C. The inoculum and agar medium is mixed thoroughly and poured into a sterile tube 20 cm long and 0.3–0.5 cm in diameter. The ends of the tube may either be closed with rubber stoppers or sealed. The anaerobes which develop are subcultured either by pushing out the column of agar, or by cutting the tube at the area of growth.

Media containing reducing substances are also used in pour plates. Very often 0.2% sodium thioglycolate or 0.1 ml of an 8% solution of ascorbic acid in a 10% sodium carbonate solution and 0.002% methylene blue are added as reducing agents to a nutrient agar. Twenty-five milliliters of the sterile medium are poured into dishes 10 cm in diameter and 40 ml into dishes 15 cm in diameter. After the medium solidifies, the center of the dish is streaked and the lid is replaced by the bottom of another dish, this second bottom having already been matched before sterilization so that when it is placed on the agar, the latter is displaced around the edge in a narrow ring. In well prepared dishes the agar in the center remains colorless; that along the edges becomes sky blue as a result of the oxidation of the methylene blue. Special lids for culturing anaerobes by this method have been designed. One of these is shown in Fig. 51.

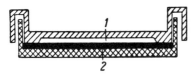

Fig. 51. Brewer cover. *1,* Air; *2,* Nutrient medium.

The disadvantages of this rapid and simple technique are: (1) pellicles may form as a result of the surface moisture; (2) growth of certain species of bacteria may be inhibited by the reducing substances.

This method may be used in another way. The medium is poured onto the bottom of the petri dish and inoculated. The medium and inoculum are thoroughly mixed and allowed to solidify. Next a layer of agar is added, into which is placed a sterile glass plate of the same diameter as the petri dish. After this layer of agar solidifies, a layer of water agar, containing 0.1% sodium thioglycolate, is added.

The following method also yields good results. An anaerobic culture is streak inoculated onto the surface of a nutrient medium in a petri dish. A cover (made from filter paper 7×4 cm) containing a freshly prepared mixture of 3 parts pyrogallic acid, 3 parts K_2CO_3, and 15 parts diatomaceous earth, is attached to the cover of the petri dish. Paraffin, melted at 54–57 C, is used as an adherent.

Matthews (1961) has developed a very simple method of culturing anaerobes. Inoculated plates, together with an indicator, are placed in a cellophane bag. A fine-pored disc from a Seitz apparatus is saturated with a 60% aqueous pyrogallol solution, after which 1 ml of a 50% aqueous solution of Na_2CO_3 is added. This disc is placed in the cellophane bag. The bag is sealed and incubated.

Special instruments have been designed (Rabotnova, 1957) for measuring the oxidation-reduction potential (Eh) during culture of anaerobes (Fig. 170).

Isolating Anaerobes from Water

If the purpose of an investigation is not a quantitative enumeration of anaerobes but a description of the species of anaerobes in a water mass, the best results are obtained with inoculations carried out with liquid media containing 0.1–0.2% agar. Dual inoculations may also be performed: the first with unheated material, the second after heating the sample (1–2 minutes in a boiling water bath or 20 minutes at 80 C).

Obtaining pure anaerobic cultures from an enrichment culture may be complicated by the presence of aerobes. Isolation is most easily accomplished from heated cultures, in which spores of the anaerobes are present. For this purpose old cultures in media containing certain carbohydrates (see above) appear to be the most suitable.

This procedure is not useful if the culture contains sporogenous aerobes because the spores of aerobic microorganisms also have high heat-resistance. In this case it is recommended that a solution of crystal violet be added to the medium in a final dye concentration of approximately 1:100,000 (anaerobes are significantly more resistant to this dye than aerobes). The dye should be added after the medium is sterilized.

A very useful development for field culture of anaerobes is the disposable Gas-pac manufactured by Bio-Tech Laboratories, Division of Becton, Dickinson Corporation, Baltimore, Maryland. With this unit, field studies of anaerobes can be made without such equipment as vacuum pumps and desiccator jars. The cardboard carrying case of the unit holds a plastic bag which can be tightly sealed. Into the plastic bag are placed the inoculated tubes and plates, and the methylene blue indicator and oxygen-removing chemicals are exposed in the plastic bag by unwrapping the metal foil containers. The plastic bag is then sealed and the Gas-pac unit is incubated.

It should be noted that, with the exception of the evacuated and substituted atmosphere methods of incubation, the anaerobes which are isolated are most frequently facultative anaerobes. Strict anaerobes are exceedingly difficult to isolate, culture, and purify.

Chapter 4

Isolating Pure Cultures
and Identifying Bacterial Species

In water masses bacteria do not occur in isolation but in distinct biocenoses. When studying microorganisms it is absolutely necessary to obtain pure cultures, that is, populations containing cells of only one species. Determination of the morphological and physiological characteristics of the species is possible only after a pure culture has been obtained.

Identification of all the microbial species in a body of water is very seldom made because of the amount of work involved in such an investigation. Usually the organisms involved as activators of biological processes are identified. Heterotrophic saprophytes are rather seldom defined to the species level, more often only to the generic level.

Difficulties arise in identifying bacterial species because of the great variability of bacteria and their extremely rapid reaction to external changes. Morphological and physiological features are related to conditions of culture such as incubation temperature, composition of the nutrient medium, and pH. Cells develop in the form and dimensions characteristic of a given species only under fixed conditions. Because of this, media of a strictly determined composition are used, the culture is grown at standard temperatures, and the morphology of the microorganisms is studied in cells of a specific age. Often only those microbial species which predominate in a water mass are identified.

METHODS OF ISOLATING BACTERIA

Pure cultures are isolated from colonies which develop when the surface of a solid nutrient medium (beef-peptone or a special agar) is inoculated with the starting material. There are two basic methods of isolating pure cultures: (1) methods based on the principle of mechanical separation of micro-organisms; and (2) biological methods. The first are the most common.

Methods of Mechanical Separation of Microorganisms

These methods are characterized by the separation of a mixture of cells and their groups into isolated cells in a nutrient medium, or on its surface. These cells develop into colonies, from which a transfer inoculation is made to the surface of a solid medium.

A pure culture may be isolated on a solid medium by inoculations into the medium by pour plating or on the surface by streak plating. In either case it is very important first to prepare dilutions of the material being studied. This is done by selecting five or six previously prepared test tubes containing sterile tap water or water from the water mass under investigation. A small amount of the inoculum is placed in the first test tube and then mixed well with the sterile water by rotating the tube between the palms of the hands. One or two loops are then transferred from the first to the second test tube. After mixing, loopfuls are placed in the third tube, and so on (Fig. 52). Pipettes calibrated to deliver volumes of liquid can also be used in preparing dilutions.

From each dilution, starting with the second or third, one to two loops are taken and placed in test tubes containing beef-peptone agar which has been melted and cooled to 40 C. The deposited material is thoroughly mixed with the agar and poured into a petri dish. Dilutions can be made directly using melted and cooled agar (in the test tubes). In this case the material under study is placed in liquid agar and thoroughly

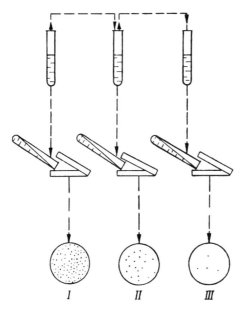

Fig. 52. Diagram of dilution procedure. *I–III,* Inoculations from subsequent dilutions.

mixed. One or two loopfuls are then transferred from this tube into the next one, and so forth. The contents of each test tube are poured into a petri dish. Labels should be placed on the petri dishes before the inoculations, usually by marking the petri dish bottom with a grease pencil or glass and plastic marking pen. When the agar in the dishes solidifies the dishes are inverted and placed in an incubator.

Of all the petri dishes inoculated from the various dilutions, those which are suitable for isolating pure cultures are the ones in which growth is not dense, i.e. where separate colonies are located at a significant distance from one another. The first dilutions may contain such a dense growth that the individual colonies may run together and form confluent growth. Consequently, the inoculations onto the agar are often done from tubes containing water, starting from the third and fourth dilutions.

The actual isolation of a pure culture is carried out as follows. The colonies in petri dishes are first observed under a magnifying lens and the colony designated for isolation is marked. The petri dish lid is lifted slightly and, without disturbing the adjacent colonies, a small portion of the designated colony is carefully removed with a sterile loop and transferred to an agar slant. Isolation can also be performed by inoculating the surface of a solid nutrient medium. In this case the nutrient agar is poured beforehand into a sterile petri dish. After the medium solidifies a drop of the inoculum (from the second dilution, for example) is placed on the surface of the medium (from the edge of the dish) with a loop. The inoculum is spread over the entire surface of the medium with a sterile glass spreader. Immediately thereafter, the same spreader—which now has on its surface microbial cells—is used to inoculate a second and third petri dish in the same way. Inoculation can also be done with a loop. In this case, the same loop—containing only one load of the inoculum throughout the procedure—is used to streak inoculate the surfaces of the first, second, and third dishes. The inoculation is performed in a specified way, as shown in Figure 53. It is important to remember that great care must be taken that the petri dish is opened as little as possible. In this method of inoculation all

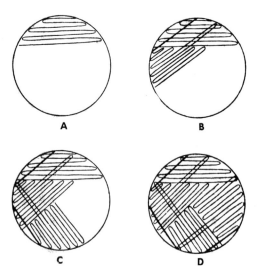

Fig. 53. Diagram of inoculations on agar surfaces. *A*, Beginning of the inoculation; *B–D*, Subsequent steps.

colonies develop on the surface of the medium; there are no deep colonies. Thus it is easier to observe the colonies under a microscope and to isolate them.

Regardless of the method used to isolate a pure culture, purity must always be tested. Dilutions are made of each culture obtained; similarity of the developing colonies is ascertained; and the process of isolation is repeated. Often the steps of purification are repeated three times before further work on identification is done.

Biological Methods of Isolating Cultures

Biological methods of isolating pure cultures are based on the diversity of microbial properties and can only be used in certain situations. Rather frequently they are used to obtain pure cultures of spore-forming bacteria occurring in association with non-spore-forming bacteria. Mixed cultures are kept in this case for a certain time to allow completion of sporogen-

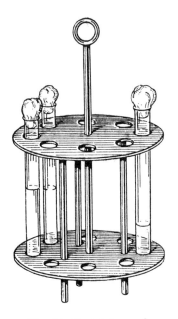

Fig. 54. Test tube rack.

esis, which is confirmed by examination under a microscope. A portion of the material is transferred to a test tube containing sterile water and is thoroughly mixed. With a special holder (Fig. 54) the test tube is placed in a water bath that contains either boiling water, in which it is kept for 2–3 minutes, or water at a temperature of 80 C, in which it is kept for 10 minutes. After the test tube is removed, it is immediately placed in water at room temperature. Inoculations are made to petri dishes containing agar. In this way, colonies in the dishes grow out only from spores; consequently, only the sporogenous species develops. This method is used, for example, for isolating *Clostridium pasteurianum.*

The biological method of isolation can be used to separate anaerobes from aerobes.

Isolating a Culture from a Single Cell

This method of isolation yields the greatest assurance of purity of a culture. For this procedure, special instruments are used—micromanipulators, which permit capture of a single cell using a microscopic pipette under a microscope (Fig. 55). When the isolation is performed, the microorganisms must be in a moist chamber, usually a hanging drop. The cell collected by the micropipette is transferred to a liquid nutrient medium, where it divides. The Perfilyev micromanipulator (Fig. 56) can also be used for this purpose. The isolation of one cell is achieved by drawing the liquid containing the bacteria into a capillary which is exactly parallel-sided. The strict parallelism of the

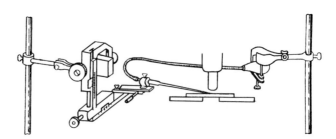

Fig. 55. Diagram of the position of the pipette when isolating one cell with a micromanipulator (Guseva, 1956).

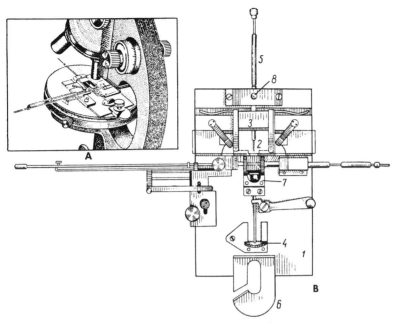

Fig. 56. Perfilyev micromanipulator (Perfilyev and Gabe, 1961). *A*, Position of the micromanipulator on the stage of the microscope. *B*, Micromanipulator apparatus: *1*, micromanipulator stage; *2*, capillary of the collector vessel; *3*, protective chamber; *4*, plunger knob; *5*, external tube of the receiving vessel; *6*, small loop for mounting on the microscope; *7*, P-form spring for securing the cover glass; *8*, spherical inlet of receiving vessel.

capillary walls makes it possible to observe it under great magnification. In the capillary, the area where only a single cell occurs is located and broken off with a special tool. The separated piece is transferred to a receiving vessel and from there into a nutrient medium.

Skerman has devised a new type of micromanipulator and microforge which is simple in design, small in size, and adaptable to use on any microscope. Working with the micromanipulator directly in a petri dish offers great advantages in single-cell isolation, particularly with water and sediment isolates which are difficult to isolate in pure culture by conventional methods.

When isolating large-celled microorganisms—such as

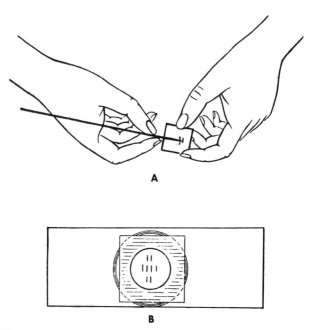

Fig. 57. A, Arrangement of drops for isolation of a pure culture of microorganisms from one cell; B, sealing the specimen.

yeast—the micromanipulator is not always necessary. The yeast
culture is isolated in the usual manner. Then successive dilutions of the culture are prepared in a sterile liquid malt medium
until one cell appears in a small drop. This is determined by
microscopic examination. When the dilution containing a single
cell is obtained, a small, flame-sterilized drawing pen is used
(Fig. 57) to place a series of small drops of the dilution on
each of several sterile cover glasses. A cover glass is then placed,
with the drops downward, on a ringed slide around which sterilized Vaseline is placed with a fine brush. The glasses are
pressed together in such a way that the Vaseline represents a
perfectly smooth surface without air channels. The droplets
are observed under a microscope at low magnification. Those
which contain only one yeast cell are marked with India ink.
The selected specimens are placed in a moist chamber (in a
Koch dish), on the bottom of which moist filter paper has been
placed, and are placed in an incubator. When the colonies
which occupy almost the entire drop develop, they are trans-

ferred to a test tube containing sterile brewer's wort or appropriate medium by means of a platinum needle or a small piece of sterile filter paper. The result is a culture representing the growth from a single cell.

IDENTIFYING BACTERIAL SPECIES

Identification of bacterial species begins with examination of the colony, followed by studies of morphological, cultural (characteristics of growth in diverse media), physiological, and biochemical features of the species, as well as behavior of the microorganism toward certain dyes and various bacteriocidal and bacteriostatic agents. Features used to characterize a given species are given on p. 94.

Only a limited number of features and testing methods are discussed by Rodina. Many more tests can and should be performed. A list of features and testing methods for aerobic, heterotrophic marine bacteria has been prepared by Colwell and Wiebe (1970). Skerman (1967) has provided a guide to the identification of the genera of bacteria and has also prepared a compendium of tests and testing methods used in bacteriology (Skerman, 1969). Rogosa, Krichevsky, and Colwell (1971) list all features used to characterize the bacteria and provide a format suitable for taxonomic data analysis by computer. Numerical taxonomy is particularly useful to aquatic microbiology. A manual on numerical taxonomy is available from the American Society for Microbiology, Bethesda, Maryland.

Colonies in dishes containing a solid nutrient medium are investigated either with the aid of a binocular magnifier or under a microscope with low-power objectives. Both the rate of growth at standard temperature and the dimensions of the colonies are recorded, and the shape and structure are studied. These features are useful in identifying species (Krasil'nikov, 1949). However, the same species of bacteria may have dissimilar colonies at different stages of growth, under different incubation conditions, and even at different inoculation densities.

Colony growth may be rapid or slow. Colonies which

Features used in characterization of bacterial species

History of species:
  Water (or sediment) sample no.
  Biotope from which species was isolated
  Date of isolation

Taxonomic data recorded for species
  1. Colony (appearance, dimensions,
     appearance of colony in outline)
  2. Cell morphology
     Cell dimensions
  3. Cell cytology: granules of metachromatin, fat and glycogen inclusion
  4. Motility
     Distribution of flagella
  5. Presence of spores
  6. Capsule
  7. Gram stain
  8. Growth in broth
  9. Growth on agar
 10. Pigment formation
 11. Growth on gelatin
 12. Growth in milk
 13. Growth on selective media
 14. Oxygen requirements
 15. Temperature requirements
 16. Formation of exchange products:
       Hydrogen sulfide
       Ammonia
       Indole
 17. Reduction of nitrates, nitrites
 18. Decomposition of fat
 19. Decomposition of starch
 20. Carbohydrate utilization
 21. Nitrogen sources
 22. Vitamin, etc., requirements
 23. Other

grow freely (not in close contact with one another) may be large (more than 4–5 mm), medium (2–4 mm), or small (1–2 mm) in dimension. Colonies of some bacteria are very small— no more than 1 mm, i.e. pinpoint colonies. The shape of colonies is, for the most part, characteristic for a species. Colonies can be distinguished as round; round with scalloped edge (irregular); round with a ridge along the edge; branching; filiform; radiating edge; spreading; rhizoid; wrinkled; irregular; concentric; and complex (Fig. 58).

The surface of colonies is also characteristic for various species. It can be smooth, radially striated, convex, wrinkled

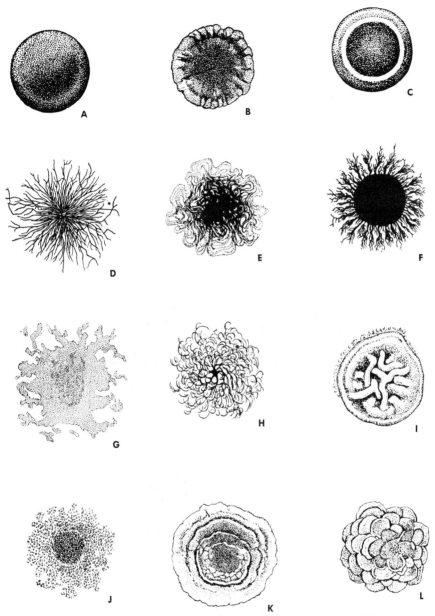

Fig. 58. Shape of bacterial colonies: *A,* Round; *B,* round with a scalloped edge (irregular); *C,* round with a ridge along the edge; *D,* branching; *E,* filiform; *F,* radiating edge; *G,* spreading; *H,* rhizoid; *I,* wrinkled; *J,* irregular; *K,* concentric; *L,* complex.

Fig. 59. Colony profiles. *A*, Flat; *B*, convex; *C*, drop-like; *D*, curved; *E*, hilly; *F*, ingrowing in agar; *G*, crateriform.

(rugose); brilliant (shiny), opaque; butyrous, leathery, and powdery.

Colony profiles for various species are distinguished above the surface of a solid nutrient medium. They are flat, convex, drop-like, low convex, sharply curved, crateriform, hilly, and ingrowing in agar (Fig. 59).

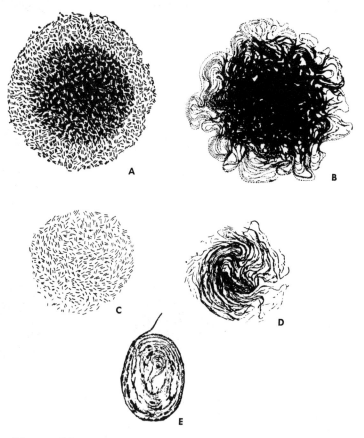

Fig. 60. Colony structures (Clifton, 1950). *A*, Grainy, *B*, friable; *C–E*, thread-like (filiform).

Each species of bacteria has its own characteristic colony structure. The internal structure may be amorphous, fine grained, heavy grained, friable, and filiform, the latter having various forms (Fig. 60). The colony edge is also characteristic. It can be smooth, wavy, scalloped, lobed, irregularly broken, ciliate, thread-like (filiform), wooly (fluffy), dendritic or branching, "hair lock"-like (Fig. 61).

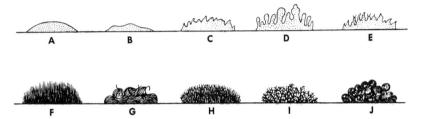

Fig. 61. Colony periphery. *A,* Smooth; *B,* wavy; *C,* scalloped; *D,* lobed; *E,* irregularly broken; *F,* ciliate; *G,* thread-like; *H,* woolly; *I,* dendritic or branching; *J,* "hair lock"-like.

A detailed study of colony formation by cutting cross sections reveals interesting information; for example, rod-shaped bacteria can be arranged vertically in the center of a colony and horizontally on the periphery.

BACTERIAL MORPHOLOGY

Bacterial species are characterized by cell shape, dimensions, and morphology under standard conditions of growth.

The following features are usually considered: (1) cell form (coccus, rod—straight and curved, spiral); (2) type of arrangement of cells (single cells, pairs, chains, filaments, small packets); (3) cell dimensions in microns; (4) shape of the ends of cells (rounded ends, square, tapered, concave); (5) presence of spores, spore shape (round, ellipsoid, oblong), dimensions, location in a cell (central, polar), type of germination (equatorial, transverse, polar); (6) motility and type of flagellation (one flagellum, a cluster, many flagella), arrangement (polar, bipolar, and peritrichous, i.e. over the entire surface of the cell); (7) presence or absence of capsules; (8) cell inclusions (metachromatin, fat inclusion, glycogen inclusions).

Cell Morphology

The best method of studying the shape of bacterial cells
is to examine them in a living state under a microscope. The
true morphology and size of the microbial cell is thus undisturbed. Drying, as has been indicated by a number of authors
(Knaysi, 1951; Timakov and Goldfarb, 1958), causes significant
deformation of bacteria (Fig. 62). When studying shape, cells
may be stained with weak dye solutions (1:500, 1:5000).
Methylene blue and neutral red are perhaps the most suitable.
Sometimes a weak solution of iodine in potassium iodide is
used. When weak dye solutions are used, the form of the living
cells is not altered significantly.

Fig. 62. Deformation of a cell as a result of drying (Knaysi,
1951). B, Bacterium in living state; B_1, in dried state; S, slide.

When preparing specimens, attention must be paid to the
size of the drop from which the specimen is prepared. There
must not be so much liquid that it extends beyond the edges
of the cover glass. To eliminate the pressure of the cover glass
on the cells, before applying the drop, thin fibrils of cotton are
placed on the slide in such a way that the edges of the cover
glass lie on them.

A more exact means of studying cell form is the suspended drop method (discussed below). Microphotography
gives an accurate representation of the cell shape of the culture under study. Polaroid photography, which is available in
the United States, is especially helpful because it enables preparation of permanent records almost instantaneously.

Length and Width of Bacteria

Measurements of bacteria (diameter for spherical forms,
length and width for others) are obtained for cultures of a

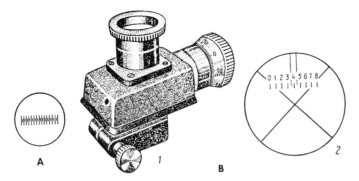

Fig. 63. Ocular micrometers. *A,* Standard; *B,* screw type: *1,* overall view; *2,* plate with cross lines and graduation lines.

specific growth stage (dimensions of bacteria and shape of the cells may change with age). Cultures 18 to 20 hours old are most often used. Measurements are taken with an ocular micrometer (Fig. 63, *A*). The values of the graduation of the micrometer must be determined. For this purpose the scale of the ocular micrometer is calibrated with that of the stage micrometer, a ruled line of which equals 1 mm, and is divided into 100 parts. Consequently, one division of the stage micrometer is 0.01 mm, or 10 μ. The calibrated stage micrometer is then placed on the microscope stage. Under the same magnifications as those used to examine the bacterial cells, the number of divisions of the ocular micrometer equalling one division of the stage micrometer is recorded. For example, if one division of the stage micrometer corresponds to five divisions of the ocular micrometer, then one division of the ocular micrometer at the given magnification and at that specific position under the microscope equals 2 μ. The diameter of a cell, then, which occupies two divisions of the ocular micrometer, equals 4 μ. Once the values of the divisions have been established, only the ocular micrometer need be used thereafter, but always at the same magnification.

A more accurate instrument for measuring bacteria is the screw ocular micrometer (Fig. 63, *B*), which has a dioptric viewing and metering mechanism. A fixed scale with a division value of 1 mm is located in the ocular, and a moveable glass slide with cross lines and a graduated index is used with the ocular. The slide with the grid is bound with an accurate micrometric screw and is moved by turning the screw, so that entire millimeters are counted along the fixed scale, whereas hundreths of a millimeter are counted off along the roller of the micrometer screw. Here also a stage micrometer is used to determine the gauging of the linear magnification of the microscope. Only after this is determined is the ocular micrometer used to measure objects. The Carl Zeiss Company, West Germany, manufactures such a micrometer which is available in the U.S.

The units of measurement for microorganisms are:

micron (μ) = 1/1,000 mm = 10^{-6} m;
millimicron (mμ) = 1/1,000,000 mm = 10^{-9} m;
Angstrom (A) = 1/10,000,000 mm = 10^{-10} m.

Methods of Fixing Bacteria and Staining Fixed Bacteria

There are different methods of staining fixed bacteria for various purposes (i.e. examination of sporogenesis; determination of the presence of capsules and flagella; and distribution of metachromatin and other cell inclusions).

Methods of Fixing Bacteria. Heat fixation is most often used in bacterial investigations (the specimen is passed, smear up, through a flame three times, for not more than 2 seconds each time). However, data concerning the disadvantage of this method are being accumulated. The cells shrink, the flagella are damaged (separated from the cells), and the cytoplasm is damaged. For these reasons, chemical fixatives are used in preference to heat fixation.

Clean, specially treated slides are used to prepare the specimen. Slides are treated in one of the following ways:

1. The slides are boiled in a 1% solution of sodium carbonate, washed with water, then with a weak solution of hydrochloric acid, and finally again with water.

2. The slides are submerged for several days in dichromate solution, rinsed with water, and then with a weak solution of sodium hydroxide, and again with water. The slides are kept

either in 95% alcohol in beakers with ground glass stoppers, or dry in beakers or Koch dishes.

The cleanliness of the slides can be checked by applying a drop of water: it should spread out uniformly.

When preparing a smear, first a drop of sterile tap water is placed in the center of the slide, and then, using a sterile loop, a small portion of the culture being studied is introduced. The culture material should be distributed uniformly and in a rather thin layer.

A thin layer of egg albumin and glycerin mixture (1:1) is sometimes placed on the slide for better fixation of the bacteria on the glass. The prepared specimens are dried and fixed. There are several methods of fixing:

1. *Fixing with ethyl alcohol.* One to two drops of absolute alcohol are placed on the specimen and allowed to evaporate. A solution of 95% alcohol can also be used, and, in this case, fixing should be for 15–20 minutes.

2. *Fixing with Nikoforov mixture.* One to two drops of ethyl alcohol and ether (1:1) are placed on the specimen and allowed to evaporate.

3. *Fixing with formalin vapors.* The specimen is placed in a Koch dish on a support. A few small pieces of cotton, wetted with strong, undiluted formalin, are placed on the bottom of the dish. The dish is closed and the bottom is gently heated. A few minutes are required for fixing.

4. *Fixing with osmic acid vapors.* The specimen is placed on a support in a wide-necked beaker with a ground glass stopper. A 1–2% solution of osmic acid is poured onto the bottom of the beaker, and the specimen remains in the fixative vapors for 5–10 minutes. This should be done in a chemical hood.

5. *Fixing with Carnoy liquid.* The specimen is submerged in the fixative—glacial acetic acid (10 ml), 96% ethyl alcohol (60 ml), chloroform (30 ml)—for 15 minutes.

Staining Fixed Bacteria.   A variety of aniline dyes is used in staining fixed bacteria. They are divided into two groups— basic and acidic:

Basic Dyes		
Reds	*Violets*	*Blues*
Neutral red	Gentian violet	Methylene blue
Safranin [1]	Crystal violet	Victoria blue
Fuchsin	Methyl violet	Toluidine blue
Pyronine	Hematoxylin	
	Thionine	
Greens		*Browns*
Malachite green		Chrysoidine
Methylene green		Vesuvin
Janus green		
	Black	
	Induline	

Acidic Dyes		
Reds and Roses		*Yellows*
Erythrosin		Aurantia
Eosin		Picric acid
Fuchsin		Fluorescin
Congo red		
	Black	
	Nigrosin	

A very simple way to determine whether a dye is acidic or basic is by use of water solutions of the dye and strips of filter paper. Filter paper has a negative charge, and when the strips are dipped into the dye solution, the dye will spread over the paper above the level of the solution only if it possesses a negative charge (i.e. the same charge as that of the filter paper), when the dye is acidic. When a basic dye is present, only water rises above the level of the liquid on the paper. The molecules of the dye will occur as static, oppositely charged particles on the paper in that portion of the paper which is dipped into the dye solution.

Saturated alcohol solutions of the dyes are usually prepared first. These solutions do not change their properties when kept for a long time. When the actual staining is done, dilute alcohol solutions are prepared. Phenol solutions are often used for preparing dyes.

The following dyes are those most often used in microbiology.

---

[1] There are many makes of safranins. The best is safranin 0.

Carbol Crystal Violet

Two solutions are prepared:

1.	Crystal violet (90% dye content)	0.4 g
	Ethyl alcohol (95%)	10 ml
2.	Phenol	1 g
	Distilled water	100 ml

The solutions (1) and (2) are mixed when used.

Ehrlich Anilin Gentian Violet

Two solutions are prepared:

1.	Crystal violet (90% dye content)	1.2 g
	Ethyl alcohol (95%)	12 ml
2.	Anilin water	2 ml
	Distilled water	98 ml

The anilin water is mixed with the distilled water, allowed to stand for several minutes, and then filtered. The two solutions (1 and 2) are then mixed.

Aniline water is prepared in the following manner: 4 g of aniline oil (*Anilinum purum*) are added to 100 ml of distilled water and shaken until a homogeneous emulsion is obtained. This is filtered through filter paper soaked in water. The filtrate must be transparent.

Ziehl Carbol Fuchsin

Two solutions are prepared:

1.	Basic fuchsin (90% dye content)	0.3 g
	Ethyl alcohol (95%)	10 ml
2.	Phenol	5 g
	Distilled water	95 ml

The solutions (1) and (2) are mixed.

Loeffler Alkaline Methylene Blue

Two solutions are prepared:

1.	Methylene blue (90% dye content)	0.3 g
	Ethyl alcohol (95%)	30 ml
2.	Solution of KOH (0.01% w/v)	100 ml

The solutions (1) and (2) are mixed.

Methylene Blue in Dilute Alcohol [2]

Methylene blue (80% dye content)	0.3 g
Ethyl alcohol (95%)	30 ml
Distilled water	100 ml

Carbol Rose Bengal

Rose Bengal (90% dye content)	1 g
Phenol (5% water solution)	100 ml
$CaCl_2$	0.01–0.03 g

The content of calcium chloride determines the intensity of the staining.

Methylene Blue

Methylene blue	2 g
Ethyl alcohol (95%)	10 ml
Phenol	2 g
Distilled water	100 ml

Preparation: The dye is ground in a porcelain mortar with phenol and alcohol. This mixture is poured into a beaker with a ground glass stopper and is allowed to stand for 24 hours. After 24 hours 100 ml of distilled water are added; the mixture is shaken well and filtered.

Solutions of Neutral Red

Water solutions of 1–1.5% are used; for vital staining, 0.5% solution in 0.85% NaCl.

Chrysoidine Solution

Chrysoidine	2 g
Boiling distilled water	300 ml

---

[2] Preparation of this and all previously listed dyes is according to the Manual of the Society of American Bacteriologists (Manual of Microbiological Methods, 1957).

The dye solution is made on filter paper, on which the weighed out dye is poured.

When staining it is necessary to have the following equipment for the dye solutions: bottles with pipettes and small rubber bulbs, water baths with slide supports, containers for submersion washings, a wash bottle, and special slide forceps.

Determining Bacterial Motility and Staining Flagella

A number of bacterial species which occur in water masses are capable of independent movement. This movement is effected for the most part with the aid of flagella—long, thread-like, wavy, curved appendages. The length of the flagella may be very great and even exceed many times the length of the cell. According to recent data, flagella have a constant thickness for the greater part of their length. It has been shown by many investigators that a flagellum, at its proximal end, passes through the cell wall and the cytoplasmic membrane, starting from the basal granule located in the cytoplasm.

According to the number of flagella and their distribution on the cells, it is possible to distinguish monotrichs—bacteria with one flagellum at a single pole; lophotrichs—bacteria having a cluster of flagella at one end of the cell; and peritrichs—bacteria with flagella over the entire lateral surface of the cells (Fig. 64). Several authors distinguish still another group, amphitrichs—bacteria having flagella at each pole; others consider them two undivided cells, each with flagella on one end.

Fig. 64. Distribution of flagella on bacterial cells. *A,* Monotrichs; *B,* lophotrichs; *C,* amphitrichs; *D,* peritrichs.

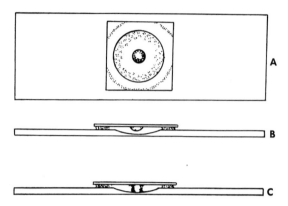

Fig. 65. Preparation of specimens for studying bacterial motility. *A,* Correctly prepared hanging drop, from above; *B,* side view of the same; *C,* improperly prepared hanging drop.

Bacterial motility is studied by means of a hanging drop of cells from young and old liquid cultures (Fig. 65). Both the character and rate of movement are noted.

A small, flat drop is placed on the surface of a cover glass in the center. A slide with a concavity is prepared beforehand by placing a layer of Vaseline or a mixture of Vaseline and paraffin around the concavity. After the drop is placed on the cover glass, the latter is turned over and placed over the concavity so that the drop hangs freely from the cover glass but touches neither the sides nor the bottom of the concavity. The cover glass is pressed tightly to the Vaseline and hermetically sealed on the outside with wax or collodion. Simply placing a drop of the culture on a slide and gently placing a cover slip on the drop is a quick, convenient method of preparation of slides for observation of bacterial morphology and motility. A drop of oil can be placed directly on the cover slip for examination of the culture with oil immersion microbiology.

Flagella can be observed and studied either in an electron microscope or by specially staining the specimens. The least equivocal representation of the distribution of flagella on bacteria can be obtained with the electron microscope. When studying flagella with the light microscope, use is made of various mordants which precipitate on the flagellum surface and increase its volume (which is small—0.002–0.003 μ in diameter); as a result it is easier to observe the flagella under high magnifications.

When preparing specimens for observation of flagella, the following requirements must be met: (1) all slides must be perfectly clean and thoroughly free from grease; (2) the bacterial culture must be in the actively motile stage (usually 15 to 18 hours old); and (3) the suspension must be prepared without mechanical damage to the flagella.

When preparing specimens, new slides (preferably heat-resistant) are used. They are soaked for 7 to 8 days in a freshly prepared dichromate solution, washed with distilled water, and dried. Neither wet nor dry slides should be handled with the hands. Dry slides are kept in a closed glass beaker. According to Leifson (1951), the slides before use are passed through the colorless part of a Bunsen burner flame with the side which will receive the smear held over the flame.

Bacteria for flagella staining are preferably grown on agar. Improper preparation of the specimen can lead to total loss of the flagella. The material for the smear is taken from an agar slant at the area where water has condensed, i.e. where there is more moisture. This water of syneresis containing motile bacterial cells is then transferred to 1 ml of sterile tap water by carefully dipping the material in water without shaking it from the loop. The process of transferring bacteria is repeated two to three times. The tube containing the bacteria is allowed to stand for 30 to 60 minutes at room temperature. A wax pencil is used to draw lines on the prepared glass slide —along the edges and across the middle—to mark off the smear. A capillary tube is used to take a drop of the culture and place it on the end of the slide. The slide is tilted and the drop is allowed to flow to the wax mark. Two to three drops of suspension are usually placed on the slide in this manner. The smear is dried at room temperature and is not fixed by any other means.

Several methods for staining flagella have been proposed. The following are a few of those which most consistently yield good results.

Leifson Methods

First method

The following is prepared:

Saturated aqueous solution of $NH_4Al(SO_4)_2 \cdot 12\ H_2O$ or $KAl(SO_4)_2 \cdot H_2O$	20 ml
Tannin (20% aqueous solution)	10 ml
Distilled water	10 ml
Ethyl alcohol (95%)	15 ml
Saturated alcohol solution of basic fuchsin	3 ml

The components are mixed in the given order. The solution is ready for use in 1 week.

Second method

Three solutions are prepared:

1.	Distilled water	100 ml
	NaCl	1.5 g
2.	Distilled water	100 ml
	Tannin	3 g
3.	Basic fuchsin	1.2 g
	Ethyl alcohol (95%)	100 ml

The solutions are mixed in equal proportions.

Staining procedure: (1) A smear is prepared; (2) the dye is poured on the smear and allowed to stand 10 minutes at room temperature; (3) it is washed with water and dried. Results: flagella are dyed red.

Morozov Method of Silver Staining

Three solutions are prepared:

1.	Glacial acetic acid	1 ml
	40% formalin	2 ml
	Distilled water	100 ml
2.	Tannin	5 g
	Liquid phenol	1 ml
	Distilled water	100 ml
3.	Solution of silver nitrate	

The silver nitrate solution is prepared as follows: 1 g of crystalline silver nitrate is dissolved in 20 ml of distilled water. An ammonia solution is carefully added dropwise to the solution until the precipitate disappears and the solution becomes lightly opalescent. It is necessary to avoid adding too much ammonia. In this case an additional small amount of silver nitrate solution can be added dropwise to the working solution until it opalesces slightly. The final solution is diluted with distilled water (1:10).

Staining procedure: (1) A smear is prepared; (2) solution 1 is poured on the smear, allowed to stand 1 minute, and decanted; the smear is then rinsed with water; (3) solution 2 is poured on the smear and heated to steaming, after which the solution is decanted and the specimen is washed with water; (4) solution 3 is poured on the smear and heated in a weak flame 1 to 2 minutes until a dark brown coloration appears, after which the smear is rinsed and dried. Results: the flagella, thickened by the silver nitrate, become clearly visible.

Gray Method

Two solutions are prepared:

1. $KAl(SO_4)_2 \cdot 12H_2O$ (saturated aqueous solution)   5 ml
    Tannin (20% aqueous solution) [3]   2 ml
    $HgCl_2$ (saturated aqueous solution)   2 ml
2. Basic fuchsin (saturated alcohol solution)   0.4 ml

Solutions 1 and 2 are mixed for 10–15 hours before being used. After combining they rapidly break down.

Staining procedure: (1) A smear is prepared; (2) the freshly filtered mixture of the solution is poured on the smear and allowed to stand 8 to 10 minutes; (3) the smear is rinsed with distilled water; (4) the fuchsin solution is poured on the smear and allowed to stand 5 minutes without heating; (5) the smear is rinsed with water; (6) it is then dried and observed.

Flagella can be detected in a fluorescent microscope

---

[3] If a large amount of tannin (tannic acid) solution is prepared, a few drops of chloroform should be added as preservative.

(Bukatsch, 1958) in the natural state. A drop of a fluorochrome
(either a water solution of auramine 1:1000, or phenolic berberine sulfate, 1:1000) is added to a drop of the culture. Phenolic berberine sulfate is prepared by adding an aqueous solution
of berberine sulfate (1:1000) to a saturated solution of phenol.
When treated with auramine, the cells and flagella fluoresce
golden yellow; when treated with phenolic berberine sulfate,
they fluoresce greenish-yellow.

The motility of bacteria can be observed in semisolid
media. The media are poured into test tubes and stab inoculation is performed with a needle. Growth of motile bacteria is
registered as a cloudy appearance around the stab. Non-motile
bacteria grow only along the stab. Various semisolid media
can be used.

Peptone Agar

Tap water	1000 ml
Peptone	10 g
Agar	5 g

The medium should be transparent. This is done by
filtering twice through thick filter paper.

Beef-Peptone Agar

Distilled water	1000 ml
Peptone	10 g
Beef extract	1.5 g
NaCl	2 g
Agar	5 g

pH is adjusted to 7.2 with a saturated solution of
$NaHCO_3$. The media are poured into test tubes in 5- to 7-ml
amounts and sterilized for 15 minutes at 110 C.

Beef-Peptone Agar with Gelatin

Two solutions are prepared:

1. Tap water	400 ml
Peptone	10 g

Beef extract	3 g
NaCl	2 g
Agar	4 g
2. Tap water	600 ml
Gelatin	80 g

When the ingredients are dissolved, the solutions are mixed together, pH is adjusted to 7.2, and the medium is filtered and poured into test tubes.

Gram Method of Staining

Gram staining of bacteria is an important diagnostic criterion. It is associated with the presence in cells of ribonucleic acid—a specific chemical compound which reacts grampositive (Bartholomew, 1962), and thus is a specific feature of the cytoplasm. All bacteria are divided into gram-positive and gram-negative groups. The former are distinguished from the latter by a series of properties (Table 3). In addition to the original Gram staining method, there are a number of modifications. The contemporary modifications of the basic method are used.

Bartholomew Modification of Gram Method
(Bartholomew, 1962)

Necessary reagents:

1. *Hucker crystal violet*. This is prepared by mixing two solutions:

A. Crystal violet (90% dye content)	2 g
Ethyl alcohol (95%)	20 ml
B. Ammonium oxalate	0.8 g
Distilled water	80 ml

Solutions A and B are mixed and allowed to stand 48 hours. The solution can then be kept for months.

2. *Burke iodine solution*. This is prepared by diluting Burke basic solution with distilled water (1:2). The basic solution is prepared as follows: 2 g of potassium iodide are placed

Table 3. Properties of gram-positive and gram-negative bacteria [1]

(Lamanna and Mallette, 1953; Timakov and Goldfarb, 1958)

Properties	Gram-Positive	Gram-Negative
Isoelectric point at pH	2–3	approximately 4–5
Behavior to action of proteolytic enzymes	resistant	sensitive
Behavior towards alkali	resistant; do not dissolve in 1% KOH	non-resistant, soluble
Sensitivity to basic dyes	very high	significantly smaller
Sporogenesis	often form spores	seldom form spores
pH of maximum growth	relatively high (alkaline)	acidic
Bacteriostatic action of iodine	growth inhibited by iodine	less sensitive to action of iodine
Nutrient requirement	complex	relatively simple (some species are autotrophs)
Permeability to stain		
In living state	more permeable	less permeable
Acid resistance	more resistant	less resistant
Solubility of lipids in solvents	resistant	less resistant
Sensitivity towards antibiotics	sensitive to gramicidin and penicillin	more sensitive to streptomycin

[1] See Stanier, Doudoroff, and Adelberg, 1970, for discussion of the significance of the gram reaction.

in a porcelain mortar; 1 g of iodine is added, and the two substances are ground with a pestle for 5 to 10 seconds; 1 ml of water is added and the mixture is again mixed by grinding; 5 ml of water are added and grinding is continued. The iodine and postassium iodide should be in solution. Ten milliliters of water are added and mixed well. The mixture is transferred to a flask for staining. The pestle and mortar are rinsed with water, which is then also transferred to the flask and brought to 100 ml in volume. This solution is stable and can be kept for months.

3. *Solution of safranin.* A 2.5% dye solution in 95% ethyl alcohol is diluted with distilled water in a proportion of 1:10.

Staining procedure: (1) A thin smear is prepared and fixed; (2) crystal violet is poured on the smear (it is best first

Table 4. Recommended time in different solutions
(Gram stain)
(Bartholomew, 1962)

Solution	Time Recommended for Differentiation (sec)	Limits of Time Required for Differentiation (sec)
n-Propyl alcohol	180	30–1800
95% n-Propyl alcohol	90	15– 360
95% Ethyl alcohol	90	15– 180
Acetone	45	5– 120

to place filter paper on the smear) and allowed to stand for 1 minute; (3) the filter paper is removed and the excess dye is washed off by submerging the slide for 5 seconds in a vessel containing 250 ml of water (the water in the vessel should constantly be replaced with a water flow of 30 ml per second); (4) excess water is removed by pouring Burke solution onto the smear and allowing it to stand for 1 minute; (5) this is washed with water for 5 seconds in the manner indicated above; (6) the glass is placed in a vessel for differential staining (n-propyl alcohol, according to Bartholomew, yields the best results; time and necessary reagents are given in Table 4); (7) this is washed again with water for 5 seconds as mentioned above; (8) the excess water is removed by 0.25% safranin solution, which is poured onto the slide; 1 minute is allowed for staining; (9) the slide is washed with water for 5 seconds, dried, and observed. Results: gram-positive bacteria appear blue; gram-negative, red.

The differentiation procedure is carried out in three vertical staining jars (75 × 25 mm). The specimen is kept in each jar for one-third of the total time. It is suggested that the slides be agitated slightly while in the jars.

Hucker Modification of Gram Method

Necessary reagents:

1. Hucker crystal violet (preparation given earlier)
2. Lugol solution: crystalline iodine (1 g); potassium iodide (2 g); distilled water (300 ml)
3. Safranin solution (preparation given earlier)

Staining procedure: (1) A thin smear is fixed in a flame, (2) stained with crystal violet for 1 minute, (3) washed with tap water for a maximum of 2 seconds, (4) submerged for 1 minute in iodine solution, (5) rinsed with water, and the water is removed from the smear with filter paper; (6) the specimen is decolorized in 95% ethyl alcohol for 0.5 minute, with slight agitation, (7) stained 10 seconds in safranin solution, and (8) washed in tap water and dried. Results: gram-positive bacteria appear blue-violet; gram-negative, red.

Basic Gram Method of Staining
(Contemporary Modification)

Necessary reagents:

1. *Carbolic gentian violet:* 1 g of dye is dissolved in 10 ml of alcohol; the solution obtained is poured into 100 ml of a 5% solution of pure phenol.

2. *Lugol solution:* Potassium iodide (2 g); crystalline iodine (1 g); distilled water (300 ml). Because the iodine is insoluble in water, in order to prepare the above solution it is necessary first to dissolve 2 g of potassium iodide in 5 ml of water and then add the 1 g of iodine to this solution. Only after it dissolves is water added to a volume of 300 ml.

3. *Counterstain:* Ziehl diluted fuchsin (1 ml of dye in 19 ml of distilled water), or 1% aqueous solution of neutral red, or 0.5–1% aqueous solution of eosin.

Staining procedure: (1) A smear is prepared from a 1-day old culture. The smear should be thin—its thickness should not exceed a single cell layer. The results of the staining can be different depending on the thickness of the smear. The smear is fixed by passing the slide through a flame. (2) Filter paper is placed on the fixed smear and the gentian violet solution is poured on and left for 1–2 minutes. (3) The paper is removed, the excess dye is poured off, and Lugol solution is poured on the specimen without rinsing the latter with water; the Lugol solution is left there for 1 minute and then decanted. (4) The specimen is placed in 95% ethyl alcohol for 0.5–1 minute. (5) After decolorizing, the specimen is rinsed with water. (6) The smear is stained again for 1–2 minutes with a strongly diluted dye solution, either fuchsin, neutral red, or safranin. Bacteria stained by the Gram method retain a basic violet color of varying intensity; however, gram-negative species acquire a red color.

To assure that the staining was performed correctly, Omelianski suggests the following. Three smears are placed on a slide in a row: one is known to be gram-positive; another gram-negative; the third is the unknown. The accuracy of staining is checked by the appearance of the two test smears.

Burke and Kopeloff-Beerman Modifications

Burke Method

1. A. Gentian or crystal violet   1 g
   Distilled water   100 ml
   B. NaHCO₃   1 g
   Distilled water   20 ml
2. Iodine solution: I, 1 g; KI, 2 g; distilled water, 100 ml
3. Solution of 2 g of safranin 0 in 100 ml of distilled water

Kopeloff-Beerman Method

1. The same basic solutions (A and B) are used
2. Iodine solution: I, 2 g; normal solution of NaOH (40.01 g per liter), 10 ml. After the iodine dissolves, the solution is increased to a volume of 100 ml with distilled water
3. Ether-acetone mixture (1:1 or 1:3)
4. Counterstain: basic fuchsin, 0.1 g; distilled water, 100 ml

Staining procedure: (1) A thin smear is prepared and dried in air without heating. (2) Solution A is poured onto the smear and 2 to 3 drops of solution B are added (more depending on the area of the smear); these are mixed on the slide and allowed to stain for 2–3 minutes. Kopeloff and Beerman suggest mixing the solutions beforehand (1.5 ml A and 0.4 ml B) and staining 5 minutes. (3) The smear is rinsed with either of the two iodine solutions given above by pouring the solution on the smear and allowing it to stand for 2 minutes. (4) The smear is rinsed with tap water, which is removed from the surface of the smear by blotting with filter paper—without drying (at this point in staining, it is necessary to be careful that the cells do not dry out). (5) The smear is rapidly decolorized in a mixture of ether and acetone. This solution should be added dropwise to the slide until there is practically no coloring in the draining drops; this usually takes less than 10 seconds (the rate of decolorizing depends on the proportion of ether and acetone: the more acetone, the more rapid the process; for slower decolorizing a mixture of ether with acetone in a proportion of 1:1 is used). (6) The smear is air-dried, (7) stained again for 5–10 seconds—preferably with the safranin solution, (8) rinsed in tap water, and dried. Results: gram-positive bacteria appear blue-violet; gram-negative are red.

When using the Gram method to stain an unknown culture, more than one procedure should be used, i.e. different methods should be applied. Moreover, the cultures can be streak inoculated onto agar containing either bee venom or gentian violet. This permits confirmation of the Gram reaction and provides an additional characteristic of the cultures.

Only gram-negative bacteria will grow on solid nutrient media containing 0.6–1 μg of bee venom per milliliter (Ortel, 1957). Adding bee venom in a concentration of 6–10 μg per milliliter to a young culture in a liquid medium (up to 20 hours old), according to Ortel's data, causes death of gram-positive bacteria, and inoculation of ordinary media with such a culture yields growth of only gram-negative bacteria.

A medium with gentian violet diluted 1:1,000,000 can also be used. Gram-positive bacteria ordinarily do not grow in agar containing this dye concentration whereas gram-negative bacteria are resistant. However, gram-positive bacteria also grow in a medium containing gentian violet in a concentration lower than 1:5,000,000. Therefore, the dye concentration must be no lower than 1:1,000,000.

Sporogenesis and Staining Spores

Spore formation is observed almost exclusively in rod-shaped bacteria; one spore-forming species of cocci, *Sarcina ureae*, is well known.

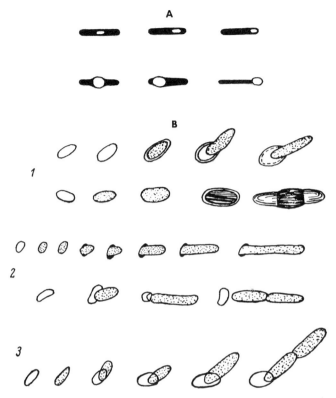

Fig. 66. A, Spore distribution in cells; B, types of outgrowth:
1, polar; *2*, equatorial; *3*, transverse.

In the identification of bacterial species according to Krasil'nikov (1949), useful diagnostic criteria include: the shape of spores, diameter of the cells, mechanism of spore formation, location of spores in the cell (whether they are formed in the center or at the poles of the cell), and presence or absence of swelling in the cells (Fig. 66). The type of spore outgrowth should be noted (polar, when the spore grows out at one end; bipolar when growth takes place from both ends; equatorial, when vegetative growth arises from the middle; transverse, with growth at the side). Note is also taken of how growth takes place—with or without discarding of the spore coat. Thus, for example, spores of the hay bacillus (*Bacillus subtilis*) have an oval form, are distributed in the cells ex-

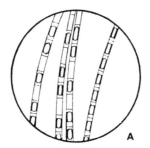 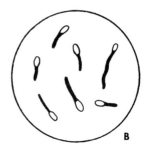

Fig. 67. Cells of spore-forming bacteria with spores. A, Bacillus subtilis; B, Clostridium pectinovorum.

centrally without strict localization, have dimensions such that deformation of mother cells is not observed (Fig. 67, A), and demonstrate equatorial outgrowth. The flax-retting bacilli (*Clostridium pectinovorum*) have round, rather large spores which are distributed terminally. The ends of the cells swell during sporogenesis (Fig. 67, B) and the spores grow out at one pole.

Spores have thick coats composed of two layers: an outer and an inner layer. Recent electron micrographs, in addition, reveal a laminated thick cortex surrounding a spore membrane. The outer layer of the coat protects the spore from external effects; the wall of the cell forms from the inner layer when outgrowth takes place. Spores are distinguished in chemical composition from bacterial cells primarily by relatively low concentration of water and large content of Ca^{++} and a roughly equivalent amount of dipicolinic acid. Owing to this and to the presence of the thick two-layered coat, spores remain viable for a long time and can withstand action of unfavorable physical and chemical factors. Thus, a temperature of 100 C usually does not kill spores. To destroy spores, steam at a temperature of 120 C is required.

Spores grow out on transfer (under optimal growth conditions). The process of germination begins with swelling when biochemical processes involving uptake of water and a loss of solid content (especially calcium dipicolinate and soluble glycopeptide) in the spore are activated. The outer coat breaks and the outgrowing cell emerges from it. The outgrowth process of a spore lasts for 4 to 5 hours.

The thick coat of the spore is responsible for the fact

that spores absorb and retain stains with difficulty. It is possible to observe spores within bacteria by simply studying an unstained specimen with the diaphragm of the illuminator constricted. They can be thus observed since they refract light significantly more than cell protoplasm. If this method is unsuccessful in determining clearly the presence of spores, staining is performed.

There are several methods of differential staining of spore-bearing cells. Strong dye solutions and heating during staining are employed to stain spores. Carbol fuchsin is most often used for this.

Old Method of Preparing Dye

Saturated solution of basic fuchsin	10 ml
5% Solution of phenol	100 ml

Contemporary Method

Two solutions are prepared:

1. Basic fuchsin	0.3 g
Ethyl alcohol (95%)	10 ml
2. Phenol	5 g
Distilled water	95 ml

The solutions are combined.

Staining procedure: (1) A smear is prepared on a slide, dried by pressing gently between two paper towels, and then covered with a small piece of filter paper. (2) The freshly filtered dye is poured onto the filter paper. (3) The specimen with the dye is heated to steaming and held for 5–10 minutes. Dye is added in proportion to the evaporation so that the specimen does not dry. (4) The excess dye is decanted and the specimen is rinsed with water. (5) The specimen is submerged in 1–5% sulfuric acid or 33% ethyl alcohol to decolorize (the time required for decolorizing varies from seconds to several minutes depending on the peculiarities of the specimen; decolorizing is interrupted at the point when the protoplasm decolorizes, but the spore has not begun to decolorize; this point is empirically determined in each case). (6) The specimen is rinsed with water. (7) Counterstaining is performed with a solution of methylene blue (100 ml of distilled water, 30 ml of a saturated alcohol solution of the dye, and 1 ml of a 1% solution of potassium hydroxide); the staining is continued for 1–2 minutes. (8) The specimen is then rinsed and dried. Results: spores are stained red; bacterial cells, blue.

The second part of the staining, beginning from point 4, may be conducted in another way. The slide is submerged momentarily in 95% ethyl alcohol, then rinsed with water. A drop of a saturated aqueous solution of nigrosin (water-soluble nigrosin, 10 g; distilled water, 100 ml) is placed on the smear; this solution is kept in a boiling water bath for 30 minutes; 0.5% formalin is added, and the solution is filtered twice through double filter paper. The dye solution is added so that distribution is uniform. The slide is then quickly dried by careful heating without preliminary rinsing. Results: spores are red, vegetative cells are unstained, and the background is gray.

Detecting the Presence of Spores

Staining procedure: (1) A smear is prepared and heat-fixed in a flame; (2) a water solution of brilliant sulfoflavin (1:400) to which 0.5 ml of liquid phenol has been added is poured on the smear and allowed to stand for 1–2 minutes; (3) the smear is rinsed thoroughly with water, (4) stained for 40 seconds with an aqueous solution of coriphosphine (1:1000), (5) washed again with water, dried in air, and observed through a fluorescent microscope. Results: spores fluoresce green-yellow; vegetative cells are red.

Staining Capsules

Some species of bacteria form mucoid capsules around the cells which consist of complex polysaccharides and glyco-proteins. Mucoid capsules are often formed by cells when a large amount of carbohydrates and a small amount of nitrogenous compounds are present in the medium.

Bacterial capsules in unstained specimens are observed as a bright border around the cells. However, when establishing the presence of capsules with stains, these bacterial formations can be more easily confused with artifacts than can other structures. Thus, for example, when staining bacteria, narrow regions, or "halos," around the cells—which themselves may be mistaken for the capsules—may remain unstained. Therefore, the best method of establishing the presence of capsules is by differential staining, in which the cell itself and the capsule are stained differently.

Tyler Modification of Anthony Method

Two solutions are prepared:

1.	Crystal violet	0.1 g
	Glacial acetic acid	0.25 ml
	Distilled water	100 ml
2.	$CuSO_4 \cdot 5H_2O$	20 g
	Distilled water	100 ml

Staining procedure: (1) A smear is prepared and dried in air, (2) stained 4–7 minutes with solution No. 1, (3) washed with solution No. 2, (4) dried by gently pressing between two pieces of bibulous or filter paper, and observed. Results: bacterial cells are dark blue; capsules are blue-black.

Omelianski Method

A solution of carbol fuchsin is prepared:

Saturated alcohol solution of fuchsin	10 ml
5% Solution of phenol	100 ml

The solution is diluted with distilled water in proportion of 1:1.

Staining procedure: (1) A drop of the water-diluted carbol fuchsin is placed on a slide; a drop of the culture is placed in the dye. (2) After 2–3 minutes a small amount of liquid India ink is added, thoroughly mixed, and spread over the slide, which is then covered with a cover slip and observed. Results: bacterial cells are red; capsules are colorless and sharply visible against a dark background. A 10% solution of collargol may be used instead of liquid India ink.

Klett Method

Dye solutions are prepared:

1.	Methylene blue	1 g
	Ethyl alcohol (95%)	10 ml
	Distilled water	100 ml
2.	Fuchsin	1 g
	Ethyl alcohol (95%)	10 ml
	Distilled water	100 ml

Staining procedure: (1) the specimen is prepared and dried in air, (2) stained with a boiling solution of methylene blue for 1 minute, (3) rinsed with water, (4) stained with the fuchsin solution for 5 seconds, and (5) rinsed with water, dried, and observed. Results: bacterial cells are blue; capsules are rose-red.

Staining Bacterial Cell Inclusions

Staining Metachromatin Granules.   Grains of metachromatin (volutin) occur in cells of many bacteria. The presence of phosphorus compounds in the medium contributes to their formation. There are many methods for determining the presence of these inclusions in cells.

Loeffler Method

A dye solution is prepared:

Saturated solution of methylene blue	30 ml
Solution of KOH (1:10,000)	100 ml

Staining procedure: (1) A smear is prepared from a culture 18–24 hours old; (2) it is then heat-fixed; (3) the dye is poured on the smear and allowed to stand 2–3 minutes; and (4) it is rinsed with water, dried, and observed. Results: the grains of metachromatin stain blue.

Neisser Method

Two solutions are prepared:

1. A.	Methylene blue	1 g
	Ethyl alcohol (95%)	20 ml
	Glacial acetic acid	50 ml
	Distilled water	1000 ml
B.	Crystal violet	1 g
	Ethyl alcohol (95%)	10 ml
	Distilled water	300 ml

Before use, solutions A and B are mixed in the proportion 2:1.

2.	Chrysoidin	1 g
	Hot distilled water	300 ml

Staining procedure: (1) A smear is prepared and heat-fixed. (2) Solution 1 is poured on the smear and allowed to stand for 5 minutes. (3) Solution 1 is decanted; the smear is rinsed with water and dried with filter paper. (4) Solution 2 is poured on the smear and allowed to stand for 1 minute. (5) The smear is rinsed with water, dried, and observed. Results: grains of metachromatin stain blue-black; cytoplasm, yellow.

Neisser Modified Method

Two solutions are prepared:

1.	Methylene blue	0.1 g
	Absolute alcohol	2 ml
	Glacial acetic acid	5 ml
	Distilled water	95 ml

The dye is dissolved in the alcohol and then added to the acetic acid solution.

2.	Chrysoidin or Bismarck brown	2 g
	Hot distilled water	300 ml

The dye is dissolved in the hot water and filtered through thick filter paper.

Staining procedure: (1) A smear is prepared, dried in air, and heat-fixed; (2) solution 1 is poured on the smear and allowed to stand for 1 minute; (3) the smear is rinsed with water; (4) solution 2 is poured on the smear and allowed to stand for 1 minute; and (5) the smear is once again rinsed with water and then dried. Results: grains of metachromatin are blue-black; cell plasma is yellow-brown.

Albert Method

Two solutions are prepared:

1.	Toluidine blue	0.15 g
	Malachite green	0.2 g
	Glacial acetic acid	1 ml
	Ethyl alcohol (85%)	2 ml
	Distilled water	10 ml
2.	I	2 g
	KI	3 g
	Distilled water	200 ml

Staining procedure: (1) A smear is prepared and heat-fixed; (2) solution 1 is poured on the smear and allowed to stand for 5 minutes; (3) the dye is decanted, but the smear is not rinsed; (4) the slide is held almost vertically as the smear is rinsed with several drops of solution 2; the smear is then covered with solution 2 and allowed to stand for 1 minute; (5) it is then rinsed with water, dried, and observed. Results: grains of metachromatin are black; cytoplasm is green.

Timakov and Goldfarb (1958) Method

The following solutions are prepared:

1. Loeffler methylene blue
2. 1% Solution of sulfuric acid
3. 4% Aqueous solution of potassium carbonate (K_2CO_3)

Staining procedure: (1) Two smears are prepared (one smear is designated by the letter A, the other by B) and heat-fixed. (2) The smears are dyed with Loeffler methylene blue for 10 minutes. (3) The specimen marked A is placed in sulfuric acid solution for 5 minutes; the B specimen is treated for the same time period in potassium carbonate solution. (4) The specimens are dried without rinsing, then (5) dyed with a 0.25% solution of light green with 2–3 drops of concentrated acetic acid. Results: In specimen A the grains of metachromatin stain cherry-red; the cytoplasm, green. In specimen B the metachromatin decolorizes and partially dissolves; consequently, unstained particles appear in its place.

Omelianski Method

The following solutions are needed:

1. Ziehl carbol fuchsin (see p. 103 for preparation)
2. 1% Solution of sulfuric acid ˙
3. Solution of methylene blue (1 ml of a saturated alcohol solution of the dye in 40 ml of distilled water)

Staining procedure: (1) A smear is prepared, dried, and heat-fixed. (2) Ziehl carbol fuchsin is poured onto the smear and allowed to stand for 0.5–1 minute. (3) The smear is then rinsed with water and decolorized for 20–30 seconds with the sulfuric acid solution. (4) The smear is rinsed with water and stained for 15–20 seconds with methylene blue. (5) It is then rinsed with water, dried, sealed in glycerin, and observed. Results: grains of metachromatin are red; cell plasma is blue.

Staining Glycogen. Glycogen is a reserve nutrient substance of a cell similar to starch. Staining with iodine to produce a red-brown coloring is characteristic for glycogen.

Necessary reagents:

1. Solution of iodine in potassium iodide:

I	7 g
KI	20 g
Distilled water	100 ml

2. Mixture of ether with alcohol (1:1)

Staining procedure: (1) A smear is prepared and (2) treated with solution 2 for 5–10 minutes. (3) The iodine solution is introduced, the cover glass is placed on the slide, and the slide is examined. Results: the small grains of glycogen stain red-brown. Treating the smear with alcohol-ether mixture is absolutely necessary, since fat yields a tint with iodine which is close to the stain of glycogen.

Staining Fat.   Fat inclusions are found in the cells of many bacteria and yeasts. There are various methods of observing them.

Omelianski Method (using chloral hydrate)

Two solutions are prepared:

1. Chloral hydrate		5 g
Distilled water		2 ml
2. Saturated alcohol solution of methylene blue		1 ml
Distilled water		40 ml

Staining procedure: (1) A smear is prepared, fixed with 40% formalin, and dried. (2) The solution of chloral hydrate is poured onto the smear and allowed to stand for 3–5 minutes. The specimen is then (3) rinsed with water, (4) stained for 1 minute with the solution of methylene blue, and (5) rinsed, dried, and examined. Results: cytoplasm stains blue. The areas in which droplets of fat occur remain unstained.

Staining Fat Inclusions with Sudan III

The following solution is prepared:

Sudan III	0.05 g
Concentrated lactic acid	10 ml

Staining procedure: (1) A thick smear is prepared. (2) A drop of formalin (40%) is added and allowed to stand for 5 minutes. (3) A drop of methylene blue (1:10) is added and allowed to stand for 5 minutes. (4) A drop of the Sudan solution is then added and also allowed to stand for 5 minutes. (5) The cover glass is placed on the slide, which is then examined under the microscope. Results: cell cytoplasm stains blue; fat droplets are orange.

Staining with Sudan Black B, Method 1

The following solutions are prepared:

1.	Sudan black B	0.3 g
	Ethyl alcohol (70%)	100 ml
2.	Safranin	1.5 g
	Distilled water	50 ml

The safranin is dissolved in boiling water and filtered upon cooling.

Staining procedure: (1) A smear is prepared from a 1-day-old culture and heat-fixed. (2) The dye is poured on the smear and allowed to stand at room temperature for 10–15 minutes. (3) The specimen is rinsed and a drop of the safranin solution is added and allowed to stand for 5 minutes. (4) The specimen is rinsed, dried, and observed. Results: protoplasm is rose-colored; fat droplets are black.

Staining with Sudan Black B, Method 2

A saturated dye solution of Sudan black B is prepared in pure ethylene glycol (boiling point, 195–197 C).

Staining procedure: (1) A drop of Sudan black is placed on a clean slide. (2) A small amount of the culture is introduced with a loop and a suspension is prepared. (3) A cover glass is placed on the cell suspension and the suspension is observed under the microscope. Results: fat droplets appear black and cell cytoplasm remains unstained.

Flemming Method

The following solution is prepared:

1% Solution of osmic acid	15 ml
Glacial acetic acid	1 ml
2% Aqueous solution of chromous acid	4 ml

Staining procedure: (1) A smear is prepared and fixed with 40% formalin. (2) A drop of the prepared solution is introduced and allowed to stand for 5 minutes. (3) A cover glass is placed on the smear and the specimen is observed. Results: fat droplets are brownish black.

Staining with a Solution of Dimethylamidoazobenzene

The following solution is prepared:

Dimethylamidoazobenzene	0.2 g
Ethyl alcohol (95%)	50 ml

Staining procedure: A drop of water and a drop of the solution are mixed on a slide. A loop containing a portion of the culture under investigation is introduced and mixed thoroughly on the slide. The cover glass is set in place and the specimen is observed. Results: the cell cytoplasm is colorless; fat droplets are yellow.

Growth in Various Media

In the majority of instances, when determining the species of heterotrophic bacteria, the usual laboratory media are employed: yeast extract-peptone agar; nutrient broth, nutrient agar, and nutrient gelatin.

Growth of each species in these media has characteristic features and the cultural properties of a species are recorded. A series of media is used to obtain data on the physiological properties of a species: presence of enzymes which break down specific compounds is determined by growing the microorganisms in a particular medium. Appearance of decomposition products during bacterial growth indicates that the enzyme does occur in the enzymatic apparatus of the bacterial species. Several media cited below are differential-diagnostic media of this type and they make it possible to demonstrate special features of bacterial physiology and metabolism.

Growth in Beef-Peptone or Nutrient Broth.   Growth of microorganisms in beef-peptone or a similar nutrient broth provides some diagnostic criteria. The following are noted: (1) presence or absence of a pellicle; turbidity; and appearance of a precipitate; (2) character of the pellicle: ring-shaped, uniform, thin, thick, smooth, wrinkled, dry, butyrous, slimy; (3) character of turbidity: homogeneous (even), flocculent (granular), in "strings"—and degree: slight, moderate, heavy; and (4) character of the precipitate: slight or scanty, abundant, heavy and friable, compact, flaky, granular, slimy.

Growth in Beef-Peptone or Nutrient Agar. Colony morphology on agar (Fig. 58) and growth characteristics on agar slants often provide useful information.

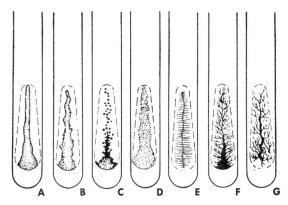

Fig. 68. Growth on solid medium (character of the inoculated growth). *A,* Continuous with even edge; *B,* continuous with wavy edge; *C,* beaded; *D,* diffuse; *E,* feathery; *F,* dendritic; *G,* rhizoid.

To determine growth characteristics, the slanted surface of a beef-peptone agar is streak inoculated; a platinum loop containing the inoculum is transferred from the lower to the upper end of the agar surface without being submerged in the condensed water. The following are noted: (1) growth: abundant, moderate, scanty; (2) characteristics of growth along the inoculation: continuous deposit (with even or wavy edges), beaded (in the form of a series of isolated colonies), diffuse, feathery, dendritic, rhizoid (Fig. 68); (3) surface of the streak: lustrous, dull, floury; (4) consistency: wet, dry, leathery, mucoid; (5) color of the deposit, color change of the medium.

Growth in Beef-Peptone or Nutrient Gelatin. A beef-peptone gelatin is inoculated by stabbing, using a medium poured into a test tube (not slanted). Incubation is usually at room temperature (about 25 C). The following are noted as growth occurs in gelatin: (1) characteristics of growth along the length of the inoculation: development on the surface and slight growth along the line of the inoculation (growth in the form of a nail)—*aerobes;* uniform growth along the entire stab (without a cap)—*facultative anaerobes;* growth only in the lower portion of the medium—*anaerobes;* (2) characteristics of growth along the inoculation (Fig. 69); and (3) color at the surface of the stab and along stab itself. Liquefication of the gelatin is noted—i.e. presence of gelatinase (p. 138). (It should be noted that the extensive characterization of the gelatin test is not employed in modern microbial taxonomy. Gelatin liquefication is recorded simply as present or absent.)

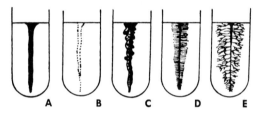

Fig. 69. Characteristics of growth along the line of inocula-tion. *A,* Filiform; *B,* beaded; *C,* uneven; *D,* hair-like; *E,* dendritic.

Growth on Potato.   (This is a classical test in bacteriology no longer done routinely in microbiology laboratories. However, since many of the older descriptions of bacterial species cite results of this test, it is included here for the sake of complete-ness.) Potato cylinders are prepared in test tubes. Firm potatoes are selected and washed with a brush under a stream of water in order to remove particles of dirt. The skin is removed with a knife; the potato is rubbed with alcohol; and the eyes are removed. The cleaned potato is once again washed with water; the tops are cut away; and, with the help of a cork borer, cylinders are cut out in lengths of about 4–5 cm, with diameters corresponding to the diameter of the test tubes selected for preparing the potato medium. Each cylinder is then cut diagon-ally into two wedges. The wedges are rinsed with water and submerged in a 1% solution of sodium bicarbonate ($NaHCO_3$)

Fig. 70. Test tube for potato.

for 1 hour, since potatoes have a slightly acidic pH. The potato wedges are placed in test tubes—either special ones with a constriction in the lower part (Fig. 70), or, when these are not available, ordinary tubes in which case a small clump of cotton, dampened with distilled water in order to keep the potato moist, is placed beforehand on the bottom of the test tube. In test tubes with constrictions, distilled water is poured in the lower part for the same purpose. The potatoes are sterilized in an autoclave for 20–25 minutes.

Inoculations are performed by smearing the inoculum onto a potato slant. When growing a culture on a potato the following are noted: the character of the growth, surface characteristics, consistency and color of the streak, and the color of the medium.

Methods of Determining Physiological Features

Relation to Oxygen. According to their requirements for oxygen, microorganisms are divided into the following groups: (1) strict aerobes, which need free oxygen for growth; (2) microaerophiles, which develop in reduced partial oxygen pressure; (3) facultative anaerobes, which are capable of using free oxygen, but are not dependent on it for growth; and (4)

obligate anaerobes, for which free oxygen is not only unnecessary, but toxic.

To determine the relation of bacteria to oxygen, a simple
test can be made when identifying bacterial species: inoculation is performed either in a liquid medium, or in an agar
medium containing glucose. The medium is poured out in 10-ml
amounts in test tubes and sterilized. Inoculations are made in
a melted and cooled medium which is thoroughly mixed after
inoculation. The medium is allowed to solidify and is placed
in an incubator.

Strict aerobes grow on the surface and in the topmost
layer of the medium; microaerophiles grow just below the surface; facultative anaerobes develop throughout the medium;
strict anaerobes are found only in the deep portion of the
medium (Fig. 71).

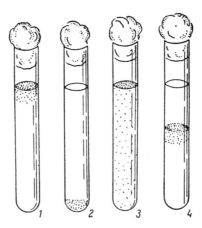

Fig. 71. Oxygen requirements. *1,* Aerobe; *2,* anaerobe; *3,* facultative anaerobe; *4,* microaerophile.

The entire diversity of oxygen requirements of bacteria, however, is not really determined in the above-mentioned test. For example, certain strict anaerobes require traces
of oxygen in the medium for growth. Moreover, the wide adaptation of bacteria to a given habitat leads to a change in type
of respiration for certain microorganisms. This is especially
often the case for facultative anaerobes, which can exist aero-
bically or anaerobically depending on conditions of the me-

dium. Growth in broth or on agar plates in a disposable Gas-pac (see Chapter 3) is a convenient test for facultative anaerobes and for isolation of anaerobes from water and sediment samples.

Therefore, the conditions of the medium, which are characterized by the oxidation-reduction potential (Eh or rH_2), reflect much more fully the direction of oxidation-reduction reactions of microorganisms.

At the present time, limit values of oxidation-reduction potentials for microorganisms have been established. The magnitude of the oxidation-reduction potential is such that low values indicate that the medium is depleted of oxygen and high values indicate that it is saturated with oxygen. Removal of oxygen from the medium by heating, followed by further measures taken to restrict entrance of oxygen to the medium, insures a low oxidation-reduction potential (and, consequently, conditions for the growth of anaerobes). When microorganisms multiply in a medium, the oxidation-reduction potential falls rapidly. Changing the oxidation-reduction potential changes the direction of metabolism in bacterial cells.

The initial rH_2 value of a nutrient medium is usually between 7.4 and 9 (Lambin and German, 1961). It depends on the composition of the medium, conditions of anaerobiosis, and the ratio of free surface of the medium to its volume. The presence of compounds which possess reducing properties (cysteine, thioglycollic acid) contributes to the decrease in rH_2. After an agar medium is sterilized in an autoclave for 20 minutes at 120 C and 15 psi, the rH_2 in the total medium mass equals approximately 7.5. After 48 hours, as a result of the penetration of atmospheric oxygen through the surface of the medium, a whole range of rH_2 values is established according to depth in the medium (Fig. 42).

Use of Carbon Sources. Many bacteria are capable of breaking down various carbohydrates and polyhydric alcohols. The ability of bacteria to utilize certain carbon compounds is a useful diagnostic criterion in species identification. In special investigations, the growth of a given strain in media supplied with various carbon sources is determined for a wide assortment of carbohydrates and alcohols. In routine studies the choice is limited to widely used and less expensive compounds.

Usually, when examining the requirement of bacteria for sources of carbon or ability to degrade these, the following are used: glucose, sucrose, maltose, lactose, galactose, glycerol, mannitol, and ethyl alcohol. However, a great variety of carbohydrates are often routinely used in identification and characterization of bacterial species. These compounds are added in a quantity of 0.5–1% either to peptone water or to a synthetic medium having 0.1–0.2% peptone. Andrade liquid medium is often used in clinical laboratories. The following media have been used to determine carbohydrate utilization by bacteria.

Beijerinck Medium

Tap water	1000 ml
$NH_4H_2PO_4$	1 g
KCl	0.2 g
$MgSO_4 \cdot 7H_2O$	0.2 g
Peptone	1–2 g
Carbohydrate	20 g

pH 7.0 is established with 1 N sodium hydroxide solution.

Andrade Medium

This medium is prepared from two solutions: (1) an indicator solution, and (2) peptone water.

1. Acidic fuchsin (0.5 g) and sterile distilled water (100 ml) are placed in a sterile flask. Added to this are 16.4 ml of 1 N NaOH, and the entire solution is sterilized in an autoclave for 5 minutes at 110 C.

2. Peptone (10 g) and sodium chloride (5 g) (ordinary table salt can be used) are added to 100 ml of distilled water; the solution is then boiled. pH is adjusted to 7.6. The solution is filtered, poured into flasks in 100-ml amounts, and sterilized in an autoclave at 115 C for 20 minutes. One milliliter of the indicator and 0.5 g of the given carbohydrate or alcohol are added aseptically to each flask. The medium is aseptically poured in 3- to 5-ml amounts into presterilized test tubes containing inverted vials. These test tubes are then sterilized by flowing steam for 30 minutes on each of 3 days. The prepared

medium should have a light straw color (without the slightest shade of rose). The presence of a rose tint either indicates overheating of the medium or a poor quality of fuchsin; if this occurs, additional NaOH is required for decolorizing. The necessary amount of NaOH is then determined experimentally, keeping in mind that the mixture should be straw yellow in color.

Fermentation of sugars and alcohols is determined by the color change of the solution and by the production of gas which causes the vials to fill with gas, displacing the liquid (Fig. 72).

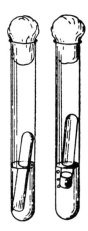

Fig. 72. Gas production in carbohydrate media.

Oxidative versus fermentative utilization of carbohydrates is now most often done with the Hugh and Leifson (1953) method, employing duplicate tubes of carbohydrate media, one overlaid with sterile mineral oil (anaerobic), the other left without an overlay (aerobic). The pH indicator is bromothymol blue. Acid production in the open and/or closed tubes indicates oxidative or fermentative attack on carbohydrates.

Semisolid media may be used for inoculations. For this purpose 0.3–0.5% agar is added to a given carbohydrate test medium before it is sterilized. The agar is melted by heating and the medium is poured into test tubes, and sterilized by flowing steam for 3 days. The inoculation is made into a tube

of agar with a needle. In such a medium, gas formation may
be observed by the appearance of gas bubbles in the medium
and rupturing or splitting of the agar.

Utilization of carbohydrates may be determined in another manner: the auxanogram method. A solid medium (synthetic medium with 1.5% agar) is prepared without carbohydrates. This medium is poured into test tubes and sterilized in
an autoclave. A heavy inoculation of the test culture is made in
a test tube, the contents of which are then poured into a sterile
petri dish. When the agar hardens in the petri dish, the plate
is divided into sectors by marking the bottom of the dish. No-
tations are made on each sector of the plate. A drop of carbohydrate is then placed in the central part of each sector
and is not allowed to extend beyond its boundaries. Growth
in the appropriate sector indicates which carbohydrate is used
by the bacteria.

A whole series of methods have been proposed for sim-
plifying the acquisition of data on the utilization of carbohydrates by bacteria. One of these methods is as follows. Small
discs of filter paper (5–10 mm in diameter) are saturated for
24 hours with 33% solutions of various carbohydrates, dried at
40 C, and stored in a sterile beaker. These discs are used as
needed. A bacteria-inoculated agar is prepared with an appropriate nutrient base which lacks a carbon source and contains
an appropriate indicator such as a solution of phenol red added
in sufficient concentration to provide a pale red color. Five discs,
saturated with various sugars, are placed on the surface of the
agar plate. Utilization of a carbohydrate is determined by
noting bacterial growth around the discs and the color change
of the phenol red to yellow (Fig. 73).

These discs may also be used in liquid media. They can
be placed in test tubes containing liquid basal medium and
phenol red as an indicator. The test tubes are first inoculated
with test cultures of bacteria. When it is necessary to determine
gas formation, strips of filter paper, saturated with carbohydrate solutions and folded in the form of the letter M, are used
instead of discs. When gas is present, the strips of paper rise
to the surface of the medium.

The use of discs saturated with carbohydrates has many
advantages compared with direct application of carbohydrates

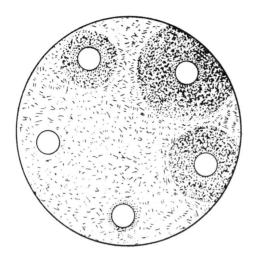

Fig. 73. Bacterial growth around discs saturated with various carbohydrates.

onto the surface of the medium: The carbohydrates do not spread along the agar surface; the possibility of their becoming mixed is eliminated; and the possibility of erroneous conclusions is avoided.

Recently the use of divided petri dishes and a multipoint inoculating device has been suggested for testing carbohydrate utilization (Lovelace and Colwell, 1968).

Nitrogen Requirements.   To determine the ability of microorganisms to utilize various nitrogen sources, a series of inoculations are made in a synthetic medium to which various nitrogen compounds are added in concentrations of 0.1–0.2%. Peptone, urea, potassium nitrate, ammonium sulfate, and individual amino acids are usually employed. Beijerinck medium may be used as a basal medium.

Beijerinck Medium

Distilled water	1000 ml
KH_2PO_4	2 g
$MgSO_4 \cdot 7H_2O$	0.5 g
$CaHPO_4$	0.5 g
Glucose	20 g

The auxanographic method is as follows. Solid medium is prepared with purified, well washed agar (2%) or Ionagar (Difco Laboratories, Detroit, Michigan), poured into test tubes, and sterilized. Discs of filter paper (5–10 mm in diameter) are saturated with solutions (1–2%) of nitrogen sources by submersion for 24 hours. They are then dried at 40 C and stored in ground glass-stoppered beakers. Test tubes containing agar are heavily inoculated with cultures of the microorganisms under investigation and then poured into petri dishes. After the medium solidifies the discs are placed on its surface. Growth will only be observed around those discs saturated with the nitrogen sources utilized by the given culture.

Reduction of Nitrates.   The ability of bacteria to reduce nitrates is tested as follows. Inoculation is usually made in beef-peptone or a nutrient broth containing 0.1% KNO_3. After a few days, gas formation is recorded. Presence of nitrites is determined by Griess reagent and presence of ammonia is detected with Nessler reagent. In either case, a drop of the culture and a drop of the reagent are mixed for the latter purpose.

Brough (1950) recommends a rapid micromethod for determining the ability of bacteria to reduce nitrates. A culture 18–24 hours old, which shows good growth, is selected. The cells are washed off, and from the resulting suspension an inoculation (of sufficient size to obtain good growth) is made in a small test tube containing 1 ml of beef-peptone or nutrient broth with 0.1% KNO_3. The medium, brought to optimal growth temperature for the given species prior to inoculation, is incubated for 15 minutes after inoculation. The nitrite analysis is then performed. If results are negative, this does not necessarily mean that the species under investigation is incapable of reducing nitrates. The investigation should be continued and inoculations made in a medium containing nitrate. It is recommended, in this case, to add 0.1–0.5% agar to the medium in order to obtain a semisolid medium.

Gelatin Liquefication.   Ability to liquefy gelatin is a diagnostic criterion of some value because the proteolytic activity of a bacterial species may be measured in a gelatin medium. The presence of hydrolysis and its characteristics are noted:

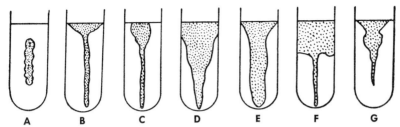

Fig. 74. Hydrolysis of gelatin along the line of stab. A, Deep; B, crateriform; C, turnip-like; D, conical; E, saccate; F, layer-like; G, funnel-like. The multiple characteristics of liquefaction as shown are those described in classical bacteriology. Modern workers simply record gelatin liquefaction as present or absent.

in layers (uniform over the width of the test tube), saccate, crateriform (Fig. 74). The presence of gelatinase can be detected by the method described by Thirst (1957). Clean slides are covered with a thin layer of 5% aqueous gelatin by submersion in melted gelatin. They are then dried in a vertical position and stored in a petri dish in a refrigerator. A drop of the broth culture of the species under study is placed on the surface of the gelatin film. The slide may be divided into sectors by a mark made on the reverse side, so that drops of several cultures may be placed on the gelatin. A drop of sterile broth may be used as control. The slides are placed in a petri dish containing moist filter paper and maintained at 22 C for 2–20 hours. After incubation the slides are placed in a mixture of formalin (1 volume) and a saturated aqueous solution of mercuric chloride (9 volumes) for 20–30 minutes, washed with water, and stained with a dilute aqueous solution of carbol fuchsin (1:100). If changes in the gelatin occur, clear zones will appear where growth of the culture occurs. This micromethod yields positive results only if there is an adequate amount of active gelatinase. Low-activity enzymes are revealed by the previously described method of inoculation in test tubes, followed by lengthy incubation, often up to 6 weeks, with 4 weeks being the minimum incubation time before cultures are recorded as negative.

Hydrolysis of gelatin may also be detected in Frazier medium. Frazier gelatin medium is prepared (see p. 255), poured into test tubes, and sterilized. One of the following solutions is prepared: $HgCl_2$ (15 g), HCl (20 ml), distilled water (100 ml); a 1% solution of picric acid; or a saturated solution of ammonium sulfate. The nutrient gelatin is poured into a petri dish. When it solidifies, the surface of the dish is divided into five sectors by marks on the outer surface of the bottom of the dish. The numbers of the cultures tested are written on the bottom of the dishes. Streak inoculations are then made of each culture in the sector designated. When the period of incubation is completed, the solution described above is poured onto the agar and allowed to stand for 5–10 minutes. The solution is then carefully decanted from the petri dish and the results are noted. The microorganisms which hydrolyze gelatin are surrounded by transparent zones. An unchanged medium is opaque.

Detection of Amylolytic Capability of Bacteria.   For detection of the enzyme amylase in bacteria (enzyme involved in the hydrolysis of starch) the liquid and solid media described in Chapter 6 (p. 196) may be used.

Detection of Ability of Microorganisms To Hydrolyze Fats.
Melted butter or a purified lipid is sterilized in test tubes. Beef-peptone or nutrient agar is prepared separately. The melted butter or lipid is poured into the bottom of a petri dish and spread in a thin layer. When the butter or fat hardens, a layer of melted and cooled beef-peptone or nutrient agar is carefully poured onto the fat surface. When the agar hardens in the petri dish it is then divided into sectors, and a cut is made in each sector with a sterile scalpel. A heavy inoculation is made in each cut with a streak from a 24-hour agar slant culture. After 18–24 hours the colonies which develop along the inoculation streaks are covered with a saturated solution of copper sulfate, and after 5–6 minutes the $CuSO_4$ is carefully decanted. The colonies of microorganisms which break down fats are stained azure, and around them a halo can be observed. When incubation is carried out for a longer period of time (3–4 days), the breakdown of fat may be determined by a decolorization of the culture. Copper sulfate solution may also be added.

Lipase may be detected by the method described by Sierra (1957) with the compounds Tween 20, 40, 60, or 80 (Atlas Chemical Industries, New Jersey).

Coagulation and Peptonization of Milk. Attack of bacteria on milk is another of the standard tests of classical bacteriology. It is a test of doubtful usefulness in bacterial taxonomy, except of course for dairy bacteriologists. However, since the tests employing milk are frequently cited, the methods are given here. Fresh skim milk is usually used for culturing bacteria. Removing the layer of fat from milk is most easily done with the aid of a siphon. The milk must form cream before it is skimmed. Skimmed milk is then divided into two parts. Bromocresol red is added to one portion, and bromophenol blue in a proportion of 1 ml of a 1.6% solution for 1 liter is added to the other. The milk and indicator solutions are mixed well, poured into test tubes in 7- to 8-ml amounts, and sterilized for 3 days in flowing steam. Inoculations of each culture are made in two test tubes containing different indicators.

Water-diluted evaporated milk or powdered skim milk may also be used. The milk is diluted with water containing an indicator (1 part milk to 4 parts water). In milk cultures, pH, coagulation, and peptonization are recorded. Two types of coagulation are distinguished: (1) acid, which results from the accumulation of acids and is associated with the change in the pH of the medium; in the presence of acid the coagulum no longer changes; and (2) rennin, in the presence of which the pH of the milk does not change, and which is usually followed by peptonization of the coagulum. Peptonization can also occur without coagulation. In Table 5 the approximate acidity of the milk is given according to changes in the color of the indicator.

Table 5. Degree of acidity of milk during growth
of microorganisms

| Acidity | Color of Milk and Its Reaction | | Approximate pH range |
	Bromocresol Red	Bromophenol Blue	
Very weak	blue to gray-green	—	6.2–6.8
Weak	subdued gray-green to greenish yellow	—	5.2–6.0
Medium	yellow; milk did not curdle	—	4.7–5.0
Strong	milk curdled	blue or green	3.4–4.6
Very strong	milk curdled	yellow	lower than 3.4

Dehydrated skim milk is available from various commercial firms which market bacteriological media. Reconstitution of skim milk is very simple—appropriate amounts of water are added. Addition of skim milk to a nutrient agar permits determination of casein hydrolysis by observation of a zone of clearing around colonies on a skim milk agar.

Litmus milk also is commercially available in prepared, dehydrated form.

Temperature Requirements.   In a medium which is optimal for a given species of bacteria, determination may be made of the minimal, optimal, and maximal temperature for growth. For this purpose the species under investigation is inoculated into a series of test tubes containing an appropriate nutrient medium. The inoculated test tubes, pre-incubated at the given temperatures prior to inoculation, are placed in an incubator at the various temperatures. Growth is then observed.

In spore-forming bacteria, the resistance of spores to high temperatures is measured. A spore-former is selected and a heavy suspension of spores is prepared in a sterile physiological solution or in water. A graduated pipette is used to place an equal amount of suspension into small test tubes containing 1 ml of a sterile medium. The inoculated suspensions are placed in a holder (Fig. 54) and kept in a boiling water bath for a specific period of time, usually 10 minutes. Inoculations are then made from the heated test tubes. The survival time of the spores in boiling water is recorded.

Methods of Determining Products of
Bacterial Metabolism

The following bacterial exchange products are most often determined: ammonia, indole, and hydrogen sulfide.

Determining the Production of Ammonia.   A large number of aquatic bacteria which use nitrogenous organic compounds are capable of decomposing them to ammonia, i.e. they possess among their enzymes the deaminases. The quantity of liberated ammonia in different species fluctuates greatly. Determination of the deaminating ability of bacteria is usually carried out in

Fig. 75. Test tube containing indicator papers.

liquid media, more often in beef-peptone broth or in peptone water. Indicator papers are suspended above the medium immediately after inoculation; the cotton stoppers are covered with rubber caps or material to prohibit gas passage; and the test tubes are sealed tightly (Fig. 75). The cultures are incubated at their optimal temperature. The following may be used as indicators: (1) papers wetted with Nessler reagent (see page 274 for preparation) and then dried; (2) red litmus paper; and (3) paper soaked in Krup reagent.

To prepare the indicator papers, filter paper is cut into strips 0.5-cm wide and 8- to 9-cm long. When treated with Nessler reagent, the color of the paper should be uniformly light gray. When ammonia is liberated by the culture, the lower end of the paper suspended above the medium acquires a brown shade. When red litmus papers are suspended, as the culture liberates ammonia, the lower ends turn blue. When using litmus paper to establish ammonia production, only freshly prepared papers should be used; commercial litmus papers are usually not sensitive enough.

Krup reagent is a mixture consisting of 1 part 3% sulfuric acid and 2 parts 1% aqueous solution of fuchsin (prepared from a saturated alcohol solution). The mixture is ready for use when it acquires a brownish tint. When the strips of filter paper wetted with Krup reagent dry, they should be colorless or slightly yellowish and should be kept in a beaker fitted with a ground glass stopper. When ammonia is present, the papers turn bright red.

When it is necessary to obtain a more accurate determination of deaminases of a culture, a quantitative analysis may be made of the liberated ammonia by distillation of the culture liquid to which magnesium oxide has been added; this is followed by titration. However, this is rarely done in routine bacteriology laboratories. Usually Nessler reagent added to 1- to 2-week-old cultures is used to determine production of ammonia.

Presence of specific amino acid decarboxylases in a bacterial species can be tested by the Møller, Falkow or Thornley methods. (See Colwell and Wiebe, 1970.)

Detecting the Formation of Indole.   The formation of indole by bacteria is associated with the presence of the enzyme tryptophanase, which breaks down the heterocyclic amino acid —tryptophan. Therefore, to test for the indole reaction, it is best to use a medium containing tryptophan, for example, a casein medium.

Casein Medium

Distilled water	1000 ml
Casein	10 g
Glucose	10 g
K_2HPO_4	1 g
$MgSO_4 \cdot 7H_2O$	0.5 g
NaCl	0.1 g
$FeSO_4 \cdot 7H_2O$	trace amounts

Casein is dissolved in 80 ml of 0.1 N solution of sodium hydroxide and the pH is adjusted to 7.4 with 1 N solution of hydrochloric acid. The remaining ingredients are dissolved in 920 ml of distilled water. The casein solution is mixed with the solution of salts, poured into test tubes in 10-ml amounts, and sterilized with flowing steam.

It is only necessary to introduce all the components of the medium if the species under investigation does not grow in the casein solution alone. A medium suitable for the indole test is commercially available in dehydrated form from several companies in the United States.

Detection of indole is made in the following manner. A culture is grown in beef-peptone or nutrient agar at its optimal temperature. The test for indole is carried out after 24 to 48 hours. A heavy inoculation is made from the beef-peptone or nutrient agar into test tubes containing a liquid casein medium, heated to 30–35 C. The inoculation is made in such a quantity that the medium becomes very turbid. The test tubes are incubated for 2 hours, and the indole is identified by the Ehrlich method. Two solutions, prepared beforehand, are used:

1. Paradimethylaminobenzaldehyde   1 g
   Ethyl alcohol (95%)   95 ml
   Hydrochloric acid (specific gravity, 1.19)   20 ml
2. Saturated aqueous solution of $K_2S_2O_3$

Five ml of the first solution are added to the liquid culture, then 5 ml of the second solution are added, and the medium is mixed by rotating the test tube between the palms of the hands. After 5 minutes the medium is colored red if indole is present.

Indole may also be detected using solution 1 and ethyl ether. In this method ether (5 ml) is added to the culture. The contents of the test tube are mixed by rotating between the palms of the hands. A few drops of solution 1 are then added along the wall of the test tube. If indole is present, the ether extract turns red.

In this manner indole and methylindole (skatole) are detected simultaneously. The same reagents may also be used to detect the presence of indole alone. For this purpose cultures are grown in a liquid medium. The stoppers of the test tubes containing cultures are replaced with stoppers of white absorbent cotton, the lower surface of which is moistened with 4–6 drops of solution 1, then with 4–6 drops of solution 2. The stopper is inserted in such a way that it is 1–1.5 cm from the surface of the culture. The test tube is placed in a holder (Fig. 54) in a boiling water bath for 15 minutes. The culture liquid should not come into contact with the stopper. Reddening of the cotton indicates the presence of indole (methylindole is non-volatile).

Indole may be detected in other ways. See also the Manual of Microbiological Methods (Society of American Bacteriologists, 1957).

Sal'kovskii Method

A culture is grown in beef-peptone broth for 7 days. It is heated first with 4 ml of 10% sulfuric acid, and then with 0.5–2 ml of a 0.05% solution of sodium nitrite. The reagents are poured down the test tube wall. The appearance of a red-violet ring indicates the presence of indole.

Oxalic Acid Strip Method

When inoculations are made, strips of filter paper—previously covered with crystals of oxalic acid—are suspended above the cultures. (These strips are soaked with a hot saturated solution of oxalic acid and dried.) A reddish stain on the lower part of the papers indicates the presence of indole.

Detecting the Formation of Hydrogen Sulfide.   The formation of H_2S by bacteria, during the decomposition of sulfur-containing compounds, is associated with the enzymatic decomposition of proteins or peptones composed of amino acids containing sulfur (cystine, cysteine, methionine). The ability of bacteria to form H_2S in protein media depends on the species, as well as the properties of the medium. Therefore, it is especially important to use strictly defined media. Usually inoculations are made in test tubes containing beef-peptone or hydrogen sulfide test broth, above which strips of filter paper— soaked with a saturated solution of lead acetate and then dried —are suspended. The presence of H_2S is based on the formation of metal sulfides. The escaping H_2S reacts with the lead acetate and reduces it; lead sulfide is formed, and the lower end of the paper turns brown or black. Instead of suspending papers, it is possible to moisten the gauze-covered test tube stoppers in the solution given above, which is sterilized beforehand; or a 0.5% solution of iron tartrate may be added to the medium. Blackening of the gauze in the first instance and blackening of the nutrient liquid in the second indicates the presence of the liberated H_2S.

The release of H_2S may also be detected in solid media. For this purpose, beef-peptone or hydrogen sulfide test agar is used. Lead carbonate or acetate is added to the medium. The lead salt used is finely ground in a mortar with gum arabic (0.5 g of gum arabic is added to 2–3 g of the salt) or a reagent-grade lead acetate is used. Blackening around the colonies or along the line of inoculation indicates H_2S production by the bacteria. Dehydrated, prepared media for detection of H_2S production by bacteria is also commercially available (Difco Laboratories, Detroit, Michigan).

Special media for detecting the liberation of H_2S are given below.

Lead Acetate–Egg Albumin Agar

Lead acetate (1%) is added to a melted beef-peptone agar (pH 7.2–7.4), which is then cooled to 45 C. Egg albumin is added to a final concentration of 2% and thoroughly mixed with the medium. The medium is then poured into test tubes and sterilized with flowing steam for 30 minutes on each of 3 days. Stab inoculation is made in an agar tube with a needle. If the bacterial species studied is capable of forming H_2S, then the medium turns black along the inoculation; this may also be due to mercaptans. The presence of mercaptans may be established with the addition of isatin-sulfuric acid (the medium turns green).

Phenol Red Agar

Tap or distilled water	1000 ml
Peptone	20 g
Lactose	10 g
Glucose	1 g
Solution A	10 ml
Solution B	10 ml
Solution C	10 ml
Indicator (phenol red)	20 ml
Agar	15 g

pH 7.0

Solution A: 2 g of iron sulfate (FeSO$_4 \cdot$ 7H$_2$O) are dissolved in the cold in 100 ml of distilled water.

Solution B: 0.8 g of sodium hyposulfite (Na$_2$S$_2$O$_3 \cdot$ 5H$_2$O) is dissolved in the cold in 100 ml of distilled water.

Solution C: 4 g of sodium sulfite (Na$_2$SO$_3$) are dissolved in 100 ml of cold distilled water.

Indicator: 0.1 g of phenol red is dissolved in 10 ml of warm 96% alcohol, and warm distilled water is added to obtain a total volume of 100 ml.

First the peptone and agar are prepared and dissolved in water. pH is then adjusted to 7.0 and the sugars, solutions, and indicator are added. The prepared medium is poured into test tubes in 3-ml amounts and sterilized in flowing steam for 20 minutes. The inoculation may be a stab done in an agar tube. The following are noted: blackening of the medium, gas formation, coloring of the surface and deep layers.

Many other tests may be performed on pure cultures of bacteria. A set of tests and testing methods are given in the series, *Identification Methods for Microbiologists*, published by the Society of Applied Bacteriology (Academic Press, 1966, 1968, 1969). Colwell and Wiebe (1970) have compiled a list of tests for aerobic, heterotrophic, marine and estuarine bacteria.

Finally, the handling of large amounts of taxonomic data can best be done by computer, utilizing the methods of numerical taxonomy. A handbook on numerical methods for bacterial taxonomy has been compiled by the Taxonomy Committee (Subcommittee on Numerical Taxonomy) of the American Society for Microbiology. The handbook, *Methods for Numerical Taxonomy*, is available on request from the American Society for Microbiology Publications Office, 4715 Cordell Avenue, Bethesda, Maryland 20014. Skerman, in *Guide to the Identification of the Genera of Bacteria*, second edition, also treats the methods of numerical taxonomy in bacterial taxonomy. Programming for the computer is discussed and a program is provided in a manual prepared by Colwell and D'Amico (1972).

Chapter 5

Quantitative Determination of Microorganisms in Water and Sediment

In investigations of water a quantitative enumeration of microorganisms is made as soon as possible after the samples are taken to avoid changes in bacterial content which inevitably arise if samples are allowed to stand. The total number of microorganisms is determined by the direct count method.

METHOD FOR DIRECT COUNT
OF BACTERIA IN WATER

This method yields a total count of bacteria which occur in water at the moment of sampling, including dead cells which have not yet undergone autolysis (however, according to results of several authors, the number of these cells is not supposed to be great). Modifications of direct count methods have been devised. The ultrafiltration (bacteria-retaining filter) method suggested by Razumov (1932) is most widely used because of its simplicity. The method consists of filtering a specific volume of a water sample through membrane filter No. 1 or No. 2 under negative pressure (characteristics of filters are discussed in Chapter 3). Bacteria occurring in a water sample are more or less uniformly distributed over the surface of the membrane filter.

When filtering water, membrane filters and an appropriate apparatus (or several apparatus, if several samples are to be filtered at one time) are needed. Various kinds of metal apparatus are often used for filtering water. However, stainless steel, glass or plastic filters are preferred (Fig. 34). Sterile plastic disposable filters are very useful in the field.

The metal stage and tube of the filter apparatus are flame-sterilized immediately before filtration. Flame-sterilization is performed with a wad of cotton soaked in alcohol and wrapped on a metal rod. All metal parts are flame-sterilized with the burning cotton wad. Two small circles of filter paper, cut to the diameter of the funnel, are placed on the surface of the lower part of the funnel; the membrane filter is placed on top of the filter papers. Before the filter is set in place, it is marked on the edge with a pencil. The number of the sample passed through the given filter is recorded, as well as the quantity of filtered water.

The membrane filter is held by the upper part of the apparatus, which is screwed to the lower part, or else both parts are joined with screw clamps. The flask, into which the funnel is placed, is partially evacuated. A water sample placed in the funnel is filtered completely.

Membrane filters should be treated before use. They are boiled several times in freshly distilled water (freshly distilled water freshly filtered through a membrane filter is preferable); the water should be changed two to three times. The purpose of boiling is to displace air retained in the pores of the filters, as well as to remove particles of solvents used in preparing the filters. When boiling filters the following requirements must be met so that the filters are not ruined: the water must be heated slowly and, when it begins to boil, it should not be allowed to boil violently.

It is necessary to make a preliminary check of each group of filters for the presence of bacterial contamination. For this purpose, after boiling, several filters must be dried, stained, and viewed under a microscope before any material is filtered. Filters contaminated with bacteria are unsuitable for counting purposes. Moreover, each filter must be carefully examined separately before use in order to reveal any defects such as perforations or cracks. Pre-sterilized filters are available from various companies. (See Plate 2.)

The amount of water sample which should be filtered depends on the type of water mass, the expected bacterial content of the water, and the diameter of the filter used. It is suggested that 50–100 ml of water be filtered when the water contains very small quantities of bacteria (i.e. artesian waters) and when the filter diameter is 2 cm; 5–10 ml (up to 25 ml) may be filtered for clean lakes, and 2–3 ml for ponds or estuaries. When water is badly contaminated, it should be diluted with water freshly filtered (but not distilled) through an ultrafilter.

After the water is filtered the filter is removed from the apparatus, placed on filter paper in a petri dish (where a piece of cotton soaked with formalin has already been placed), and dried. Filters may be stored in this way over a period of time or else be subjected immediately to further processing. Filters must be handled very carefully during transport, so that they are not damaged in any way and are well protected from dust. Filters dried in formalin vapors should be transported in special cartons made of thick cardboard with round holes cut according to the diameter of the filters. The cartons should be provided with a tightly fitting lid.

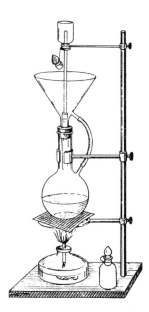

Fig. 76. Treating bottles with steam.

If it is impossible to filter samples under field conditions, they may be fixed with formalin and filtered later. The following are needed for fixation: (1) bottles with tightly fitting ground glass stoppers prepared as follows: the bottles are washed out with dichromate cleaning solution, flushed with distilled water, allowed to stand in a stream of steam (Fig. 76) for 10 minutes, dried immediately in a desiccator, and closed tightly; and (2) formalin filtered through a membrane filter. When a water sample is taken, a specific volume (e.g. 10 ml) is placed in the bottle with a pipette. Formalin is then added to a final concentration of 3%. The bottle is closed with the stopper and shaken well. A label is attached, on which are recorded the date, sample number, place of collection, and volume of water placed inside. The mouth and stopper of the bottle are covered with Mendeleyev putty (a hermetic sealer) or paraffin, tightly covered with a rubber cap, and finally bound. The samples can thus be transported safely to the laboratory. After the fixed water is filtered, the bottle is rinsed with water which has been pre-filtered through a membrane filter. This water, too, is in turn filtered.

The filter is further treated by staining. Staining is accomplished with a 5% aqueous solution of erythrosin in 5% phenol water as follows: filter papers, cut into small circles with a diameter slightly larger than the filters, are placed in a petri dish and moistened with the stain. Filters with the bacterial filtrates are placed on the dye-treated filter paper with the side containing the suspended bacteria facing up. The petri dish is then closed. Staining lasts for 2–3 hours. When it is completed, the excess dye is washed off in the same manner: by transposing the filters from one petri dish containing filter paper soaked in distilled water to another dish of the same type. The washing procedure is considered finished when the filter is a weak rose color. The washing procedure is very important for counts. Care must be taken in washing filters. When a filter has been properly washed, its surface retains a rose coloring; the bacteria, however, are bright red in color.

Membrane filters may also be stained by methods of combined staining which, according to a number of authors, in-increases the contrast of the image.

1. The membrane filter is stained for 1 hour with pheno-

lic erythrosin in the usual manner and then washed until it is a weak rose color. The washed filter is partly dried in an incubator and then dyed with freshly prepared Pfeiffer fuchsin for 1–2 minutes. The excess dye is washed off by transfer to water-soaked filter paper. A 0.5% solution of brilliant green may be added for staining.

2. The membrane filter is stained for 1 hour on filter paper soaked with a 1% aqueous solution of methylene blue. Excess dye is washed (by transfer to a water-soaked filter paper) until a sky-blue color is attained. The membrane is then partly dried and again stained with either a freshly prepared, filtered solution of Pfeiffer fuchsin or a 5% solution of erythrosin for 2–5 minutes, after which it is washed off and dried in formalin vapors.

Immediately before the count is made, the dry layer or, more appropriately, a portion cut out of it, e.g. half of the filter, is placed on a slide on which a drop of immersion oil has

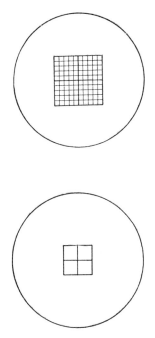

Fig. 77. Ocular micrometer grids.

already been placed. Another drop of immersion oil is placed on the layer. The specimen is covered with a thin cover glass, and the bacteria precipitated on the filter are counted. The calculation may be performed with a 15× ocular and 90× objective, using an anoptral or a phase-contrast apparatus, which gives high image contrast and allows the bacteria to be easily distinguished from non-organic particles. Microorganisms are enumerated by measuring the central portion of the field of vision with an ocular grid micrometer (Fig. 77). The dimensions of the squares of the grid should not be too small (1 × 1 mm is best). In each field of vision four central squares are counted. The fields of vision in which a count is made should not be less than 30. The ocular grids can be machined, if necessary, if graduating machines are available. Homemade preparations are usually not satisfactory, however, because of the large coefficients of error at the higher microscopic magnifications. Very small deviations in the lines of grids can lead to substantial errors in counts.

When counting bacteria the specimen is moved diagonally with the aid of an x-shaped stage. Rod-shaped forms, cocci, yeast cells, nitrogen-fixing-bacteria-like cells,[1] iron bacteria, and sulfur bacteria are counted separately, registering the number of cells of each group in a separate tally. Calculation by morphological groups makes it possible, on one hand, to characterize the microbial flora of the water mass and, on the other hand, to calculate the biomass of microbes from the data.

Bacterial numbers are calculated according to the following formula:

$$x = \frac{SN}{sV} \; ;$$

where S is the filtering area of the instrument (in μ^2); s is that area of the field of vision in which a count is being made (in μ^2); N is the mean number of microbe cells in the field of vision; and V is the volume of filtered water.

The area of the field of vision which is bordered by a

---

[1] The use by Rodina of the description, "nitrogen-fixing-bacteria-like cells," is retained since what is meant is the morphology of those cells capable of fixing nitrogen, i.e., the Azotobacteraceae and Rhizobiaceae.

specific portion of the micrometer grid is determined with a stage micrometer. The sides of that portion of the micrometer grid in which a count is made are calibrated several times with the graduations of the stage micrometer. The dimensions of a side of the square and then the area of the counted portion of the field of vision are thus calculated. The filtering area of the instrument is determined by measuring the inner diameter of the instrument according to the formula πr^2.

When the same magnification of one microscope and a single filtering apparatus are used, the proportion S/s is a constant value. Calculating a coefficient, K, makes it possible to simplify subsequent calculations. In this case the formula becomes:

$$x = \frac{KN}{V}.$$

To avoid false accuracy, results are usually rounded off for the direct counts. The following ranges for rounding off figures are used:

At Magnitudes of:	Round to the Nearest:
10,001–50,000	100
50,001–100,000	500
100,001–1,000,000	1,000
1,000,001–10,000,000	10,000
> 10,000,000	50,000
> 1,000,000,000	100,000

Thus, for example, the number 10,889,138 in 1 ml would be expressed as 10,900,000 in 1 ml; 10,810,138 would be 10,800,000 and so forth.

METHODS FOR DIRECT COUNT
OF BACTERIA IN SEDIMENT

Direct count of bacteria in sediment may be made in various ways, which are based on the method developed by

Winogradsky (1952) for soils. Filtration is the primary method for direct bacterial count in sediments.

Filtration

One gram of sediment is weighed out into a small sterile flask and covered with 25 ml of a 0.0004 N solution of sodium hydroxide previously filtered through a membrane filter; the flask is closed with a stopper and shaken for 20 minutes. The sediment mixture is allowed to stand for 1–2 minutes to allow large particles to precipitate. Then 1–2 ml of the suspension are placed in the funnel of the filter apparatus fitted with a No. 2 membrane filter. Two or three milliliters of water, filtered through a membrane filter, should be placed on the filter beforehand in order to provide a more uniform distribution of the bacteria. Once all the liquid is filtered, the membrane filter is processed as described earlier for making direct counts of bacteria in water.

Counting Bacteria with a Fluorescent Miscroscope
(Modified method of Pochon and Tardieux, 1962)

Preparation is made of a 0.1% solution of sodium pyrophosphate ($Na_4P_2O_7$) in water filtered through a membrane filter, and an aqueous solution of acridine orange (0.1% solution). One gram of sediment is weighed out in a small flask, to which 50 ml of the sodium pyrophosphate solution are added; the flask is then shaken for 20 minutes. Ten milliliters of the liquid are added to each of three sterile test tubes. The acridine orange solution is added to each: 0.5 ml to the first, 1.5 ml to the second, and 3 ml to the third.

Specimens are prepared from the sediment suspension: a specific volume is placed on a slide and covered with a cover glass. The edge of the specimen is sealed with Vaseline. The bacteria which show up bright green are counted in a fluorescent microscope.

The dimensions of the fields of vision in which counts are taken may be determined with an ocular micrometer. The area of the specimen is determined in the usual manner. Calculation of the number of bacteria is reported on a per gram of dry sediment basis.

DETERMINING THE BIOMASS OF BACTERIA
IN WATER AND SEDIMENT

Direct counts indicate the quantity of microbes in a specific volume of water or in a specific weight of sediment. In many cases, especially those involving the productivity of water masses, the biomass of bacteria must be determined as one of the components of the living material supported by the water mass. It is, moreover, a very important component in the cycling of matter and, at the same time, a fundamental nutritional source for planktonic and benthic invertebrates.

Bacterial biomass may be calculated from data obtained on total (direct) counts of distinct morphological groups—rods, cocci, nitrogen-fixing-bacteria-like cells, yeast, and sulfobacteria—with cell shape and size as identifying features. Microbial biomass is usually expressed in weight units per volume of water or weight of sediment (in milligrams per liter, grams per liter, and milligrams per gram).

To determine the volume of one cell, the formula for the volume of a sphere is used for cocci, that of a cylinder for rods, ellipsoid for yeast cells (in a body of water, the majority of yeast is represented by the species of the Torula group, the cells of which are generally oval), and so forth.

Measurement of cells is necessary in determining the volume of any species. To calculate the total biomass of bacteria, it is necessary to use one set of measurements (mean) for all cocci, bacilli, and so forth in the water mass. However, since not only the number, but also the dimensions of bacteria in a water mass vary depending on such environmental factors as temperature, depth, oxygen content, and nutrient concentration, it is first necessary to calculate separately the mean dimensions of the rods, cocci, and nitrogen-fixing bacteria cells which occur in the given water mass.

The following procedure can be used to determine the

mean dimensions of the bacterial cells: a significant amount of a water sample is partially filtered through a membrane filter. The filtration is stopped when about 0.5 ml of water, in which a large number of bacteria are concentrated as a result of the filtration process, remains above the surface of the filter. A pipette, thoroughly washed with a dichromate cleaning solution and steamed out well, is used to transfer a small amount of the water to the surface of a clean, degreased slide. Calculation of the dimensions of the bacteria is made in the usual manner, with an ocular micrometer. To simplify calculations, the bacteria can be lightly stained with a weak iodine or methylene blue solution. (The dimensions have to be calculated more than once because the measurements change with the environment.)

When obtaining mean dimensions of cells of each type (cocci, rods, and so forth) it is best to collect the largest possible number of measurements. It must be taken into consideration that calculations according to the arithmetic mean yield smaller figures for the diameters of cocci and widths of rods and, accordingly, smaller volume calculations. Therefore, the cubic mean should be used in determining the mean diameter of cocci; for the mean width of the rods, the root mean square is used. To simplify computations, the results of the measurements can be grouped into variation ranges, and the total number of measurements expressed as percentages or fractions of the total number of cells of each type. When calculating the mean by ranges, the means of the ranges are used. Below are a few examples of calculating mean volume and root mean square.

Calculating the Diameter of Cocci (D)

Diameter of Cocci (μ)	Mean of the Range	Number of Cases (n)	D^3	D^3n
0.1–0.5	0.3	6	0.027	0.162
0.5–1.1	0.8	34	0.512	17.408
1.1–1.5	1.3	29	2.197	63.713
1.5–1.9	1.7	16	4.913	78.608
1.9–2.5	2.2	9	10.648	95.832
2.5–3.1	2.8	6	21.952	131.712
		100		387.435

Accordingly, the mean diameter of the coccus used for mean volume is expressed as: [2]

$$\overline{D} = \sqrt[3]{\frac{387.435}{100}} = \sqrt[3]{3.87435} = 1.5706 \, \mu,$$

whereas the arithmetic mean yields 1.305 μ.

The mean volume of the coccus is

$$\overline{V} = \frac{1}{6} \cdot 3.14 \cdot 3.87435 = 2.0276 \, \mu^3$$

(according to the arithmetic mean calculation of the diameter, the mean volume is: $\frac{1}{6} \cdot 3.14 \cdot 1.305^3 = 1.165 \, \mu^3$, i.e. the decrease constitutes more than one-third of the volume).

When calculating the volume of rod-shaped bacteria, calculation of the root mean square (rms) of the width of the rods (d) is as follows:

Width of Rods (μ)	Mean of the Intervals (d)	Number of Cases (n)	d^2	d^2n
0.2–0.6	0.4	24	0.916	3.84
0.6–1.0	0.8	46	0.64	29.44
1.0–1.4	1.2	15	1.44	21.60
1.4–1.8	1.6	11	2.56	28.16
1.8–2.2	2.0	4	4.00	16.00
		100		99.40

The mean value of the width is

$$\overline{d}_{rms} = \sqrt{\frac{99.4}{100}} = 0.997 \, \mu.$$

This calculated value is also used when computing the volume of rods. (The arithmetic mean would be 0.9 μ.) If the length of the rods is 2 μ, then the mean volume is:

$$\overline{V} = \frac{1}{4} \cdot 3.14 \cdot 0.994 \cdot 2 = 1.56 \, \mu^3$$

(if the width calculated by arithmetic mean were used, the mean volume would be 1.27 μ^3, i.e. it would be one-fifth less).

---

[2] The line above the algebraic values indicates means.

The mean length of rod-shaped bacteria, unlike the mean width, must be determined by the arithmetic mean method.

Once the mean dimensions of cells of each group are determined, the biomass of the bacteria can be determined by means of the direct count data. The direct count yielded, for example, 200,000 cocci and 400,000 rod-shaped cells in 1 ml of water. The biomass of the bacteria is determined separately for each group.

Calculating the Biomass of Cocci

The mean diameter of cocci in a given water mass is, for example, 1 μ. Then the volume of one cell, using the formula for the volume of a sphere $V = \frac{1}{6}\pi D^3$, is

$$V = \frac{1}{6} \cdot 3.14 \cdot 1^3 = 0.52 \ \mu^3.$$

Since the volume of one coccus is 0.52 μ^3, and since the number of cells in 1 ml of water is 200,000, the biomass of the cocci in 1 ml is

$$0.52 \cdot 200,000 = 104,000 \ \mu^3$$

or

$$\frac{104,000}{1000^3} = 0.0001 \ mm^3.$$

The biomass of the bacteria in 1 liter equals 0.1 mm³. Thus, to determine (in mm³) the volume of bacterial biomass contained in 1 liter of water by direct count taken for 1 ml, if the volume of a cell is 0.52 μ^3 and there are 200,000 cells in 1 ml, the following calculation is made:

$$\frac{0.52 \cdot 200}{1000} = 0.104 \ mm^3.$$

Taking into consideration the fact that the specific gravity of bacteria may be assumed to equal that of water, i.e. 1, the weight of 1 mm³ of the biomass of the bacteria is 0.001 g, or 1 mg.

Therefore, to determine the cell biomass expressed in milligrams in 1 liter of water, the volume of one cell is multiplied by the number of thousands of cells observed by direct count in 1 ml; the result is then divided by 1000.

Calculating the Biomass of Rod-Shaped Cells

If in a given water mass the mean width of rods is 1 μ, and the mean length 1.78 μ, then the volume of one cell (V), using the formula for the volume of a cylinder, $V = \frac{1}{4}\pi d^2 h$, where d is width and h is height, is determined as follows:

$$V = \frac{1}{4} \cdot 3.14 \cdot 1^2 \cdot 1.78 = 1.397 \ \mu^3 \cong 1.4 \ \mu^3.$$

If the number of rods in 1 ml of water is 400,000, then the biomass in 1 liter is expressed as:

$$\frac{1.4 \cdot 400}{1000} = 0.56 \text{ mg.}$$

Calculating the Biomass of Yeast Cells

To determine the volume of one yeast cell (and any other cell having an oval shape, for example, certain species of nitrogen-fixing bacteria and certain species of *Chromatium*), the formula for the volume of an ellipsoid, $\frac{1}{6}\pi D d^2$, where D is the length of a cell and d is the width (along the short axis), can be used.

When the length of the yeast cell is 7 μ and the width is 5 μ, then the volume of one cell is $\frac{1}{6} \cdot 3.14 \cdot 7 \cdot 5^2 = 91.58 \ \mu^3$.

Further calculations of the biomass of yeast cells are carried out on the basis of direct count data by the method described above.

If in the water mass under study the yeast cells have a different shape, then calculations of biomass are made separately for each group: for example, the formula of a sphere is used for round yeast cells, the formula for an ellipsoid for oval cells. The total biomass of bacteria is obtained by adding together biomasses of the cells of each group.

The bacterial biomass for sediments is usually given

per gram; therefore, to obtain values expressed in milligrams, it is necessary to multiply the volume of one cell by the number of cells in 1 g, and divide the result by 1000^3.

In Tables 6 and 7 are given biomasses of bacteria in 1 liter of water (Table 6) and in 1 g of sediment (Table 7) for several cell dimensions and for various sample sizes.

DETERMINING GENERATION TIME OF BACTERIA AND PRODUCTION OF THE BACTERIAL BIOMASS

The total ultraplankton (ZoBell, 1946) observed in a water mass at any time is a measure of reproduction of the bacteria and their consumption by zooplankton and dying off. Therefore, it is essential that these measurements be made to describe the actual production of microorganisms in a water mass.

The time of one generation, i.e. the period of time for division of a number of cells, is considered an indicator of bacterial reproduction rate.

Two methods of determining generation time are employed in aquatic microbiology. The first method, suggested by Razumov (1948), is determination of the generation time in isolated samples of water which have been freed of zooplankton by filtration and submerged in a water mass for a specified length of time. The second method, developed by Kriss and Rukina (1952) is by calculation of dividing bacteria on submerged slides.

When calculating generation time, it is assumed that the number of dividing bacteria increases according to a geometric progression with a coefficient of progression equal to 2; that is, if the original number of bacteria, b, was equal to 10, then after the first division there should be 20 cells, after the second 40, after the third 80, and so on.[3] If t is the time of observation of change in number of bacteria, and g is time of one generation,

_____

[3] A number of authors take 1.8 as the coefficient, acting on the hypothesis that 90% of the cells are capable of division.

Table 6. Biomass of microbes in water *(In milligrams per liter)*

Cell Species	Volume of One Cell (μ^3)	Quantity of Cells in One Milliliter (in thousands)								
		50	100	150	200	300	400	500	1000	5000
Cocci	0.178	0.0089	0.0178	0.0267	0.0356	0.0534	0.0712	0.089	0.178	0.89
	0.265	0.0128	0.0256	0.0384	0.0512	0.0768	0.1024	0.128	0.256	1.28
	0.52	0.026	0.052	0.078	0.104	0.156	0.208	0.260	0.52	2.6
	1.72	0.086	0.172	0.258	0.344	0.516	0.688	0.860	1.72	8.6
	2.07	0.1035	0.207	0.3105	0.414	0.621	0.818	1.035	2.07	10.35
Rod-shaped cells	0.4	0.02	0.04	0.06	0.08	0.12	0.16	0.20	0.4	2.0
	0.6	0.03	0.06	0.09	0.12	0.18	0.24	0.30	0.6	3.0
	0.8	0.04	0.08	0.12	0.16	0.24	0.32	0.40	0.8	4.0
	1.0	0.05	0.10	0.15	0.20	0.30	0.40	0.50	1.0	5.0
	1.2	0.06	0.12	0.18	0.24	0.36	0.48	0.60	1.2	6.0
	1.4	0.07	0.14	0.21	0.28	0.42	0.56	0.70	1.4	7.0
	1.6	0.08	0.16	0.24	0.32	0.48	0.64	0.80	1.6	8.0
	2.0	0.10	0.20	0.30	0.40	0.60	0.80	1.00	2.0	10.0
Bacilli	2.4	0.12	0.24	0.36	0.48	0.72	0.96	1.20	2.4	12.0
	2.7	0.135	0.27	0.405	0.54	0.81	1.08	1.35	2.7	13.5
	2.9	0.145	0.29	0.435	0.58	0.87	1.16	1.45	2.9	14.5
	3.5	0.175	0.35	0.525	0.7	1.05	1.4	1.75	3.5	17.5
	4.8	0.240	0.48	0.72	0.96	1.44	1.92	2.40	4.8	24.0
	5.9	0.295	0.59	0.885	1.18	1.77	2.36	2.95	5.9	29.5
	16.0	0.80	1.60	2.40	3.20	4.80	6.40	8.0	16.0	80.0
Nitrogen-fixing-bacteria-like cells	3.52	0.176	0.352	0.528	0.704	1.056	1.408	1.76	3.52	17.6
	4.07	0.2035	0.407	0.61	0.814	1.221	1.628	2.035	4.07	20.35
	12.08	0.604	1.208	1.812	2.416	3.624	4.832	6.04	12.08	60.4
	17.6	0.88	1.76	2.64	3.52	5.28	7.04	8.80	17.60	88.0
	24.5	1.225	2.45	3.675	4.90	7.35	9.8	12.25	24.50	122.5
	35.3	1.765	3.53	5.295	7.06	10.59	14.12	17.65	35.30	176.5
Yeast	52.0	2.60	5.20	7.80	10.40	15.6	20.8	26.0	52.0	260.0
	91.6	4.58	9.16	13.74	18.32	27.48	36.64	45.8	91.6	458.0

Table 7. Biomass of microbes in sediments (*In milligrams per gram*)

Cell Species	Volume of One Cell (μ^3)	Quantity of Cells in One Gram (*in thousands*)				
		100	200	500	1000	1,000,000
Cocci	0.178	0.0000178	0.0000356	0.000089	0.000178	0.178
	0.256	0.0000256	0.0000512	0.000128	0.000256	0.256
	0.52	0.000052	0.000104	0.00026	0.00052	0.52
	1.72	0.000172	0.000344	0.00086	0.00172	1.72
	2.07	0.000207	0.000414	0.001035	0.00207	2.07
Rod-shaped cells	0.4	0.00004	0.00008	0.0002	0.0004	0.4
	0.6	0.00006	0.00012	0.0003	0.0006	0.6
	0.8	0.00008	0.00016	0.0004	0.0008	0.8
	1.0	0.0001	0.0002	0.0005	0.001	1.0
	1.2	0.00012	0.00024	0.0006	0.0012	1.2
	1.4	0.00014	0.00028	0.0007	0.0014	1.4
	1.6	0.00016	0.00032	0.0008	0.0016	1.6
Bacilli	2.4	0.00024	0.00048	0.0012	0.0024	2.4
	2.7	0.00027	0.00054	0.00135	0.0027	2.7
	2.9	0.00029	0.00058	0.00145	0.0029	2.9
	3.5	0.00035	0.0007	0.00175	0.0035	3.5
	4.8	0.00048	0.00096	0.0024	0.0048	4.8
	5.9	0.00059	0.00118	0.00295	0.0059	5.9
	16.0	0.0016	0.0032	0.008	0.016	16.0
Nitrogen-fixing-bacteria-like cells	3.52	0.000352	0.000704	0.00176	0.00352	3.52
	4.07	0.000407	0.000814	0.002035	0.00407	4.07
	12.08	0.001208	0.002416	0.00604	0.01208	12.08
	17.6	0.00176	0.00352	0.0088	0.0176	17.6
	24.5	0.00245	0.00490	0.01225	0.0245	24.5
	35.3	0.00353	0.00706	0.01765	0.0353	35.3
Yeast	52.0	0.0052	0.0104	0.026	0.052	52.0
	91.6	0.00916	0.0183	0.0458	0.0916	91.6

then the number of generations is $K = t/g$. The number of bacterial cells (B) at the end of the experiment is

$$B = b2^K = b2^{t/g}.$$

After the logarithm is taken, the following is obtained:

$$\log B = \log b + \frac{t}{g} \log 2; \qquad \log B - \log b = \frac{t}{g} \log 2;$$

$$g = \frac{t \log 2}{\log B - \log b}.$$

This formula is used for a single determination of generation time in a flask containing a water sample. It must be noted that conditions in a flask differ from those in nature; because of the removal of plankton and organic particles, water does not circulate freely in a flask; exchange products accumulate; antagonisms occur; and bacteria attach to the walls of the flask. Therefore, application of data on division of bacteria in flasks to an entire water mass must be done with great care and reservation. In particular, it is advisable to carry out rather short-term experiments (4–6 hours or, in any case, less than 8 hours).

The method of counting dividing bacteria on submerged slides, developed by Kriss and Rukina, is outlined as follows: the slides are placed in the water mass in the usual way, kept there 1, 2, or 3 days, then removed, lightly dried, and stained with a solution of erythrosin. Because bacterial cell division depends to a large extent on temperature, during the exposure time of the slides the temperature should be determined three times a day or monitored continuously. The stained slides are examined under high magnification, and in an area of 4 cm² (8 cm² according to Kriss and Rukina) all fields are observed one after another. All solitary cells, all microcolonies, and the number of cells in each colony are counted.

According to the method of random counting it is more expedient to calculate 100 fields of vision at specific intervals, moving along the diagonal. This is verified by data analyses done by Rodina.

The generation time in this case is calculated as a mean (according to the division of different species in microcolonies).

It is assumed that each microcolony is formed from one cell, that is, $b = 1$, $\log b = 0$.

In that case the formula for calculating the generation time is simplified and the following is obtained:

$$g = \frac{t \log 2}{\log B} = \frac{0.301\, t}{\log B} \qquad (\log 2.0 = 0.30103).$$

The value of B, the number of cells in each colony, is 1 or greater. Therefore, instead of $\log B$, the mean value should be substituted in the formula, i.e.

$$\bar{g}_1 = \frac{t \log 2}{\log B} \qquad \text{and} \qquad \bar{g}_2 = \frac{t \log 2}{\log B}.$$

The mean may be calculated in different ways: either as the mean logarithm of B ($\overline{\log B}$) or as the logarithm of the mean of B ($\log \bar{B}$). Below is given an example of the calculation. It is assumed that the duration of the experiment is 8 hours.

No. of cells in colony (B)	log B
1	0.00000
4	0.60206
5	0.69897
10	1.00000
20	2.30103

The mean number of cells in the colonies is:

$$\bar{B} = \frac{1 + 4 + 5 + 10}{4} = \frac{20}{4} = 5.$$

The logarithm of the mean number of cells is:

$$\log \bar{B} = \log 5 = 0.69897.$$

The mean logarithm of the number of cells is:

$$\overline{\log B} = \frac{0 + 0.60206 + 0.69897 + 1.00000}{4} = \frac{2.30103}{4} = 0.57526.$$

The mean generation time using the log of the mean number of cells is:

$$\bar{g}_1 = \frac{0.30103 \cdot 8}{0.69897} = 3.4454 \text{ hours.}$$

The mean generation time using the mean logarithm of the number of cells is:

$$\bar{g_2} = \frac{0.30103 \cdot 8}{0.57526} = 4.186 \text{ hours.}$$

To explain the significance of these two computed mean generation times, the means may be substituted in the formula

$$1 \times 2^{t/g} = B.$$

The substitution $\bar{g_1} = 3.4454$ yields

$$1.2^{8/3.4454} = 1.2^{2.3222};$$

$$2.3222 \log 2 = 2.3222 \cdot 0.30103 = 0.69897 = \log 5$$

(logarithm of the mean number of cells). Thus, after calculating the mean generation time and after substituting it in the formula, we obtain the exact same mean number of cells, 5, and sum of cells, 20, as in the original data.

The substitution $\bar{g_2} = 4.186$ yields

$$1.2^{8/4.186} = 1 \times 2^{1.91133}; \qquad 1.9111 \log 2 = 0.57526$$

(mean logarithm of the number of cells).

If, according to this mean logarithm, we find the number to which it corresponds, then it is not 5, but 3.766. Therefore, evidently, the first method of computation is preferred when determining the mean generation time.[4]

In contrast to the simplified numerical example presented for calculating the mean generation time, the number of cells on the submerged slides may vary in greater ranges, and the number of microcolonies may be very large. To simplify an awkward calculation, an interval variation series can be constructed and a weighted mean calculated according to the means of the intervals with weighting according to the frequencies of occurrences of each group. It is understood that, when calculating the mean $\bar{g_2}$ (by the method of the mean logarithm of the number of cells), the grouping and construction of the variation series must be derived according to the logarithms of the numbers of cells in a colony.

---

[4] This question requires further analysis.

The following example illustrates this:

Number of cells in colony (B)	1	2	3–5	6–10	11–15	16–25	26–50
Number of colonies (n)	421	240	125	92	80	183	200

The mean number of cells in the colonies is therefore:

$$\overline{B} = \frac{1 \times 421 + 2 \times 240 + 4 \times 125 + 8 \times 92 + 13 \times 80 + 20.5 \times 183 + 38 \times 200}{421 + 240 + 125 + 92 + 80 + 183 + 200}$$

$$= \frac{14,529}{1,341} = 10.834.$$

The number of colonies may be expressed as a percentage of the total number, e.g. 1,341 is taken to be 100 or, in fractions, 1,341 is 1. The mean number of cells in a colony does not change from this.

If it is assumed that the experiment ran for 48 hours; $\log \overline{B} = 1.03479$. The mean time of one generation is then

$$\overline{g} = \frac{0.301 \times 48}{\log 10.834} = \frac{14.448}{1.03479} = 13.96 \text{ hours.}$$

Absolute growth is important for calculating increase of bacterial biomass in regard to its nutritive significance. It is possible to consider that the bacterial biomass in a body of water is maintained at a specific level for a more or less long time. It is then possible to assume that the increase is used as nutrient for zooplankton and to compensate for bacterial lysis. Therefore, it appears possible to assume that the absolute increase of bacterial mass is not greater than the mass calculated by direct count, that is, the absolute increase is 100% of biomass for the period of time equal to a generation time.

In this case, for 1 day this increase t/g would be 24 hours/ g hours and would be expressed as number of generations per day (the hours would cancel). In the first example presented, when the generation time is 3.4454 hours per day, then there will be

24/3.4454 = 6.97 generations of bacteria. In the second example, when the generation time is 13.96 hours, the number of generations per day will be 24/13.96 = 1.72.

When this method of calculation is used, it is assumed that one and the same, on average, bacterial growth mass is doubled. Therefore, it is sufficient to multiply this same increase by the number of generations. Such an assumption and calculations are possible only in the case of a water mass where the number of bacteria can be considered constant over a more or less long period of time, when the increase or decrease of the size of the biomass need not be considered. Such a state is not true in flask culture and on submerged slides, where the numerical change of bacteria is considered.

The increase and decrease of microbial mass can occur in any water mass and given dynamics must be taken into consideration, e.g. seasonal and daily changes. In all cases in which the dynamics as well as the rates of reproduction cannot be ignored, the calculation of growth increases and decreases must be made with formulas for geometric progression (because here development proceeds on the basis of what has already been reached—daughter cells participate along with the original mass in creation of new increase).

The possibility of accepting bacterial biomass in a water mass as constant over a more or less lengthy period of time is assumed in the formulas for calculation provided by Ierusalimskii (Kriss and Rukina, 1952) and Ivanov (1955). These formulas differ only in appearance, not in substance. Ierusalimskii defines t/g in times per day. To calculate absolute growth increase according to Ierusalimskii's formula, the bacterial biomass must be multiplied by t/g. In Ivanov's formula, absolute growth and bacterial production are not calculated for 24 hours, but in a general form for any number of hours. Multiplication by the number of hours according to either of the formulas assumes that the real biomass remains at the same level, and that the increase is consumed and covers the decrease from lysis.

Generation time, growth increase, and biomass must be determined again for altered conditions of bacterial growth. In particular, it is important to measure these indicators repeatedly to calculate monthly, seasonal, and yearly characteristics. It should be clear that it is necessary to investigate special features of individual zones of a water mass when calculating total mass.

DETERMINING NUMBERS OF DISTINCT PHYSIOLOGICAL GROUPS OF BACTERIA

In determinations of the total number of bacteria it is not possible to judge the character of biogenous processes which occur within the water mass. The energy of the transformations of various compounds of nitrogen, phosphorus, and other elements is determined not by the total number of bacteria, but by presence of bacteria of definite physiological groups, and by the activity of these microbes.

Among microbiological sampling procedures there are not many methods for determining composition of different groups in a sample. Winogradsky's method of enrichment media is the principal method at the present time. Presence of bacteria of a certain group is observed by growth of these bacteria in enrichment media most favorable for growth of each group and according to chemical changes effected in these media.

Quantitative content of bacteria of different groups is determined in two ways: by inoculating known aliquots of water or sediment onto solid enrichment media and then counting the colonies which grow, for example, counting all or only selected colonies of bacteria; or by inoculating various volumes of water and sediment into liquid enrichment media (the method of titer or extinction dilution).

Enumerating Bacteria by Inoculating Solid Nutrient Media

In this approach to determining the number of bacteria of a certain physiological group, two methods of inoculating the same solid medium may be used: deep or surface inoculation. When agar and gelatin nutrient media are used, the first method is usually preferred, because it is simpler. When silica gel is used, the second method is followed.

Regardless of which method is used, the amount of water and sediment used for inoculations must be determined beforehand. Different amounts of inoculum are added depending on both the physiological group marked for investigation and the character of the water mass. When inoculations are made to estimate protein-degrading bacteria in water taken from uncontaminated water masses, 0.5–1.0 ml is used. When water samples are taken from clearly contaminated bodies of water, 0.01–0.0001 ml is used. Water from contaminated water masses is diluted with water which has been sterilized in an autoclave in small sterile flasks. Dilutions may be made in various ways. One method is shown in Fig. 78.

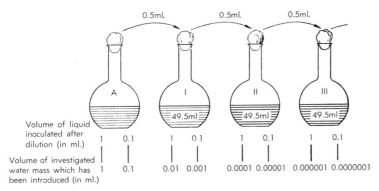

Fig. 78. Water dilution flow chart. *A, I, II, III,* original and subsequent dilutions. *Upper volumes,* volume of liquid inoculated after dilution (in milliliters); *lower volumes,* volume of original water sample inoculated (in milliliters).

It is necessary to have a minimum of two previously designated dilutions from each water sample to be used for inoculations. Inoculations in petri dishes must be made from each dilution at least in duplicate.

When dilutions are prepared, the following precautions must be taken. The pipette used to collect the water being investigated must be introduced into the bottom layer of the water. Before transferring the water from the pipette into the test tube containing sterilized water, the edges of the pipette must be flame-sterilized.

If the entire quantity of water collected with the pipette is to be delivered into the test tube, then the pipette is blown out to the end, and the last drop is removed by touching the wall of the test tube at the surface of the water.[5] While the water is delivered from the pipette into the test tube, the pipette should be placed no more than 3 mm under the surface of the water (to avoid washing bacteria from the outer surface of the pipette). The test tube should be held at an angle to prevent bacteria from entering the tube from the air. The edges of the test tube are flamed before the cotton stopper is reinserted or the cap replaced. The cotton stopper or test tube cap should be held so that the part which was in the test tube faces down and does not touch the hand (this is usually done whenever inoculations, i.e. standard bacteriological aseptic techniques, are made).

Pipettes of 1 or 2 ml with 0.01-ml graduations are used for inoculations. They must be plugged at the wide end with cotton stoppers, thoroughly wrapped in paper, and sterilized. They are unwrapped directly before inoculations are made; and the ends are sterilized by passing through a burner flame. Sterile disposable plastic or glass pipettes are very useful in the field.

A separate pipette is used for each dilution. As an exception to this, when there is a lack of pipettes one pipette may be used for all inoculations, provided that the higher dilutions are transferred first, proceeding to lower ones.

The following method is employed when total bacterial content of the water mass is very small. A water sample (100 ml) is filtered through a membrane filter, which is then placed in a petri dish on the nutrient agar surface and allowed to grow. Bacteria grow on the surface of the filter at the expense of those minimal quantities of nutrient substances which rise up through the pores of the filter. The colonies which develop on the filter are stained with erythrosin and are counted.

Because of the abundance of microorganisms, when sediments are sampled, inoculations are made from dilutions. There are several dilution methods. One method is illustrated in Fig. 79. To perform a series of sediment dilutions it is necessary to use sterile glassware: small flasks with a 150-ml capacity (pref-

---

[5] The procedure may vary, depending on whether Mohr measuring pipettes or serological (measured to deliver) pipettes are used.

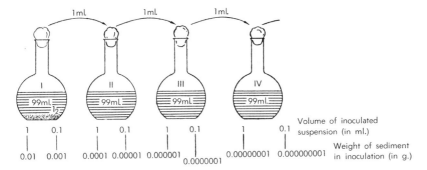

Fig. 79. Soil dilution chart. *I–IV,* original and subsequent dilutions. *Upper volumes,* volume of inoculated suspensions (in milliliters); *lower volumes,* weight of original soil inoculated (in grams).

erably flat-bottomed), test tubes, pipettes of 100, 10, and 1 ml (the latter with 0.1-ml graduations), and sterile water (tap water or water from the investigated source) or a sterile physiological saline solution.

One gram of sediment sample is weighed out and placed in a small sterile flask, and sterile conditions are maintained at all times. The sediment sample is covered with 99 ml of sterile tap water and thoroughly shaken for 10–20 minutes. In this way the original suspension is obtained. The shaking serves both to break up clumps of bacteria in the sediments and to separate bacteria from sediment particles. In sediments, as in soils, many bacteria are strongly adsorbed by particles of benthic deposits. In preparing a suspension for inoculation, many bacteria are not washed from the sediment particles and thus cannot be counted when inoculations are made. To loosen thoroughly the bacteria from sediment particles, small, sterile, glass balls or glass beads are placed in a flask containing a specific suspension of sediment, thus permitting a more accurate count of bacteria after shaking the suspension with the beads. It is even better to use a special shaking apparatus which produces greater uniformity in particle size by breaking up bacterial clumps in different sediment samples. The suspension is shaken using the same number of revolutions throughout. When working with a shaking apparatus, a larger sediment sample may be used. The better dispersed the suspension of sediment, the less the amount of error in enumerating the bacteria. If no shaker is available, shaking should be done by hand for 10–20 minutes.

Inoculations are made from suspensions of the following amounts of sediment: 0.01, 0.001, 0.0001, 0.00001, 0.000001 g; for smaller inocula it is necessary to resort to greater dilutions. The dilutions are made as follows. A 1-ml quantity is taken from the central layer of the original suspension with a sterile pipette and placed in a sterile test tube containing 9 ml of sterile water (in this way, a dilution for inoculations of 0.001 g is prepared). The same pipette is used to take 1 ml of the suspension just obtained and place it in a flask containing 99 ml of sterile water (in this way, a dilution for inoculations of 0.00001 g is prepared). After the flask is shaken thoroughly, 1 ml of solution is placed in a test tube containing 9 ml of sterile water, another milliliter is placed into a flask containing 99 ml of sterile water, and so on.

After the dilutions are prepared, inoculations are made. The test tube and flasks containing the sediment dilutions are shaken for 5–10 minutes. A sterile pipette is then used to inoculate two petri dishes for each dilution. Agar is quickly added, mixed well with the inoculum, and allowed to solidify. After solidification, all inoculated dishes are inverted and placed in an incubator. Surface plate counts are made by depositing 0.1 ml of a given dilution on the surface of a nutrient agar plate (predried by incubation overnight in a 37 C incubator) and dispersing the inoculum over the entire surface of the plate with a sterile bent glass rod or glass "hockey stick."

Dilutions may be prepared in another way. One or ten grams of sediment are weighed out and placed in a round-bottomed flask containing corresponding 99 [6] or 90 ml of sterile water. After thorough shaking in a shaker or by hand (20 minutes in this case), a sterile pipette is used to remove 1 ml from the central layer of the suspension and place it in a second, identical flask. This is shaken for 5 minutes, after which 1 ml is

---

[6] A 100-ml sterile volumetric pipette is used to place 100 ml of sterile water in a sterile flask, after which 1 ml is removed with a graduated pipette.

Fig. 80. Automatic colony counter.

transferred to a third flask, and so forth. The inoculations are performed as stated earlier.

When different physiological groups are sought, colony counts are made at different times for each group. When protein-degrading bacteria are enumerated, counts are made at 48 hours.

The most accurate means of determining the number of colonies is by direct count of the entire agar plate. The petri dish is held upside down and the count is made with a magnifying lens. As the count is made, the colonies are marked off

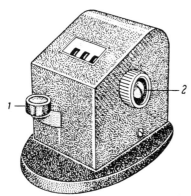

Fig. 81. Single-key calculator. *1,* Key, *2,* clearing knob.

on the bottom of the dish with a brush pen or a wax pencil. In most microbiology laboratories, the count is made with a magnifying lens or under a dissection microscope.

The simplest counting method is with an automatic counter (Fig. 80). Use of this instrument enables the worker to focus his entire attention on the colonies. When an automatic calculator is not available, one of the following may be used: in Russia, a single-key calculator (Fig. 81) (such as those produced in the Kievan Medical Apparatus Factory), and in the United States, a simple hand-operated counter. These calculators are designed for counting any type of similar objects. Colony counters, devices with graduations in squares (Fig. 82) or in sectors, may be used when the number of colonies is large and it is impossible to count them directly.

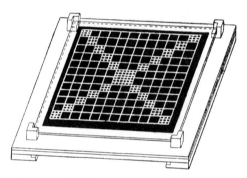

Fig. 82. Wolfhügel counting instrument.

In place of colony counters, which are not always available, a simpler device may be used for counting colonies: an ordinary transparent celluloid ruler with a centimeter scale in which a window 1-cm square is cut out. This window is placed on various locations on the dish and colonies are calculated over an area of 10 cm². The mean number of colonies is determined for 1 cm², and is converted into a number for the entire area of the dish. From all the inoculations a mean is computed for 1 g of wet sediment. The final conversion of the number of bacteria must be made for 1 g of dry sediment. To do this the moisture content of the sediment must be determined. Small weighing beakers washed out with dichromate cleaning solution are dried in a desiccator at 110 C until a constant weight is obtained. A given quantity of sediment is placed in the weighing beaker, which is then weighed with the sediment. The sediment is dried at 100 C until a constant weight is reached. When the weight is constant, the moisture content can be calculated according to the weight loss.

Determining Numbers of Bacteria by Titer
(Most Probable Number—MPN)

In computing titers or MPN, selective media for a particular physiological group of bacteria are inoculated with various quantities of water and sediment. Usually, the quantity of inoculum selected for each successive inoculation is 10 times greater than the preceding one; however, in a number of instances it is necessary to use other ratios. In bacteriology a titer is referred to as the smallest quantity of inoculum (of water or sediment) which yields growth of the desired microorganisms when a nutrient medium is inoculated. A titer may be converted to quantity of bacteria in a specific volume of water (1 ml, 1 liter). The presence of the bacterial group sought is determined by observing the inoculations, by microscope survey, or by chemical reaction.

Both the selection of the medium and its nutrient qualities have great significance in the results of this approach to estimation of bacterial populations. The best possible selective medium must possess high nutritive qualities for the specific group of bacteria.

The accuracy of quantitative determinations of bacteria by the titer method depends on the number of parallel inoculations made: the more made, the greater the accuracy. It is appropriate to make 10 parallel inoculations with each given quantity of water; however, when several physiological groups are being studied, such a number of inoculations greatly complicates the work. Therefore, in many instances the number of parallel inoculations has to be limited (but not to less than 3). It is not recommended that the number be reduced when only one group is being enumerated.

Usually inoculations are run to determine the number of protein-degrading bacteria.

Inoculations are made with sterile pipettes with increasing quantities of inoculum; the quantities differ according to the physiological group under investigation. For example, when determining the titer of ammonifying bacteria in water, inoculations of 0.00001, 0.0001, 0.001, 0.01, and 0.1 ml are prepared. When inoculating for nitrogen-fixing bacteria in a water mass, it is better to select large volumes of water: 0.1, 0.5, 1.0, 3.0, 5.0, 10.0 ml, and so on.

Quantities of water beginning with 0.1 ml and greater are inoculated without dilution by means of graduated pipettes. Smaller quantities are inoculated after appropriate dilutions of the water under investigation with sterile water.

Sediment inoculations are made from those dilutions prepared for bacterial counts in beef-peptone agar.

When sampling any group of bacteria by inoculation of liquid media, incubation is also made of control flasks or test tubes containing an uninoculated medium, in which the same chemical reactions are carried out as in the inoculations. The control eliminates the possibility of ascribing purely chemical reactions to biochemical transformations observed in the inoculated media.

When growth is obtained in a series of dilutions, it is necessary to calculate the number of bacteria of the selected group in 1 ml of water or 1 g of sediment or both. In extinction dilutions growth is sometimes not detected in all parallel inoculations. In this case (when the dilution is 10×), Mc-Cready's tables are used to calculate the number of microorganisms. These tables are based on mathematical statistics methods (Table 8). To use these tables, a "numerical" characteristic is constructed. This is a number with a threefold meaning: the first digit is the number of test tubes containing the last dilution which yielded growth in all test tubes; the second and third digits are the number of test tubes in which bacterial growth is observed in the next two following dilutions. In the latter case, if there was growth in subsequent higher dilutions, the number of such test tubes is added to the last digit.

Once the numerical characteristic has been determined, the corresponding probable number of bacterial cells in 1 ml of the dilution is found in the table. After this, a conversion to 1 ml of water or 1 g of sediment is computed. For example,

Table 8. Table for analyzing results of quantitative count of bacteria by dilution method (MPN)

Numerical Characteristic	Probable Number in Presence of Parallel Test Tubes			Numerical Characteristic	Probable Number in Presence of Parallel Test Tubes			Numerical Characteristic	Probable Number in Presence of Parallel Test Tubes		
	3	4	5		3	4	5		3	4	5
000	0.0	0.0	0.0	222	3.5	2.0	1.4	433	—	30.0	—
001	0.3	0.2	0.2	223	4.0	—	—	434	—	35.0	—
002	—	0.5	0.4	230	3.0	1.7	1.2	440	—	25.0	3.5
003	—	0.7	—	231	3.5	2.0	1.4	441	—	40.0	4.0
010	0.3	0.2	0.2	232	4.0	—	—	442	—	70.0	—
011	0.6	0.5	0.4	240	—	2.0	1.4	443	—	140.0	—
012	—	0.7	0.6	241	—	3.0	—	444	—	160.0	—
013	—	0.9	—	300	2.5	1.1	0.8	450	—	—	4.0
020	0.6	0.5	0.4	301	4.0	1.6	1.1	451	—	—	5.0
021	—	0.7	0.6	302	6.5	2.0	1.4	500	—	—	2.5
022	—	0.9	—	303	—	2.5	—	501	—	—	3.0
030	—	0.7	0.6	310	4.5	1.6	1.1	502	—	—	4.0
031	—	0.9	—	311	7.5	2.0	1.4	503	—	—	6.0
040	—	0.9	—	312	11.5	3.0	1.7	504	—	—	7.5
041	—	1.2	—	313	16.0	3.5	2.0	510	—	—	3.5
100	0.4	0.3	0.2	320	9.5	2.0	1.4	511	—	—	4.5
101	0.7	0.5	0.4	321	15.0	3.0	1.7	512	—	—	6.0
102	1.1	0.8	0.6	322	20.0	3.5	2.0	513	—	—	8.5
103	—	1.0	0.8	323	30.0	—	—	520	—	—	5.0
110	0.7	0.5	0.4	330	25.0	3.0	1.7	521	—	—	7.0
111	1.1	0.8	0.6	331	45.0	3.5	2.0	522	—	—	9.5
112	—	1.1	0.8	332	110.0	4.0	—	523	—	—	12.0
113	—	1.3	—	333	140.0	5.0	—	524	—	—	15.0
120	1.1	0.8	0.6	340	—	3.5	2.0	525	—	—	17.5
121	1.5	1.1	0.8	341	—	4.5	2.5	530	—	—	8.0
122	—	1.3	1.0	350	—	—	2.5	531	—	—	11.0
123	—	1.6	—	400	—	2.5	1.3	532	—	—	14.0
130	1.6	1.1	0.8	401	—	3.5	1.7	533	—	—	17.5
131	—	1.4	1.0	402	—	5.0	2.0	534	—	—	20.0
132	—	1.6	—	403	—	7.0	2.5	535	—	—	25.0
140	—	1.4	1.1	410	—	3.5	1.7	540	—	—	13.0
141	—	1.7	—	411	—	5.5	2.0	541	—	—	17.0
200	0.9	0.6	0.5	412	—	8.0	2.5	542	—	—	25.0
201	1.4	0.9	0.7	413	—	11.0	—	543	—	—	30.0
202	2.0	1.2	0.9	414	—	14.0	—	544	—	—	35.0
203	—	1.6	1.2	420	—	6.0	2.0	545	—	—	45.0
210	1.5	0.9	0.7	421	—	9.5	2.5	550	—	—	25.0
211	2.0	1.3	0.9	422	—	13.0	3.0	551	—	—	35.0
212	3.0	1.6	1.2	423	—	17.0	—	552	—	—	60.0
213	—	2.0	—	424	—	20.0	—	553	—	—	90.0
220	2.0	1.3	0.9	430	—	11.5	2.5	554	—	—	100.0
221	3.0	1.6	1.2	431	—	16.5	3.0	555	—	—	180.0
				432	—	20.0	4.0				

three test tubes are inoculated with each of 1 to 4 dilutions. The results are:

Dilution	No. of Tubes in which Growth Is Obtained
1 (1:100)	3
2 (1:1000)	2
3 (1:10,000)	1
4 (1:100,000)	0

The numerical characteristic is 321. The probable number according to the table is 15. Accordingly, in 1 ml of the first dilution there are 15 bacteria of the specified physiological group sampled and in 1 ml of water from the water mass there are 1500 bacteria of the group.

Chapter 6

Methods of Studying the Carbon Cycle and Microorganisms Involved

The cycle of carbon transformations in water masses is extremely complex. A whole series of physiological groups of microorganisms participates and each group breaks down specific compounds and forms different compounds as intermediate and end products. Methods of detecting yeast and bacteria which effect the decomposition of starch, hemicellulose, cellulose, pectins, salts of organic acids, and hydrocarbons, and bacteria involved in the butyric fermentation are discussed in this chapter, as well as the oxidation of hydrogen and common methods of observing oxidation of hydrocarbons which yield gas.

ISOLATING YEAST AND YEAST-LIKE FUNGI
IN WATER MASSES

An investigation of a water mass should include the enumeration of yeast and yeast-like fungi, especially in determinations of microbial biomass.

Yeast cells are considered separately when direct counts are made because of their large cell size. They are enumerated by the pour plate method as are bacteria. This method usually yields lower numbers of yeast since yeasts in water masses usually occur in aggregates. The aggregates are difficult to separate by shaking to disperse cells. When an inoculation is made from an aggregate, one colony develops.

Inoculations are usually made onto a solid medium in a

petri dish. The amount of inoculum should be 0.5 to 1–2 ml of water and platings should be in duplicate for each volume of water sampled. Sediment inoculations are made from dilutions prepared for inoculation to beef-peptone agar. Plates are prepared as for nutrient or beef-peptone agar: sterile nutrient agar, cooled to 42 C, is poured into the petri dishes, thoroughly mixed with the inoculum, and allowed to solidify. The plates are inverted and incubated at 22–23 C.

Counts are made first after 5 days and then after 10 days. All the colonies which appear are examined under the microscope; yeast colonies are noted and counted.

The following are the most appropriate media for yeast from water samples.

Wort and Saccharose Medium for Isolating Yeasts

Saccharose	150 g
$(NH_4)_2 HPO_4$	2 g
KH_2PO_4	1 g
$MgSO_4 \cdot 7H_2O$	1 g
Unfermented wort	10 ml
Sodium propionate	0.5 g
Streptomycin	up to 10 mg
Tap water	1000 ml
Agar	20 g

Glucose Medium for Yeasts

Glucose	60 g
KH_2PO_4	1 g
K_2HPO_4	0.1 g
$MgSO_4 \cdot 7H_2O$	0.7 g
NaCl	0.5 g
$Ca(NO_3)_2$	0.4 g
Yeast broth	50 ml
Yeast autolysate	50 ml
Tap water	900 ml
Agar	20 g

The yeast broth is prepared by boiling a 200-g yeast cake in 1 liter of tap water or water from the area sampled.

Normally a combination of low pH (4.0) and antibiotics such as: 2 ppm actinomycin and 50 ppm aureomycin; 50 units/ml penicillin and 100 units/ml streptomycin; or 20 ppm chloramphenicol, 20 ppm streptomycin and 100 ppm chlortetracycline, is used in preparing yeast isolation media to depress or inhibit bacterial growth. Some useful information on isolation of non-pathogenic yeasts is given by F. W. Beech and R. R. Davenport, 1969, Soc. Applied Bacteriol. Tech. Ser. No. 3, Academic Press, N.Y., pp. 71–88.

Wort Agar

The starting material is unfermented malt wort (this can be obtained from breweries or prepared in the laboratory from malt). Industrial unfermented wort usually contains 12–16% sugar. For medium preparations, the wort must be diluted with water. The wort is diluted to half strength. To prepare a solid medium, purified agar of a high quality (20 g per liter) is selected. The weighed agar is dissolved in 500 ml of water, combined with 500 ml of undiluted wort, stirred well, and brought to a boil. The medium is then poured into test tubes and sterilized at 110 C for 30 minutes. Streptomycin and sodium propionate must be added aseptically to the medium.

When preparing wort in the laboratory, 1 liter of water is heated to 48–50 C and stirred constantly while 250 g of ground malt are added. The flask with the malt is kept at 48–50 C for 30 minutes, after which the temperature is increased to 55–58 C; the flask is held at this temperature until the starch is completely broken down as determined by an iodine test (on testing, the medium should not yield a blue color). The coagula of protein precipitates are filtered off and the sugar content of the transparent wort is determined. Wort containing 6–8% sugar is used for preparing the medium.

Yeast colonies from the media listed above are picked and transferred to a slant agar. After the purity of the culture is checked (for yeast cultures it is best to perform single-cell isolation streaking on solid media), the culture is identified to genus and species.

Morphological characteristics and physiological capabilities are useful in classifying yeasts.

Morphological Characteristics

Important morphological features (Lodder and Kreger-Van-Rij, 1952) in classification of yeast include: (1) characteristic vegetative reproduction; (2) shape and dimensions of cells; (3) formation of ascospores and their morphology; (4) culture characteristics.

The characteristic vegetative reproduction has taxonomic significance. Vegetative reproduction is studied in wort and wort agar. Yeast may reproduce either by budding (e.g. *Torulopsis, Rhodotorula*) (Fig. 83, *A*); or by division (*Schizosaccharomyces* (Fig. 83, *B*). Both modes of reproduction may be encountered together; vegetative reproduction begins with formation of a bud, which is then separated from the mother cell by a wall (the *Saccharomycetaceae* family) (Fig. 83, *C*).

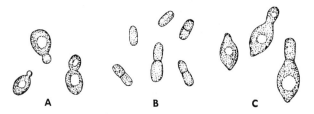

Fig. 83. Methods of yeast reproduction. *A*, Budding; *B*, division; *C*, budding with separation by wall.

When yeast cells which reproduce by budding are encountered, the characteristics of this process should be noted, i.e. location of bud development, whether on various parts of the surface of the cell, on the tip, or on both poles (Fig. 84).

The shape and size of the cells are examined in young cultures in wort and wort agar. Yeast cells are round, oval, elongated-oval, cylindrical, lemon-shaped, flask-shaped, three-cornered, or drop-shaped (Fig. 85).

In addition to the shape and size of yeast cells, notation is also made of the formation of a pseudomycelium, which is often encountered in aquatic yeast, or a true mycelium. A pseudomycelium is referred to as branching composed of elongated cells that develop through budding (Fig. 86, *A*). In the early stages the cells which form filaments are of equal size.

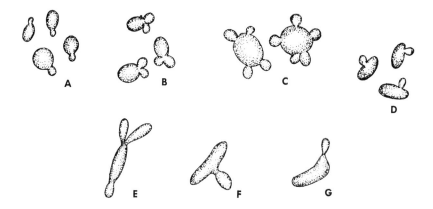

Fig. 84. Location of bud development on yeast cells. *A,* At the constricted end of a cell; *B,* at one of the ends of the cell; *C,* at various parts of the cell; *D,* at the side of the cell; *E,* on both ends of the cell; *F,* in the middle of the lateral part of the cell; *G,* at the edge of the cell.

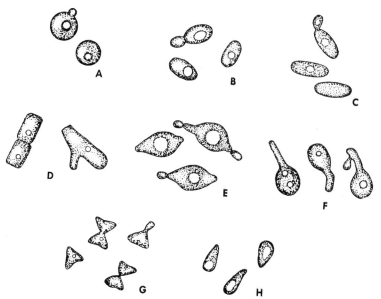

Fig. 85. Shape of yeast cells. *A,* Round; *B,* oval; *C,* elongated-oval; *D,* cylindrical; *E,* lemon-shaped; *F,* flask-shaped; *G,* three-cornered; *H,* drop-shaped.

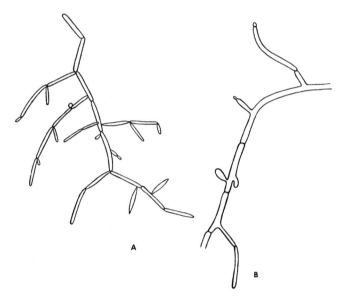

Fig. 86. Pseudomycelium *(A)* and true mycelium *(B)* of yeast.

Later the pseudomycelium branches by budding, and blasto-spores, which are round or oval cells, develop on the pseudo-mycelia. The branching of a true mycelium proceeds by a lateral outgrowth of the cells (Fig. 86, *B*). True mycelium consists of syncytial and non-septate hyphae (Fig. 87).

The forms of pseudomycelium and the characteristic blastospore formation vary (Fig. 88) in different genera. Blasto-spores can collect into compact spherical clusters around the top of pseudomycelial cells. They may also be located in free branching whorls, or in small chains. Blastospores distributed in fine verticils are present in certain genera in a highly branching pseudomycelium. Blastospores are often distributed symmet-rically in pairs. Blastospores may also develop in a way remi-niscent of the formation of conidia in *Penicillium.*

Many genera of yeast appear in dormant forms called arthrospores. According to Kudryavtsev (1954), arthrospores are distinguished from vegetative cells by the absence of vacu-oles, and the presence of a reinforced coat, more sharply light-refracting plasma, and a great quantity of the reserve nutrient substances glycogen and fat.

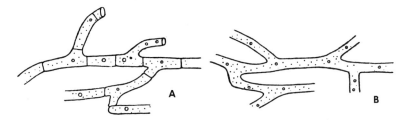

Fig. 87. Syncytial (A) and non-septate (B) mycelium (Clifton, 1950).

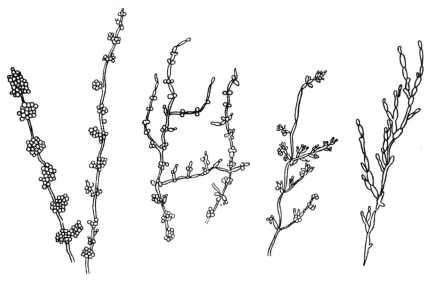

Fig. 88. Various distributions of blastopores on the pseudomycelium.

The formation of ascospores and their shape are always taken into account when identifying species of yeast. The following specific conditions are necessary for yeast sporulation: (1) vigorously growing culture in active vegetative reproduction; (2) high fat content of the inoculated cells; (3) sudden transition from rich to minimal medium and transfer of the yeast to media in which vegetative reproduction cannot proceed because of a lack of nutrient; and (4) temperature favorable to sporulation.

To fulfil the first condition, several transfers (two to three in 2 days) are made in a fresh nutrient medium (wort or a similar rich medium). Under these conditions a culture is obtained with cells rich in the reserve nutrient substances necessary for spore formation in a minimal medium.

Other media suggested for culturing yeast before transfer to a sporulation medium are described below.

Yeast Sporulation Medium

Glucose	5.0 g
KH_2PO_4	0.2 g
$CaCl_2 \cdot 6H_2O$	0.05 g
$MgSO_4 \cdot 7H_2O$	0.05 g
$FeSO_4 \cdot 7H_2O$	0.001 g
$(NH_4)_2SO_4$	0.5 g
Distilled water	1000 ml

Beet Extract Medium

Extract from beet leaves	10 ml
Extract from beet roots	20 ml
Apricot juice	35 ml
Grape juice	16.5 ml
Dry baker's yeast	2 g
Glycerin	2.5 ml
Agar	3 g
Chalk	1 g
Tap water	up to 100 ml

Demonstrating the ability of yeast to sporulate is not always easy. Some species express this process only weakly and in some cases various media must be used.

The following are suggested sporulation media.

Gorodkowa Agar

Beef extract	1.0 g
Peptone	1.0 g
NaCl	0.5 g
Glucose	0.25 g
Agar	1–2 g
Tap water	100 ml

Lodder and Kreger-Van-Rij Modification of Gorodkowa Agar

Glucose	1 g
Peptone	10 g
NaCl	5 g
Agar	30 g
Tap water	1000 ml

Gypsum Blocks. Gypsum blocks are prepared by mixing 2 parts baked gypsum with 3 parts water. The mass obtained is mixed thoroughly, transferred to small paper forms (without a bottom and no more than 2 cm high), placed on a smooth glass surface, and allowed to harden. The compact blocks are placed in Koch dishes with the smooth surface up. Water is added to half the height of the gypsum blocks and the plates are sterilized at 120 C for 30 minutes. A loop is used to place a drop of a dense culture of a healthy young yeast on the smooth surface of the cooled block.

Spores usually form on gypsum blocks after 28–30 hours at about 25 C. The number of spores in the asci of different species varies from 2 to 8. Thus the number and shape of the spores are noted.

Yeast sporulation can be obtained in a medium prepared from washed agar (2%) and distilled water. Spores form in this medium after a few days. A temperature of about 25 C seems to be most suitable for sporulation. However, Kudryavtsev notes that for yeast of the genus *Nadsonia*, sporulation occurs sooner at a lower temperature (18 C).

The following characteristics are often used to identify yeasts: spore shape and method of spore formation and germination.

Ascospores may be spherical, ellipsoidal, hemispherical (depressed on one side), crescent-shaped, bean-shaped, needle-shaped, spindle-shaped, hat-shaped, or saturn-like (with circumscriptions like the planet Saturn) (Fig. 89). The surface of ascospores may be smooth or warty, or may possess a "collar." The shape of the ascospore should be studied carefully with a microscope. The method of formation of the ascus which contains spores is determined only after repeated examination of the yeast in a sporulation medium under a microscope.

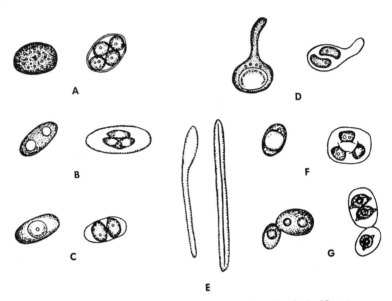

Fig. 89. Formation of ascospores in yeast cells. *Left,* Vegetative yeast cells; *right,* asci containing ascospores. Ascospore shape: *A,* spherical; *B,* ellipsodial; *C,* hemispherical; *D,* beanshaped; *E,* needle-like; *F,* hat-shaped; *G,* saturn-like.

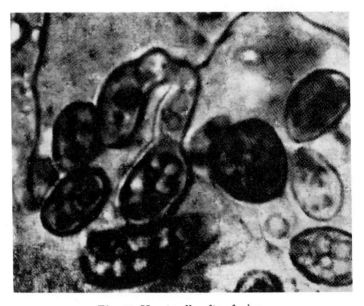

Fig. 90. Yeast cells after fusion.

The method of sporulation differs for various species of yeast. For some species, spores develop apogamously, i.e. directly from vegetative cells; for others, spores develop by sexual means, as a result of the process of cell fusion. When the latter occurs, two cells unite by means of projections which form on their surfaces. At the point of fusion of these projections, a membrane opens and the cell contents flow together.

Isogamous fusion (when cells of the same size fuse) may be differentiated from heterogamous fusion (when cells of different dimensions are joined). The result of the fusion process is usually a dumbbell-like form (two cells with a fusion bridge) which is easily observed in the specimen (Fig. 90).

Culture characteristics of yeasts are determined by a variety of means. The description of a streak of a young culture on wort agar after 1 week is usually used. A description of a giant colony is seldom given because of the great variability of this criterion.

Physiological Features

Identification of yeasts is made according to the following physiological features: (1) formation of a pellicle; (2) utilization of sugars; (3) fermentation of sugars; (4) utilization of nitrogen sources; (5) use of alcohol as the sole source of carbon; (6) decomposition of arbutin; (7) formation of carotenoid pigments; (8) formation of starch-like compounds.

1. The formation of a pellicle, a ring, and/or precipitate are noted in liquid wort culture after 2–3 days of incubation at 25 C and after 1 month. The characteristic color and thickness of the pellicle are noted in classical descriptions of yeast species.

2. Carbohydrate utilization is usually determined by the auxanograph method. Behavior with respect to carbohydrates (sucrose, maltose, lactose, glucose, galactose, and raffinose) is tested. A basal medium containing no carbon source is used. Lodder and Kreger-Van-Rij medium is of this type.

Lodder and Kreger-Van-Rij Medium

$(NH_4)_2SO_4$	0.5%
KH_2PO_4	0.1%
$MgSO_4 \cdot 7H_2O$	0.05%
Washed agar	2%

A 100× concentrated vitamin solution (0.05 ml) is added to the medium:

Biotin	2 μg
Calcium pantothenate	400 μg
Inositol	2000 μg
Niacin	400 μg
Paraaminobenzoic acid	200 μg
Pyridoxine hydrochloride	400 μg
Thiamine	400 μg
Riboflavin	200 μg
Tap water	10 ml

The vitamin solution may be substituted by a yeast water or yeast extract solution (see p. 193).

The sterile medium is melted in a water bath, cooled, and poured into petri dishes containing about 2 ml of a thick suspension of the yeast species under investigation and 1 drop of sterile yeast extract or vitamin solution. The agar is thoroughly mixed with the inoculum. After solidification of the agar, the dishes are dried in an incubator at 25 C with the lid barely open to ensure a dry surface. Next, a star cut from filter paper, with each ray of the star extending to the edge of the dish, is soaked in a carbohydrate solution and gently pressed onto the agar surface. Small discs of filter paper, soaked in carbohydrate solutions and dried, can also be used. The latter are distributed on the surface of the agar. Another method is to divide the dishes into sectors by drawing lines on the bottoms with a brush pen or wax pencil. Notation is made of which carbohydrate is added to each sector, after which a given amount of that carbohydrate is then placed on the surface of the agar. In each of the above methods, yeast growth is observed after 1–2 days' incubation (only in those areas in which carbohydrate is utilized by the yeast).

Utilization of carbohydrates may also be determined in liquid media. A basal medium is prepared in large quantity and aliquots are poured into a series of flasks, each of which contains one of the carbohydrates under study. When the carbohydrate dissolves, the media are poured into test tubes in 5-ml quantities and sterilized by flowing steam for 3 days. The basal medium can be autoclaved and filter-sterilized carbohydrate added before inoculation. Test tubes containing different carbohydrates are inoculated simultaneously with a given cell suspension. Observations are made after 24, 48 hours and 1–2 weeks.

Assimilation of a particular carbon source is determined by the density of yeast growth appearing in the test tubes.

Kudryavtsev (1954) recommends a special medium for determining utilization of carbon sources by yeast.

Kudryavtsev Medium

Carbohydrate	60 g
$MgSO_4 \cdot 7H_2O$	0.7 g
NaCl	0.5 g
$Ca(NO_3)_2$	0.4 g
KH_2PO_4	1 g
K_2HPO_4	0.1 g
Yeast broth	50 ml
Yeast autolysate	50 ml
Tap water	1000 ml

Three milliliters of a 0.04% alkaline solution of the indicator bromophenol blue are added to the medium, and 2 g of agar are added per 100 ml of medium. The pH of the medium is 5.8–5.9.

3. Carbohydrate fermentation by yeast is detected in yeast water or yeast extract containing 2% added carbohydrate. To prepare yeast water, 1 liter of water is poured over a 200-g yeast cake and sterilized in an autoclave at 120 C for 15 minutes. The mixture, while hot, is filtered twice through a folded filter. Carbohydrate is added to the filtrate, which is then poured into Dunbar tubes (Fig. 91) and sterilized with flowing steam. (This method is no longer used routinely, however, since yeast

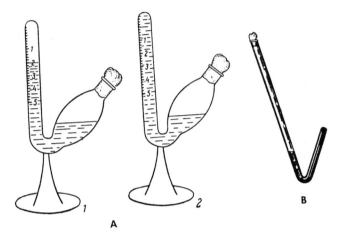

Fig. 91. Apparatus for detecting carbohydrate fermentation (Carter and Smith, 1957). *A,* fermentation tubes: *1,* presence of gas; *2,* absence of gas. *B,* Dunbar tube.

extract is commercially available.) The medium is poured into test tubes which contain an inverted vial (for detection of gas) and sterilized as above.

Yeasts which ferment carbohydrates liberate carbon dioxide. The carbon dioxide collects in the closed bend of the tube. Observation is made daily.

4. The auxanogram method is most often used to detect utilization of nitrogen sources by yeast. The following medium is used in establishing this utilization:

Tap water	1000 ml
Glucose	20 g
KH_2PO_4	1 g
$MgSO_4 \cdot 7H_2O$	0.5 g
Washed agar	20 g

The medium is sterilized for 15 minutes at 110 C.

When the inoculated agar solidifies in the petri dishes, either a pattern cut from filter paper (each ray of which has been saturated with the appropriate nitrogen source) or filter paper discs saturated with solutions of the nitrogen sources and then dried are placed on the agar. KNO_3, $(NH_4)_2SO_4$, peptone, and asparagine are usually used.

The above-mentioned medium may also be used in liquid form. The prepared medium is divided into parts: KNO_3 is added to one part, $(NH_4)_2SO_4$ to the second, peptone to the third, and asparagine to the fourth. The prepared solutions are poured in 5-ml quantities into test tubes and sterilized with flowing steam. After sterilization, 1 drop of vitamin mixture is added aseptically to each test tube.

5. The use of alcohol as the sole source of carbon can be demonstrated in Kudryavtsev or Lodder and Kreger-Van-Rij medium. Alcohol is added aseptically, to 1%, after sterilization (before the agar solidifies). The agar is mixed well and the test tube is placed in a slanted position for solidification of the agar.

6. The decomposition of arbutin, according to Lodder and Kreger-Van-Rij, can be used to characterize yeast (for separating the genus *Hansenula* (arbutin-positive) from the genus *Pichia* (arbutin-negative)). Agar in yeast water containing 0.5% arbutin is used for this purpose. One drop of a solution of $FeCl_3$ is placed in a petri dish and agar is poured over it. Inoculations are made after the agar solidifies. A positive reaction is the appearance of dark-brown coloring around the colonies in the medium which occurs after 2, 4, or 6 days.

7. Formation of carotenoid pigments is noted in cultures of yeast in wort agar. Many aquatic yeasts form different pigments, for example, rose, red, orange, and black pigments.

8. Formation of starch compounds, according to Lodder and Kreger-Van-Rij, is helpful in identifying some yeast species. A synthetic medium containing 2% agar (p. 192) is used for this purpose. One drop of yeast extract or a vitamin mixture is placed in each dish, which is then inoculated. Agar is added and the inoculated plates incubated.

After 1–2 weeks, the presence of starch compounds is determined by treating the surface of the agar with Lugol solution. Appearance of blue coloring indicates the formation of starch compounds by the yeast.

Many other tests are useful in identifying and classifying yeast species. It is best to consult the literature for additional test methods as those given here are only a minimum for identifying yeasts.

ISOLATING AMYLOLYTIC BACTERIA

When aquatic vegetation in a water mass dies off, amyloid substances become available for bacterial decomposition. Starch belongs to the class of compounds easily decomposed by bacteria. Many aquatic bacteria contain amylase, the enzyme by which breakdown of starch can be effected. Usually the process proceeds in two steps: hydrolysis occurs, and as a result of this, dextrin is formed and then split.

The action of bacteria on starch can be a useful diagnostic characteristic. Starch agar is usually used to detect amylolytic bacteria; pour plate or streak inoculation of starch agar is usually done, although liquid media may also be used, by means of the extinction dilution method. The pour or streak plating method yields more accurate results.

In starch agar around the colonies of amylolytic bacteria zones of hydrolysis are obtained which do not show a reaction with iodine and which vary in diameter, depending on the activity of the bacteria. When the colonies on the starch agar are flooded with an iodine solution, the medium turns blue, except for the zones of hydrolysis around the colonies of amylolytic bacteria. These zones either remain colorless, or stain reddish brown. The color of the zones around the colonies depends on the degree of hydrolysis of the starch: if it is hydrolyzed to the stage of dextrin, then the zones are reddish brown; if the breakdown has gone further, they are colorless. The colonies with zones of hydrolysis are counted.

Various starch media may be used:

Starch Agar

Tap water	1000 ml
KH_2PO_4	0.5 g
K_2HPO_4	0.5 g
$MgSO_4 \cdot 7H_2O$	0.2 g
$(NH_4)_2SO_4$	0.2 g
Starch	10 g
Agar	15 g

pH 7.0

Imshenetskii Medium

20% Potato agar	1000 ml
Peptone	5 g
Insoluble starch	30 g
Chalk	1 g

Agar is added to 15 g per liter for a solid medium.

Pochon Medium

Winogradsky standard salt solution, with the following composition, is used:

Tap water	1000 ml
K_2HPO_4	5 g
$MgSO_4 \cdot 7H_2O$	2.5 g
NaCl	2.5 g
$Fe_2(SO_4)_3 \cdot 9H_2O$	0.05 g
$MnSO_4 \cdot 5H_2O$	0.05 g

The medium is prepared by combining two solutions:

1. Winogradsky standard solution diluted 1:20	750 ml
NH_4NO_3	1 g
Agar	15 g
Aqueous solution of phenol red (0.02%)	15 ml
2. Rice starch	1.5 g
Tap water	250 ml

Both solutions are mixed. The medium should be pinkish (rose-tinted). The first counts on this medium are made after 5 days: the colonies surrounded by a zone of yellowish agar are counted. The plates are then treated with Lugol solution (diluted five-fold) and the colonies with zones of clearing around them are counted.

Potato Agar

Potato broth (10%)	1000 ml
Chalk	1 g
Agar	15 g

Wheat Bran or Flour Agar

Wheat bran or flour extract (5%)	1000 ml
Agar	15 g

Potato Starch Agar

Tap water	1000 ml
Potato starch	0.2 g
Peptone	0.5 g
K_2HPO_4	trace
$MgSO_4 \cdot 7H_2O$	trace
$FeCl_3 \cdot 6H_2O$	trace
Agar	15 g

Beef Extract Agar

Tap water	1000 ml
Beef extract	3 g
Soluble starch	10 g
Agar	20 g

pH 7.4

To determine the biochemical activity of starch-hydro-lyzing (amylolytic) bacteria, Imshenetskii medium (cited above) is poured into small conical flasks to form a layer of 1.5–2 cm. After sterilization, inoculations are made with water or sediment by the extinction dilution method. The inoculated flasks are placed in an incubator at 23–25 C. After 24 hours, tests are made in the cultures for accumulated carbohydrate break-down products.

Breakdown of starch by bacteria in liquid media is judged by the disappearance of the blue coloring characteristic for starch when iodine is added.

HEMICELLULOSE DECOMPOSITION

Hemicelluloses (pentosans—arabans, xylans, araboxy-lans; methylpentosans; and hexosans) form a part of plant tissue. They therefore undergo decomposition together with cellulose in water masses. This process is occasionally tested for in water masses.

Decomposition of hemicelluloses proceeds most intensively under aerobic conditions. Different groups of microorganisms—fungi, actinomycetes, and bacteria—induce decomposition.

The process is judged by the total activity of the microflora capable of breaking down hemicelluloses (Pochon and Tardieux, 1962). Basically, this method is described as follows: a selective medium for bacteria which degrade hemicellulose is inoculated with various dilutions of sediment or other sample material. The inoculations are observed daily; the dilutions are examined and records are kept of initiation and completion of the process. The beginning of hemicellulose degradation (phase I) is recorded with the appearance of nitrites, which is established with Griess reagent (see p. 273 for preparation). The end of the process (phase II) is determined by reaction with a solution of 1,3,5-benzenetriol in pure hydrochloric acid (0.5 g of 1,3,5-benzenetriol in 1000 ml of HCl). Curves are constructed on the basis of the data obtained (Fig. 92). By means of the curve which shows variation with time, estimation is made of both the intensity of growth of bacteria capable of breaking down hemicelluloses and the energy fluctuations of the process in relation to the number of bacteria.

The test is carried out as follows: each day 0.5 ml is taken from each test tube, placed in micro-test tubes, and

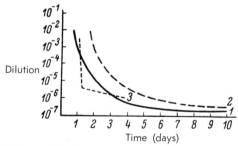

Fig. 92. Determination of decomposition of hemicellulose (Augier, 1956). 1, Curve at beginning of decomposition; 2, curve at end of decomposition; 3, difference. Abscissa, time in days; ordinate, dilutions.

heated; 1 drop of Griess reagent is then added. The appearance of red or rose-yellow coloring indicates the presence of nitrites (it is necessary to wait 2 minutes for the reaction to appear). The test tubes in which the presence of nitrites was indicated are marked positive; test tubes with samples which yield a very pale color are marked "weak" ($\pm$); those with no color reaction are marked negative. One milliliter of 1,3,5-benzenetriol solution is then added. The test tubes are placed in a boiling water bath for 15 minutes. If no hemicelluloses are present, the solution appears yellow and transparent. Where 30 to 100% of the hemicelluloses remains the reaction is brown, with a noticeable blackish precipitate. The results are recorded as follows: tubes with a yellow liquid as $+$ and those with black precipitate as $-$.

At this point, it is necessary to mention that nitrites and nitrates yield a red color with 1,3,5-benzenetriol and this color reaction masks the hemicellulose reaction.

Augier medium is a selective medium containing hemicellulose as the sole source of carbon and potassium nitrate as a source of nitrogen.

Augier Medium

Winogradsky standard salt solution	
(see p. 197 for preparation)	50 ml
Soil extract	3 ml
KNO$_3$	0.3 g
Hemicellulose	1 g
Trace elements solution	1 ml
CaCO$_3$	1 g
Distilled water	up to 1000 ml

Five-milliliter amounts of the medium are poured into test tubes and sterilized in an autoclave for 20 minutes at 110 C.

If a commercial preparation is not available, hemicellulose may be prepared as follows: 100 g of oak sawdust are selected and treated by a method involving four steps.

1. Removal of substances soluble in water. Water (0.5 liter) at 70 C is poured on the sawdust and the mixture is placed in a mechanical agitator or on a shaker for 15 minutes.

Plate 1. Water analysis kit for field work. Reagents for testing oxygen, carbon dioxide, methyl orange alkalinity, phenolphthalein alkalinity and temperature. (Courtesy Wildlife Supply Company, Saginaw, Michigan.)

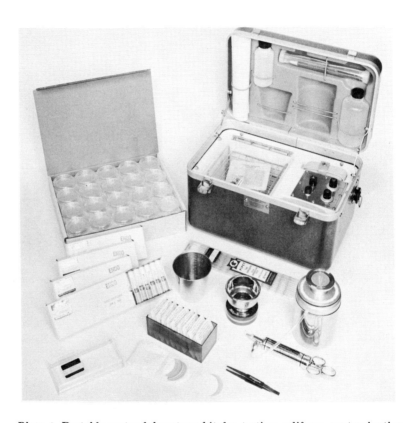

Plate 2. Portable water laboratory kit for testing coliform contamination in water by the bacteria-retentive filter method, using pre-sterilized 47-mm filters. The incubator is a portable incubator fitted with a rack for petri dishes. (Courtesy Millipore Corporation, Bedford, Massachusetts.)

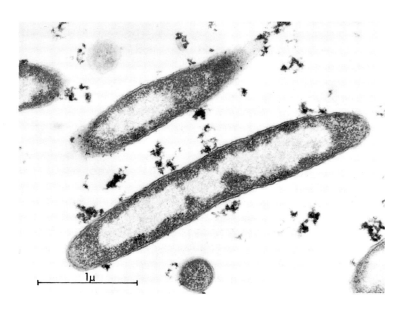

Plate 3. Ultrastructure of deep-sea isolate, *Pseudomonas bathycetes* strain C6M. The ultrathin sections reveal the typical gram-negative structure similar to that of the terrestrial gram-negative bacteria. A distinct nuclear region, granular cytoplasm with numerous ribosomes, plasma membrane, and cell wall can be seen. × 27,000. (From Kettling, Colwell, and Chapman, 1971, Proc. Conf. High Pressure Aquaria. Univ. North Carolina Press.)

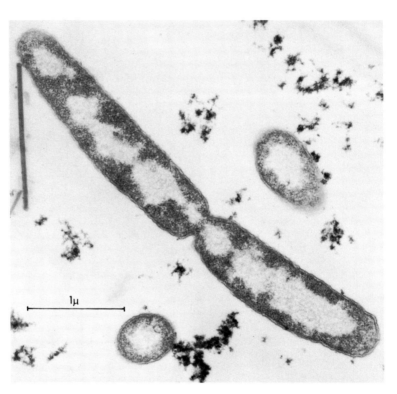

Plate 3B

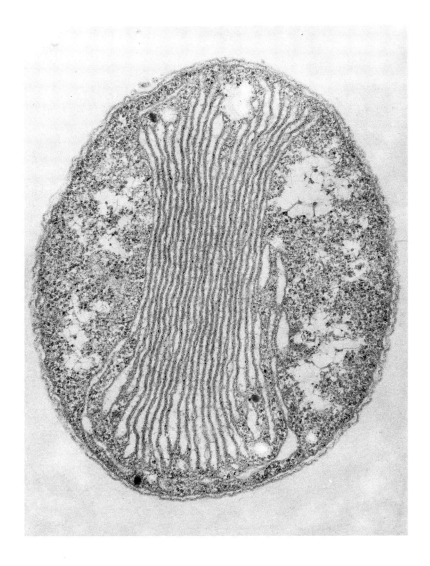

Plate 4. Section of *Nitrosocystis oceanus* showing the cell wall, plasma membrane, cytomembranes, and nucleoplasm. × 36,720. (From Watson and Remsen, 1970, J. Ultrastruct. Res. *33*:148–160. Courtesy of Academic Press, Inc.)

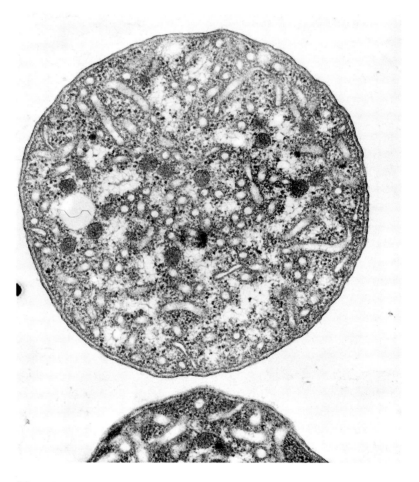

Plate 4B

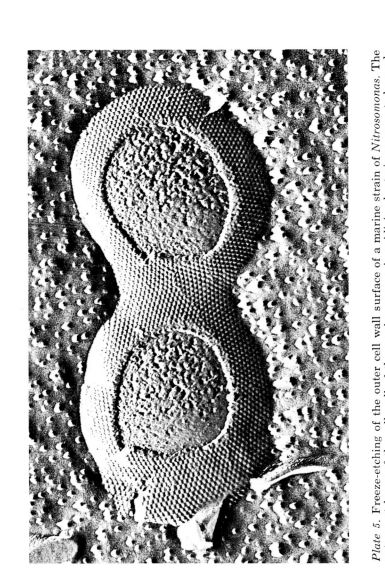

Plate 5. Freeze-etching of the outer cell wall surface of a marine strain of *Nitrosomonas*. The outermost layer of the cell wall of the marine ammonia-oxidizing bacteria appears to be made up of protein subunits arranged in a regular manner and linked together through metal-oxygen bonds. The large subunits measure approximately 150 A. ×76,500. (From Watson and Remsen, 1969, Science *163*:685–686. Courtesy of the American Association for the Advancement of Science.)

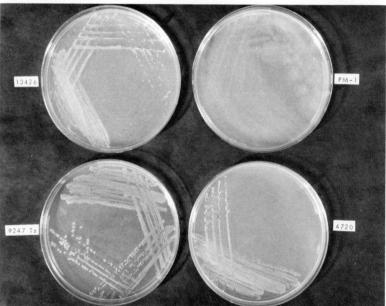

Plate 6. Bacterial colony growth showing pigmentation (*top left*), spreading (*bottom left*), mucoid (*top right*), and variations in colony size (*bottom right*). Numbers given are culture identification: 12542 = *Chromobacterium violaceum*; 12473 = *Chromobacterium lividum*; 6918 = *Chromobacterium viscosum*; 12540 = *C. violaceum*; 13426 = *C. violaceum*; 9247 Tz = *Proteus morganii*; PM-1 = *Proteus mirabilis*; 4720 = *Agrobacterium*

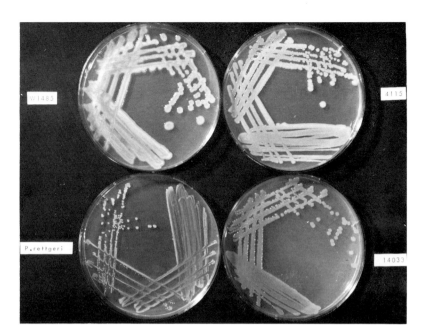

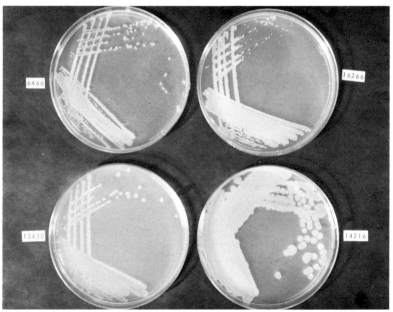

tumefaciens; W 1485 = *Escherichia coli* K12; *Proteus rettgeri;* 4115 = *Enterobacter aerogenes;* 14033 = *Vibrio cholerae;* 6466 = *Agrobacterium radiobacter;* 13430 = *Pseudomonas fluorescens;* 16266 = *Pseudomonas fluorescens* var. *antirrhinastri;* 14216 = *Pseudomonas aeruginosa.* (From Moffett and Colwell, 1968, J. Gen. Microbiol. *51*:245–266.)

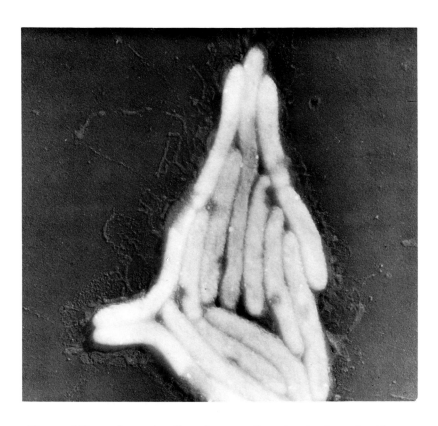

Plate 7. Micro-colony of a *Cytophaga* sp. Strands of polysaccharide can be seen surrounding the cells. The bending and twisting of the bacterial cells is demonstrated. × 9360.

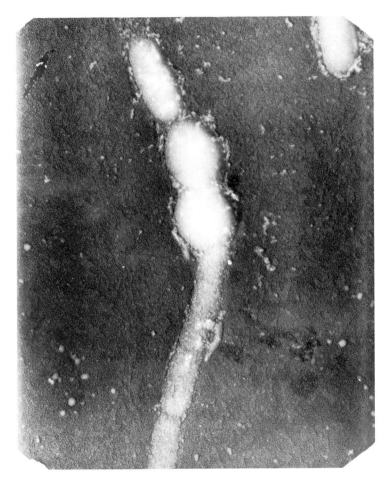

Plate 8. Rounded structures observed in *Flavobacterium aquatile* strain 11947, presently recognized to be a *Cytophaga* sp. × 9360.

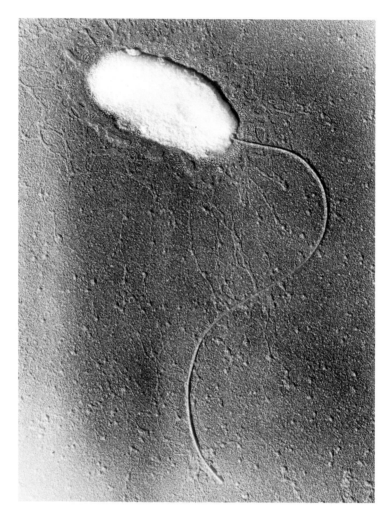

Plate 9. Polarly flagellated marine *Vibrio* sp. strain 328. × 18,000. (From Colwell and Gochnauer, 1968, Proc. Tenth I.A.M. Symp. 1968. *Taxonomy of Microorganisms,* pp. 113–131.)

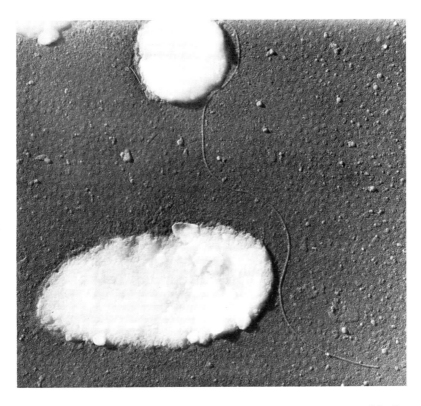

Plate 10. An agar-digesting marine bacterium, *Vibrio* sp. 859. Motile round body forms seen in wet mount preparations are confirmed by electron microscopy. × 15,600.

Plate 11. The J-Z water sampler is shown after a successful water sample cast.

Plate 12. The Niskin water sampler shown is being attached to a hydrographic wire prior to sampling.

Plate 13. Right, A sediment sampler useful for sampling in estuaries. *Below,* Retrieving the sediment sampler containing estuarine mud for bacteriological analysis. The inner liner of the sampler, made of an autoclavable, inert plastic, can be removed aseptically and the mud sample retained for study.

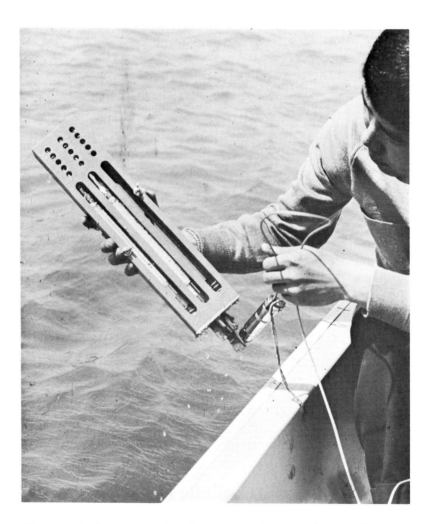

Plate 14. A simple reversing thermometer for near-shore and estuary studies.

The water is decanted and the sawdust is dried thoroughly in a small linen bag or filtered in a Büchner funnel. This step is repeated four times. After the last extraction the sawdust is dried.

2. Extraction of hemicelluloses. Sodium carbonate solution (40 g of Na_2CO_3 in 1 liter of water) is poured on the sawdust and allowed to stand for 2 days. From time to time it is shaken. First it is filtered through cloth, then through a Büchner funnel. The Na_2CO_3 solution is also used to precipitate the hemicelluloses. The sawdust is rinsed on the funnel with 150 ml of distilled water.

3. Precipitation of the hemicelluloses. Acetic acid is poured into the Na_2CO_3 solution and mixed with 150 ml of rinse water to obtain a pH of 3.0. The solution acquires a reddish or light brown coloring. An equal volume of 95% alcohol is added. A very delicate precipitate forms; this is allowed to settle, and the supernate is decanted with a siphon. The precipitate is centrifuged and collected. A minimum amount of alkaline water is poured in.

4. Purification of the hemicelluloses. Fifty milliliters of 95% alcohol are poured over the precipitate, which is then poured into a Büchner funnel, allowed to drain, and rinsed with 200 ml of absolute alcohol, followed by a mixture of ether and alcohol, and finally with ether. The precipitate on the filter is removed, dried at 60 C, and then ground into a fine powder in a mortar (hemicellulose yield from 100 g of sawdust is approximately 5 g).

CELLULOSE DECOMPOSITION

When aquatic plants in water masses die off, much of the plant residue, which is rich in cellulose, undergoes decomposition. An especially large amount of cellulose enters the water in the autumn. Decomposition of cellulose can proceed under either aerobic or anaerobic conditions.

Aerobic Decomposition

The presence and total number of aerobic bacteria which decompose cellulose can be determined in two ways: (1) by the extinction dilution (titer) method; (2) by inoculations onto solid nutrient media (surface spread plates) and counting the colonies which develop.

Many media have been devised for culturing aerobic cellulose bacteria. Several are listed below.

Hutchinson Medium

K_2HPO_4	1 g
$CaCl_2 \cdot 6H_2O$	0.1 g
$MgSO_4 \cdot 7H_2O$	0.3 g
NaCl	0.1 g
$FeCl_3 \cdot 6H_2O$	0.01 g
$NaNO_3$	2.5 g
Distilled water	1000 ml

pH 7.2–7.3

Winogradsky Medium

KH_2PO_4	1 g
$MgSO_4 \cdot 7H_2O$	0.5 g
NaCl	0.5 g
$MnSO_4$	0.01 g
$Fe_2(SO_4)_3 \cdot 9H_2O$	0.01 g
$CaCO_3$	20 g
KNO_3	4 g
Distilled water	1000 ml

The pH of the medium is adjusted to 7.2 by adding a 10% solution of potassium carbonate as necessary.

Fåhraeus Medium

$(NH_4)_3PO_4$	2.5 g
$MgSO_4 \cdot 7H_2O$	0.5 g
$FeSO_4 \cdot 7H_2O$	0.01 g
KCl	0.5 g

$CaCl_2 \cdot 6H_2O$	0.02 g
$MnSO_4 \cdot 5H_2O$	0.001 g
Distilled water	1000 ml

pH is adjusted to 7.4 with a solution of NaOH.

McBeth Medium

K_2HPO_4	1 g
$MgSO_4 \cdot 7H_2O$	1 g
Na_2CO_3	1 g
$(NH_4)_2SO_4$	2 g
$CaCO_3$	2 g
Distilled water	1000 ml

Dubos Medium

$NaNO_3$	0.5 g
K_2HPO_4	1 g
KCl	0.5 g
$MgSO_4 \cdot 7H_2O$	0.5 g
$Fe_2(SO_4)_3 \cdot 9H_2O$	trace
Distilled water	1000 ml
pH 7.5	

Cellulose must be included in some form in all of these media.

The prepared medium is poured into small conical flasks or test tubes. Because the best conditions for growth of aerobic cellulose bacteria depend not only on the composition of the medium, but also on the conditions of aeration, the medium is poured out in a thin layer (no more than 1 cm thick). Into each flask is placed a filter folded in creases with the wide side down, and the apex of the cone up (Fig. 93, A). Filter paper folded in creases and laid on the bottom of the flask (Fig. 93, B) may be used in place of conical filters. When test tubes are used, a small amount of the medium (in a layer no thicker than 2 cm) is added, and several narrow strips of filter paper are placed into each test tube so that part of the strip is above the medium (Fig. 94, A). Flasks and test tubes containing media are ster-

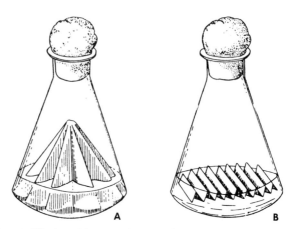

Fig. 93. Flasks with a nutrient medium for culturing cellulose bacteria. *A,* With a conical-shaped folded filter; *B,* with a crimped filter.

ilized in an autoclave. Inoculated media are incubated at a temperature of about 25 C.

Growth of cellulose-degrading bacteria is observed on the filter paper, first at the border of the medium where favorable oxygen as well as nutrient conditions prevail (Fig. 94, *B*). The filter paper is gradually broken down completely by the bacteria and settles to the bottom of the flask (Fig. 95). Often it becomes slimy, i.e. gelatinizes. If cellulose-decomposing myxobacteria develop, moist mucilaginous spots of varied color will appear on the filter at the edge of the medium. However, in a number of cases growth of aerobic bacteria does not proceed very rapidly under such conditions of aeration. Imshenetskii (1953) recommends methods for providing more ideal growth conditions which provide good aeration of the culture, e.g. culturing in flasks placed on shakers. The culture is poured into round flasks so that it is distributed in a thin layer over the walls of the flask while the flask is being shaken. Air can be bubbled through the medium to improve aeration.

All of the media discussed may be used in solid form, in which case cellulose (5 g per liter), scalded with boiling water and ground into very small fibers, and 1.5% thoroughly washed or purified agar are added to the media.

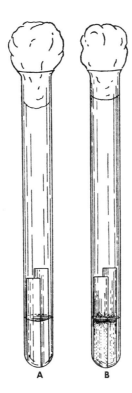

Fig. 94. Test tubes with filter paper for growth of aerobic cellulose bacteria. *A,* Before inoculation; *B,* beginning of growth.

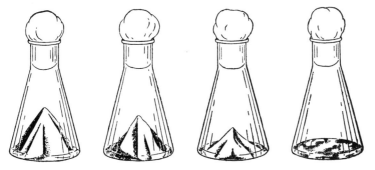

Fig. 95. Gradual decomposition of filter paper by bacteria (Imshenetskii, 1953).

Once the presence of cellulose bacteria in the water mass has been established, the bacteria responsible for the process of cellulose degradation should be identified. Because cellulose may be broken down by various bacterial species, the pieces of filter paper undergoing decomposition should first be examined under a microscope. To identify the species of the bacterial agents involved, the investigator should have some familiarity with those bacteria which cause aerobic decomposition of cellulose. Difficulties arise because of the fact that associated bacteria grow with the cellulose-decomposing bacteria. Aerobic cellulose bacteria, capable of growing in the mineral media cited above, can be divided into a group which utilizes only cellulose as a carbon source, and another which is capable of growing in ordinary media as well. Representatives of the following genera are most frequently encountered in natural bodies of water: *Cytophaga; Sporocytophaga; Cellvibrio; Chromobacterium;* and *Pseudobacterium. Cytophaga* is a representative of the first group, but *Cellvibrio, Chromobacterium,* and *Pseudobacterium* are capable of assimilating other organic nutrients in addition to cellulose. Certain actinomycetes, fungi, and myxobacteria of the genus *Sorangium* also decompose cellulose. The vegetative cells of *Cytophaga* Winogradsky have a very characteristic morphology: slightly curved rods with pointed ends (Fig. 96), 0.3–0.5 × 4–6 μ, very motile, gram-negative. Morphologically similar to this genus are representatives of the genus *Cellfalcicula* Winogradsky.

Cells of *Sporocytophaga* Stanier go through several stages in their life cycle: young, curved, rod-shaped bacteria with pointed ends gradually change, first turning into elongated

Fig. 96. Cytophaga Winogradsky.

Fig. 97. Growth cycle of *Sporocytophaga* Stanier (Imshenet-skii, 1953).

rods with a cylindrical shape, then becoming more and more shortened, finally yielding round or oval microcysts (Fig. 97).

Cells of *Cellvibrio* Winogradsky are bow-shaped, curved rods with rounded ends, and are gram-negative, with dimensions of $0.3–0.5 \times 2–4.5\ \mu$ (Fig. 98).

When a microscopic observation is made, a small piece of decomposing filter paper is transferred with a needle into a drop of water on a slide and then torn into separate fibrils with the needle. Bacteria are found on the fibrils. Several specimens are prepared. Some are examined under the microscope without prior staining. The others are examined after staining.

To obtain a more complete representation of the bacteria which have developed, specimens should be prepared from different parts of the spot, since older cells are found in the center and younger ones on the periphery.

To elucidate the motility of aerobic cellulose bacteria, Imshenetskii (1953) suggests that agar film and cellophane strip methods should be applied in addition to the hanging drop method (see Fig. 65) since cell movement of myxobacteria often occurs only when they are located on the surface of a firm substratum. It is easiest to use cellophane. In this case, the cellulose bacteria are placed on a small quadrangular piece of cellophane set in the center of a cover glass. A small drop of a liquid nutrient medium and some of the original sample is placed on the cellophane. The cover glass containing the cellophane is placed on a well slide. The transparency of the cellophane permits observation of cell movement and the growth stages of the bacteria.

A special medium is needed to provide optimal conditions for culturing *Cellvibrio*. Transfers into this medium must be done to isolate this organism.

Cellvibrio Medium

$MgSO_4 \cdot 7H_2O$	0.5 g
$NaNO_3$	2 g
KCl	0.5 g
$Fe_2(SO_4)_3 \cdot 9H_2O$	0.01 g
KH_2PO_4	0.14 g
K_2HPO_4	1.2 g
Yeast extract	0.5 g
Agar	7.5 g
Distilled water	1000 ml
Filter paper	10 g

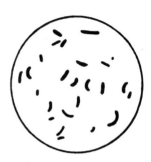

Fig. *98. Cellvibrio* Winogradsky.

Filter paper is either finely ground and added to the medium, or it is cut into pieces of the diameter of petri dishes and placed on the surface of a medium poured into a petri dish. The filter paper must be sterilized beforehand. In order to grind up the filter paper, it is first wetted and then divided into separate fibrils with a scalpel or glass rod. The shredded paper is gathered into a mortar, a minimal quantity of hot water is poured in, and the mixture is ground into a homogeneous mass.

According to Allen (1957), Norman and Fuller cellulose-dextrin agar medium is most favorable for growth determination and maintenance of cellulose-decomposing bacteria.

Norman and Fuller Agar

$NaNO_3$	1 g
K_2HPO_4	1 g
KCl	0.5 g
$MgCl_2 \cdot 6H_2O$	0.5 g
Yeast extract (10%)	10 ml
Non-water-soluble cellulose dextrin	10 ml
Agar	15 g
Distilled water	1000 ml

pH 7.0

The non-water-soluble cellulose dextrin is prepared as follows:

1. A large porcelain mortar and 100 ml of 72% sulfuric acid are placed in a refrigerator or on ice for several hours.

2. The cold mortar is surrounded by finely crushed ice, and a minimal amount of water is added. The acid is poured into the mortar and 20 g of finely ground filter paper are added. The solution is mixed thoroughly. The temperature of the suspension at this time should not exceed 10–12 C. When the filter paper is completely disintegrated (a non-water-soluble dextrin will be obtained) the mixture is allowed to remain in the cold for 1–2 hours.

3. The mixture is poured into 600 ml of ice water containing several pieces of ice and is allowed to stand for 15–30 minutes.

4. The dextrin is filtered onto a Büchner funnel through several layers of thin fabric or cheesecloth.

5. The dextrin precipitate is transferred from the funnel to a folded filter-paper filter, washed several times with water to remove the acid, and neutralized with dilute alkali. The precipitate is again washed with water to remove all soluble salts and finally is used to prepare the medium.

According to Allen, cellulose-dextrin agar has many advantages: good growth of different species of cellulose bacteria, identification of active forms by formation of a halo, and retention by cellulose-decomposing bacteria of their enzymic activity when maintained on this medium over a long period of time.

Rivière agar may also be used for detecting and isolating cellulose bacteria.

Rivière Agar

Powdered cellulose	5 g
$NaNO_3$	1 g
$NA_2HPO_4 \cdot 2H_2O$	1.18 g
KH_2PO_4	0.9 g
$MgSO_4 \cdot 7H_2O$	0.5 g
KCl	0.5 g
Yeast extract	0.5 g
Hydrolysate of casein	0.5 g
Agar	10 g
Tap water	1000 ml

The powdered cellulose is either obtained commercially or is prepared by soaking filter paper in 1 N hydrochloric acid for 12 hours in the cold. The obtained mass is transferred onto a Büchner funnel, and the precipitate is washed to free the acid and then dried.

The medium is sterilized at 120 C for 20 minutes and poured into dishes in such a way that the cellulose is evenly distributed in the medium. The dishes are lightly dried. Inoculations are made by placing the inoculum on the surface of the medium and distributing it over the surface with a sterile spatula or bent glass rod. The medium is incubated for 8 days at 25 C. This medium is suitable for the majority of cellulose bacteria.

To isolate *Cytophaga* and *Sporocytophaga* in pure culture, Rivière recommends this medium with addition of antibiotics (2.5, 5, and 10 mg per liter).

Inoculation to solid media may be made in petri dishes containing gel soaked in one of the above media. Discs of filter paper are cut according to the diameter of the dishes and sterilized separately in an oven. The medium is also sterilized. The filter paper discs are placed on the surface of silica gel which has been partially dried under sterile conditions; 2–2.5 ml of the medium are placed on the discs. Aqueous agar (1.5–2%) prepared from thoroughly washed agar may be used in place of the gel. A filter paper disc is placed on the surface of agar in a petri dish and overlaid with nutrient medium.

Inoculations are made somewhat differently by the silica gel method (p. 276). The gel is prepared in petri dishes for inoculation, and then overlaid with 2 ml of the following medium:

Tap water	1000 ml
K_2HPO_4	10 g
NH_4NO_3	3 g
$MgSO_4 \cdot 7H_2O$	5 g
NaCl	1 g
$CaCl_2 \cdot 6H_2O$	1 g
$FeCl_3 \cdot 6H_2O$	1 ml

The gel is maintained at 56 C until the medium is well dried. The inoculum is distributed evenly on the surface of the dish. Each inoculated dish is covered with a sterile disc of filter paper which should lie on the surface of the gel plate.

When water samples are inoculated onto the surface of dishes containing gel, water is placed on the surface of the medium dropwise from a graduated pipette. When sediment inoculations are made, the inoculum may be introduced from various dilutions or the sediment may be distributed in small clumps as is done for nitrifying bacteria (p. 277).

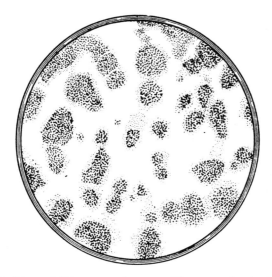

Fig. 99. Zones of filter paper decomposition around colonies of cellulose bacteria.

The inoculated plates are placed in an incubator at 24–25 C. After 4–5 days the beginning of the growth of cellulose-decomposing bacteria may be observed: zones of decomposing cellulose, often colored, appear on the filter paper (Fig. 99). The coloring of the colonies and the coloring of the paper are characteristic features of the growth of a particular species. Colonies surrounded by such zones are counted. A microscopic investigation of zones of cellulose decomposition often aids in identifying species of the cellulose bacteria.

Pigment Formed by Various Species of Cellulose Bacteria

Cellvibrio ochraceus	ocher-yellow
C. flavescens Winogradsky	yellow-brown
C. vulgaris Stapp and Bortels	light yellow
C. fulvus Stapp and Bortels	yellow-brown, then brown
C. violaceus Castalli	violet
Cellfalcicula viridis Winogradsky	green
C. mucosa Winogradsky	pale yellow (straw-colored)
C. fusca Winogradsky	brownish pale yellow
Sporocytophaga congregata Fuller and Norman	yellow
S. myxococcoides (Krzemieniewska) Stanier	light yellow
S. ellipsospora (Imshenetskii and Solntzera) Stanier	orange
Cellulomonas ferruginea (Rullmann) Bergey et al.	brownish yellow, brown
C. rossica (Kellerman and McBeth) Bergey et al.	bright yellow
C. biazotea (Kellerman et al.) Bergey et al.	lemon yellow
C. folia Sanborn	bright yellow
C. gilva (McBeth)	yellow lemonish
C. aurogenes (Kellerman et al.) Bergey et al.	bright yellow
C. liquata (McBeth and Scales) Bergey et al.	yellow
C. flavigena (Kellerman and McBeth) Bergey et al.	yellow
C. idonea (McBeth) Bergey et al.	yellow
Cytophaga hutchinsonii Winogradsky	yellow
C. aurantiaca Winogradsky	orange
C. rubra Winogradsky	rose and red

C. tenuissima Winogradsky	olive-green
C. deprimata Fuller and Norman	yellow-orange
C. lutea Winogradsky	bright yellow
C. krzemieniewskae Stanier	rose-tinted or light red
C. diffluens Stanier	rose, then red
Bacillus cytaseus McBeth and Scales	very pale yellow

A number of species do not form pigments. Also it should be clearly noted that species identification cannot be made on the basis of pigmentation alone. A variety of other tests should be run. See Skerman (1967).

Anaerobic Decomposition

There are various nutrient media which can be used for detecting cellulose-decomposing anaerobic bacteria in water masses.

Imshenetskii Medium No. 1

Tap water	1000 ml
$NaNH_4HPO_4 \cdot 4H_2O$	1.5 g
KH_2PO_4	0.5 g
NaCl	0.1 g
K_2HPO_4	0.5 g
$MgSO_4 \cdot 7H_2O$	0.4 g
Peptone	5 g
$CaCO_3$	2 g
$MnSO_4 \cdot 5H_2O$	1 drop of a 1% solution
$FeSO_4 \cdot 7H_2O$	1 drop of a 1% solution
Filter paper	15 g
pH 7.0–7.4	

Imshenetskii Medium No. 2

Tap water	500 ml
Beef-peptone broth	500 ml

| Chalk | 2 g |
| Filter paper | 15 g |

pH 7.0–7.4

Omelianski Medium

Tap water	700 ml
Beef-peptone broth	300 ml
$(NH_4)_2HPO_4$	1 g
K_2HPO_4	1 g
$MgSO_4 \cdot 7H_2O$	0.5 g
NaCl	trace
$CaCO_3$	2 g
Filter paper	30 g

Fig. 100. Test tube with nutrient medium and filter paper for culture of anaerobic cellulose bacteria.

Filter Paper Medium

Tap water	900 ml
Yeast water	100 ml
Glucose	0.5 g
Peptone	5 g
Cellulose	3 g

pH 7.4

Filter paper, cut in thin strips, is placed either in tall test tubes or in bottles, and the medium is added (Fig. 100). To provide more complete anaerobiosis, after inoculation the

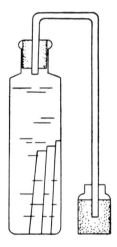

Fig. 101. Apparatus for a culture of anaerobic cellulose bacteria (Omelianski, 1940).

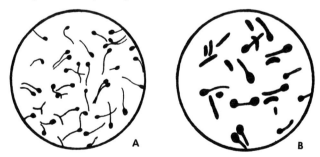

Fig. 102. A and B, Different morphological types of *Bacillus fermentationis-cellulosae* Omelianski.

bottles are filled to the top with additional sterile nutrient medium and closed with stoppers with vent tubes, the ends of which are dipped in liquid paraffin (Fig. 101).

Decomposition of the filter paper and the corresponding liberation of gas are indications of cellulose breakdown. The causative agent is *Bacillus fermentationis-cellulosae* Omelianski, a long, thin, sometimes curved, motile, rod-shaped bacterium which produces a round, terminal spore (Fig. 102). During sporogenesis, the cells swell at the pole and appear as drumstick-shaped bacteria. A characteristic of the species is the absence of granulose in the cells. *B. fermentationis-cellulosae* hydrolyzes cellulose with formation of organic acids (acetic acid, butyric acid) and gas (carbon dioxide and hydrogen).

Other species capable of decomposing cellulose have been described: *Clostridium omelianski* (Henneberg) Spray and *Clostridium cellulosolvens* Cowles and Rettger.

The cell morphology of all the species is similar: rod-shaped bacteria with a terminal spore.

PECTIN DECOMPOSITION

The rate of mineralization of plant residues in water masses depends on the decomposition of pectins (intracellular) which are insoluble in water and which form the basis of plant cells, and are thereby contained in significant amounts of plant residues. Pectins are derivatives of carbohydrates and their chemical composition differs in various plants.

Bacteria capable of breaking down pectins are widely distributed in water masses and are significant in the cycle of matter. They are also important economically, because of their involvement in the retting of bast-fibrous plants such as flax.

Pectins are broken down under natural conditions by anaerobic bacteria, aerobic bacteria, and fungi.

A variety of nutrient media may be used for observing anaerobic pectin fermenters. A selective medium containing water and small bundles of treated flax fibers is often used. The bundles of flax fibers are boiled several times in water to remove extractable material.

Cultures of pectinolytic bacteria may also be successfully obtained with the Bychkovskaya medium. Small bundles of nettle stems are used instead of flax fibers. After it has blossomed, nettle is gathered and dried. Small bundles of nettle are boiled with several changes of water and placed in test tubes. Yeast water or yeast extract solution, to a final concentration of 5–10%, and precipitated chalk are added to water. The medium is poured into the test tubes and sterilized in an autoclave at 120 C.

A mineral medium, suggested by Barinova, may also be used.

Barinova Medium

Tap water	1000 ml
K_2HPO_4	0.5 g
$(NH_4)_2HPO_4$	2 g
$MgSO_4 \cdot 7H_2O$	0.1 g
NaCl	0.1 g
Chalk	20 g

A variety of sources of pectin may be used—potato, flax fibers, soluble beet pectin, or dry sugar beet (which contains 20% pectin). According to Barinova's data, the most suitable source is sugar beet, added in the amount of 5 g per 100 ml of medium. The medium is improved by addition of yeast extract (1 g per liter).

A medium of potato broth containing yeast extract and a piece of carrot or potato yields good results. The broth is prepared by boiling 500 g of cut potato in 1000 ml of water for 30 minutes. The liquid is decanted; 5 g of yeast extract are added; and the medium is sterilized for 30 minutes at 115 C. The pieces of carrot or potato are sterilized in narrow test tubes and covered with potato broth which is added aseptically. A reducing substance is added to enhance conditions of anaerobiosis. Test tubes are filled to the top with nutrient medium; this provides the most favorable conditions for growth of anaerobes. Omelianski suggests placing inoculated test tubes in vacuum desiccators, from which air is evacuated and replaced with an inert gas. This is not mandatory, however.

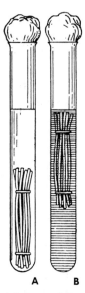

Fig. 103. Test tubes containing nutrient medium for culturing pectinolytic bacteria. A, Before inoculation; B, during growth of pectinolytic bacteria.

Anaerobic fermentation of pectins develops after 2–3 days and the small bundles are carried to the surface of the liquid by the gases liberated during the process (Fig. 103, B). Glass tubes may be used to secure the bundles to prevent them from being borne up. Fermentation of pectins is accompanied by the formation of acetic and butyric acids.

On the first few days after inoculation, growth of associated bacteria occurs. These bacteria utilize the carbohydrate material of the plants. To screen out extraneous microflora and enrich for the pectinolytic bacteria, a series of transfers is made. Localization of pectin-decomposing bacteria in plant tissues can be observed by examining the tissue under a microscope. The plant tissue sample (obtained by excising a small section of bast fiber with a scalpel or razor) is placed in 1 drop of a solution of iodine in potassium iodide. Pectinolytic bacteria are distinctly visible owing to the coloring of the intracellular granulose.

Many bacteria are capable of breaking down pectins in water masses. Common forms such as *Bacillus subtilis*, *B.*

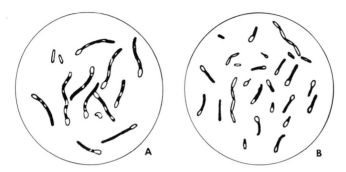

Fig. 104. Pectinolytic bacteria. *A, Clostridium pectinovorum* (Störmer) Donker; *B, C. felsineum* (Carbone and Trombolato) (Bergey et al.).

polymyxa, and species of the *P. fluorescens* and *E. coli* groups may also be pectinolytic to a slight degree. However, very active breakdown of pectins occurs in the group of bacteria which includes the *Clostridium* species (Table 9 and Fig. 104).

Pure cultures of anaerobic pectin fermenters can be isolated by the Omelianski method. An agar medium is prepared with a carrot broth base. Ground-up fresh carrot (200 g) is covered with 1 liter of tap water. A small quantity of chalk is added to neutralize the acids, and the solution is boiled for 1 hour. The liquid obtained is filtered and agar is added to the filtrate.

A potato agar medium with yeast extract or a medium with peptone can also be used.

Peptone Medium for Pectinolytic Bacteria

Tap water	950 ml
K_2HPO_4	1 g
$MgSO_4 \cdot 7H_2O$	0.25 g
Peptone	2 g
Galactose	5 g
Yeast water	50 ml
Agar	20 g
Pectin	8 g

Isolation is as follows: A small amount of liquid from a culture in which sporogenesis has occurred is transferred to small

Table 9. Pectin-decomposing bacteria
(Raynaud, 1949)

Species	Characteristics of Species	Formation of Pigments	Liquefication of Gelatin
Clostridium pectinovorum (Störmer) Donker (Fig. 104, *A*)	large rod-shaped bacteria up to 15 μ in length; gram-positive; motile; terminal spores; during sporogenesis the cells accumulate granulose	none	+
C. felsineum (Carbone and Trombolato) Bergey et al. (Fig. 104, *B*)	rod-shaped, motile, 3.5×0.9 μ, subterminal spores; during sporogenesis cells are clostridial-shaped and accumulate granulose	yellow-orange and yellow-rose	+
C. haumanni (Soriano) Prévot	rod-shaped, motile, gram-postive, $3–12 \times 0.8$ μ, terminal spores; during sporogenesis cells are plectridial, do not accumulate granulose or accumulate very little	canary-yellow or red	0
C. roseum McCoy and McClung	rod-shaped, motile, 3.9×0.8 μ gram-positive, subterminal spores; during sporogenesis cells are clostridial-shaped	rose and rose-red	+
C. corallinum Prévot and Raynaud	rod-shaped, $3–4 \times 0.8$ μ, motile, gram-positive, subterminal spores; during sporogenesis clostridial-shaped cells	coral-red	+
C. saturnirubrum Prévot	straight, rod-shaped, $4–5 \times 0.8$ μ, motile, gram-positive, subterminal spores; during sporogenesis clostridial-shaped cells	red	0
C. aurantibutyricum Hellinger	rod-shaped, very variable in dimensions: $2.8–17 \times 0.5–0.6$ μ, motile, gram-positive, subterminal spores	rose-orange or orange-red	+

Milk Coagulation	Starch Decomposition	H_2S Formation	Indole Formation	Products of Fermentation							
				Acids				Alcohols			
				Butyric	Acetic	Formic	Lactic	Butyl	Ethyl	Iso-propyl	Acetone
+	+	0	0	+	+	± trace amts	+	+	+	+	+
+	0	+	+	+	+	± trace amts	+	+	+	± or 0	+
?	0	?	?	?	?	?	?	?	?	?	?
+	+	0	0	+	+	+	+	+	+	0	+
+	+	0	0	+	+	+	+	+	+	0	+
0	+	0	0	0	0	0	+	?	?	?	?
+	+	0	+	+	+	+	+	+	+	± trace amts	+

sterile test tubes and placed in a water bath for 10 minutes at 80 C. With a sterile pipette, a small amount of the heated liquid is transferred to a test tube containing melted agar. The two are mixed well and immediately collected into sterile Pasteur pipettes which are placed in an incubator for germination after the agar hardens.

Colonies of pectin-decomposing bacteria grow rather rapidly in agar in capillaries and growth may be visible after 1–2 days. The agar around the colonies is broken as a result of the liberation of gas. Isolation is carried out in the usual manner by notching and then breaking off the capillary near the colonies. This is usually repeated several times. Isolated colonies may also be picked from the surface of agar poured into petri dishes; anaerobic culture conditions are maintained during this procedure.

It should be noted that the addition of peptone or beef extract creates less favorable conditions for pectinolytic bacteria.

Omelianski medium can be used for enriching for aerobic pectinolytic bacteria:

Omelianski Pectin Medium

Tap water	1000 ml
$(NH_4)_2SO_4$	0.5 g
K_2HPO_4	0.5 g
Chalk	20 g

The medium is poured in a thin layer into small Erlenmeyer flasks. Pectin is added to each and the solution is sterilized in an autoclave. The inoculated flasks are placed in an incubator at 25 C.

The activity of aerobic pectinolytic bacteria may be determined in a special medium.

Medium for Pectinolytic Bacteria

Two solutions are prepared:

1.	Distilled water	1000 ml
	Agar	15 g
	$CaCl_2 \cdot 6H_2O$	5 g

The solution is filtered; pH is adjusted to 7.3 with a dilute solution of sodium hydroxide; and the solution is dispensed into test tubes in 3-ml amounts and sterilized for 20 minutes at 120 C.

2.	Distilled water	900 ml
	Winogradsky standard solution	50 ml
	Soil extract	50 ml
	$NaHCO_3$ solution (10%)	5 ml
	NH_4NO_3	2 g
	Pectin	8 g

When the potato pectin is added to the medium, it must be remembered that it mixes poorly with water. The pectin is either added directly to the salt solution, heated to 80 C, and mixed with it until a homogeneous mixture is obtained, or it is placed in an absolutely dry vessel and mixed with 15 ml of 96% ethyl alcohol to which the salt solution, heated to 80 C, is added and mixed thoroughly. The pH of the solution is adjusted to 8 by adding 5 ml of sodium carbonate solution. The medium is dispensed in 15-ml amounts and sterilized in an autoclave. The temperature rise of the autoclave is stopped as soon as 120 C is reached. In this way, the destruction of the pectin molecule is minimal. After sterilization the pH drops to 7.2.

Two milliliters of the solution containing pectin is dispensed with a sterile pipette into each test tube containing agar. The test tubes are then inoculated. When 1 ml of sediment dilution or 1 ml of water is inoculated, the concentration of pectin is 5.33%. The action of microorganisms on the pectin is detected by breakdown of pectin as shown by liquefication.

Barinova suggests a special medium for fungi which decompose pectins.

Barinova Medium for Pectin-decomposing Fungi

Tap water	1000 ml
NH_4NO_3	3 g
KH_2PO_4	0.6 g
$MgSO_4 \cdot 7H_2O$	0.3 g

$FeCl_3 \cdot 6H_2O$ trace

$ZnSO_4 \cdot 7H_2O$ trace

Pectin   10 g

The pectin decomposition process proceeds through action of a series of enzymes. The first stage, primary maceration of plant tissues, takes place under influence of the enzyme protopectinase, which converts insoluble pectins to a soluble state. The enzyme pectase then hydrolyzes the ester bonds of the pectins. Hydrolysis of pectins is completed by pectinase and polygalacturonidase.

The activity of pectin-decomposing bacteria can be determined by the speed of action of the indicated enzymes: protopectinase, by the maceration of potato and carrot tissues (this can be crudely estimated by noting the time for the breaking off of a thin potato flake under strictly determined experimental conditions; see Seliber, 1962); pectase, by the change in the color of a medium containing lemon pectin; polygalacturonidase, by measuring relative viscosity of a pectin solution before and after bacterial action (Saissac, Brugiere, and Raynaud, 1952).

BUTYRIC ACID FERMENTATION BACTERIA

When plant and animal remains decompose in a water mass and in sediment, various carbohydrates and alcohols (such as mannitol and glycerol) and their compounds (glycerides, and so forth) decompose, and butyric acid is formed in large amounts. This process, termed butyric acid fermentation, is carried out by anaerobic spore-forming bacteria which are widely distributed in nature, including water masses. The bacteria ferment the above compounds with abundant production of gas (H_2 and CO_2) and formation of buytric acid and a number of by-products (fatty acids, alcohols, acetone). Butyric acid fermentation reaches its greatest development in those biotopes in which nitrogen-free organic substances accumulate. The classification of this group has not yet been adequately defined. Butyric acid fermentation is induced by large spore-forming rod-shaped bacteria which reach 10–12 μ in length. The majority are motile. At sporogenesis cell configuration changes either to a spindle shape (if the spore is centrally located) or a drumstick shape (if the spore is terminally located). Butyric acid-fermenting bacteria accumulate intracellular granulose (a reserve nutrient). As a result, the cell cytoplasm stains blue with iodine.

Various media are used to observe bacteria which cause butyric acid fermentation: potato medium; peptone water containing glucose; beef-peptone broth, to which are added carbohydrate (glucose or saccharose) in the amount of 2% and chalk for neutralizing the acid formed; and various mineral media containing glucose. Enrichment for butyric acid bacteria with specialized functions, e.g. the ability to fix nitrogen, requires media specified for that species of bacteria.

Media which contain sugar are suitable for most of the diverse forms. Thus, to culture butyric acid bacteria which are strict anaerobes, the microorganisms must be grown under conditions which ensure anaerobiosis and the media should be hot when inoculated. Non-spore-forming species are destroyed and spore-forming butyric acid bacteria are preserved.

Inoculations are made in one of the following media or any of the several commercially available media for anaerobes.

Potato-Chalk Medium

Wide, tall test tubes are filled to one-quarter or one-half with fine pieces of raw potato. Chalk (about 0.2 g) and tap water are added to two-thirds the volume of the test tube. The medium is sterilized in an autoclave.

Glucose Beef-Peptone Broth

A deep layer of beef-peptone broth containing glucose is poured into test tubes, the bottoms of which contain a small amount of chalk. Small inverted vials are placed in the test tubes for gas collection (Fig. 105). The medium is sterilized either with flowing steam or in an autoclave at 110 C for 20 minutes. During sterilization the small vials fill completely with the medium.

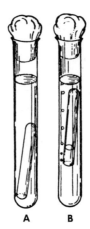

Fig. 105. Test tubes for culturing butyric acid-fermenting bacteria. *A,* Before inoculation; *B,* after growth.

Glucose Peptone Water

Tap water	1000 ml
Peptone	20 g
Glucose	20 g
Chalk	precipitate

A deep layer of the medium is poured into test tubes containing inverted vials and chalk. Sterilization is performed with flowing steam or in an autoclave at 110 C for 20 minutes.

Neronova Mineral Medium

Distilled water	1000 ml
$(NH_4)_2HPO_4$	1 g
K_2HPO_4	0.5 g
$MgSO_4 \cdot 7H_2O$	0.01 g
Glucose	20 g
Chalk (added to each test tube separately)	10–20 g
Crystalline biotin	0.001 mg
Ascorbic acid	100 mg

A deep column of the medium is formed by pouring it into tall, thin test tubes, on the bottom of which has been placed a small quantity of chalk. Sterilization is accomplished with flowing steam.

Ethyl Alcohol Medium [1]

Distilled water	1000 ml
Ethyl alcohol (96%)	8 ml
$CH_3COONa \cdot 3H_2O$	8 g
KH_2PO_4–Na_2HPO_4 buffer, 2 M, pH 7.0	25 ml
Na_2CO_3	0.1 g
$(NH_4)_2SO_4$	0.5 g
$MgSO_4 \cdot 7H_2O$	0.2 g
$CaSO_4 \cdot 2H_2O$	10 mg
$FeSO_4 \cdot 7H_2O$	5 mg
$MnSO_4 \cdot 5H_2O$	2.5 mg
$NaMoO_4 \cdot 2H_2O$	2.5 mg
Biotin	3 μg
Paraaminobenzoic acid	50 μg
Sodium thioglycolate	0.5 g

Sodium thioglycolate can be replaced with 0.2 g of $Na_2S \cdot 9H_2O$; growth factors can be replaced with 0.5–1 g of yeast extract. The alcohol is introduced aseptically into the sterilized medium.

Beef Extract Medium

Tap water	1000 ml
Yeast extract	10 ml
Beef extract	5 g
NaCl	3 g
$CH_3COONa \cdot 3H_2O$	3 g
Soluble starch	1 g
Dextrose	5 g
Cysteine hydrochloride	0.5 g
Agar	0.5 g

pH 7.2

---

[1] *Manual of Microbiological Methods*, 1957.

In all media, growth of butyric acid bacteria can be detected after 1–2 days. The turbidity of the medium is noted and a mass of gas bubbles appears which can be seen on the surface of the medium. Vials (placed in the tubes prior to growth of the culture) fill with gas and rise to the surface. Finally, the cultures produce a strong odor of butyric acid.

The bacteria are examined under a microscope before fermentation in the inoculations terminates. Morphologically, butyric acid bacteria may be distinguished by the features cited above (Fig. 106).

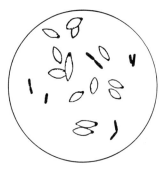

Fig. 106. Butyric   acid-fermenting   bacteria—*Clostridium butyricum.*

To demonstrate that butyric acid fermentation (during which hydrogen and carbon dioxide are formed) and not alcohol fermentation (during which only CO_2 is formed) occurs during inoculations of broth containing glucose, the following procedure may be used (Omelianski). A small piece of pure sodium hydroxide is taken up with a long forceps having curved edges, and is placed under one of the overturned vials filled with the liberated gases. The sodium hydroxide reacts with the CO_2, and the liquid enters the vial. The volume occupied by hydrogen does not fill with the liquid.

A fermented brewer's medium is used for detection of butyric acid by means of qualitative reactions.

1. Reaction that demonstrates formation of ethyl butyrate, a compound possessing a characteristic pineapple odor. For this reaction, 2–5 ml of medium and 1–2 ml of concentrated

sulfuric acid are combined. The mixture is heated and the characteristic odor of ether will be detected if butyric acid is present.

2. Reaction with $FeCl_3$. Several milliliters of the liquid are taken from the culture. Several drops of a concentrated solution of ferrous chloride are added and the mixture is heated. If neutral salts of butyric acid are present, a brown color is observed.

Gases liberated by bacteria during butyric acid fermentation can be collected and identified. For this purpose the culture is handled in the following way. Erlenmeyer flasks are filled three-quarters full with beef-peptone broth containing sugar and chalk. Rubber stoppers with holes through which bent glass tubes are passed are fitted to the flasks. The tubes with stoppers are wrapped in paper and sterilized separately from the flasks. The flasks, closed with cotton stoppers, are sterilized for 20 minutes at 110 C. Before inoculation the flasks are heated and inoculations are made into the hot medium. After the liquid cools, the flasks are filled to the top with additional medium, the cotton stoppers are replaced by the rubber ones, and the ends of the gas-conducting tubes are connected to reservoirs for gas collection (Fig. 107).

The best media for isolating butyric acid bacteria from the aquatic environment are a potato medium containing chalk,

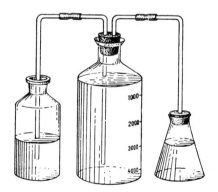

Fig. 107. Apparatus for studying butyric acid fermentation (Fedorov, 1951).

and a mixture of equal parts of potato and carrot agars. Potato agar is prepared according to one of the following methods:

1. Agar (7.5 g) is dissolved in 500 ml of water; 100 g of hot boiled potato are ground up in the 500 ml of warm agar, which is added gradually. Chalk (2.5 g) is added, and the entire mixture is blended well and poured into test tubes in 10-ml amounts. These are sterilized in an autoclave.

2. A potato is skinned, washed, cut into slices, and boiled in water (10 g per 100 ml of water). The broth is filtered and 1.5–2% agar is added to the filtrate. The agar is melted in a water bath and, if necessary, the medium is again filtered; chalk (2.5 g per 1000 ml) is added and the mixture is poured into test tubes. The medium is shaken constantly during the dispensing. The test tubes are sterilized in an autoclave.

Carrot agar is prepared in the same manner as the potato agar.

Isolation of a pure culture is achieved as follows: Several dilutions are prepared from a medium in which anaerobes have grown. A small amount of inoculum from each dilution is placed in a series of test tubes containing melted agar, mixed thoroughly, and collected into Pasteur pipettes. The ends of the pipettes are sealed, and the pipettes are incubated in a standing position in an incubator. Colonies of butyric acid bacteria develop after 1–2 days. The pipettes are sawed with a glass file and broken where isolated colonies can be seen. Sterile conditions are maintained at all times and single colonies are transferred to liquid nutrient medium.

Isolation of these bacteria may also be achieved with slices of potato and carrot, the surfaces of which are rubbed with chalk. The slices are then sterilized in an autoclave. Inoculation is made from dilutions onto the surfaces of the slices by rubbing in the drop of inoculum with a sterile spatula. The inoculated slices must be placed in a vacuum desiccator or anaerobic incubator. After 5–6 days, colonies of butyric acid bacteria appear and these are surrounded by gas bubbles. These are then transferred to liquid medium. Isolation is repeated. The cultures so obtained are examined under a microscope for the presence of contaminating bacteria.

FAT-HYDROLYZING BACTERIA

Fats are among the constant components of animal and plant cells. When either die, the fats undergo decomposition by specific microorganisms. Significant amounts of various kinds of fats enter water masses contaminated with sewage (domestic, slaughterhouse sewage, sewage from wool-washing plants, cloth plants, soap works, and other plants). Decomposition of fats occurs constantly in different water masses.

The total number of fat-decomposing bacteria may be determined by the pour plate method with solid media and by extinction dilution with liquid media. It must be noted that there is no generally accepted procedure for fat decomposition and there are no generally accepted media. Therefore, the formulas of several different media are cited below.

Beef-Peptone Agar with Fat

The following procedure is often employed in determining the number of fat-decomposing bacteria. Melted beef or pork lard, sterilized previously, is poured on the bottom of sterile petri dishes and immediately decanted. A thin layer of fat remains on the bottom of the dish. Melted butter can also be used. Inoculations are made in these dishes. The inoculum is introduced into the dishes in the usual manner, after which meat-peptone agar is added. The agar must be just barely warm, otherwise the fat will dissolve and the uniform layer on the bottom of the dish will be disturbed. The inoculum is mixed with the medium by carefully tilting the plate. The colonies of fat-decomposing bacteria are counted after 10 days. The colonies are easily discernible because the fat under them becomes white and opaque. Addition of a sterile solution of Nile blue to the medium facilitates recognition of colonies of fat-decomposing bacteria, because the medium around the colonies turns blue if free fatty acids are present; a sterile medium, containing neutral fats, remains rose-tinted. The following method may be used (in a medium without Nile blue): when the period of incubation is over, a saturated solution of copper sulfate is poured over the dishes and after 5–6 minutes the $CuSO_4$ is carefully decanted from the agar. Colonies of fat-decomposing bacteria stain azure and halos may be observed around them.

Seliber Medium

Distilled water	1000 ml
K_2HPO_4	1 g
$MgSO_4 \cdot 7H_2O$	0.3 g
$CaCl_2 \cdot 6H_2O$	0.1 g
$(NH_4)_2HPO_4$	2 g
NaCl	0.1 g
Agar	20 g

Heated melted butter or heated oil (0.5%) is added to the medium. The pH of the medium should be 6.7–7.0. The medium is dispensed into test tubes in 12- to 15-ml amounts and sterilized in an autoclave at 120 C for 15 minutes. Before the medium is inoculated, several drops of an alcohol solution (0.04%) of the indicator, bromothymol blue, are added to each test tube of melted agar until a bright-green coloring is obtained. The agar is then poured over the inoculum in a petri dish. The indicator changes color around colonies of fat-decomposing bacteria as a result of the acidification of the medium. Thus the lipolytic bacteria are easily distinguished.

When determining the number of fat-decomposing microbes by the titer or extinction dilution method, Seliber medium (without agar) may be used if sunflower or olive oil (1–2 ml per 50 ml of medium) is added instead of heated oil or melted butter.

Tauson Medium

Distilled water	1000 ml
$NaNO_3$	2 g
KCl	0.5 g
$MgSO_4 \cdot 7H_2O$	0.5 g
K_2HPO_4	1 g
$FeSO_4 \cdot 7H_2O$	0.001 g
Agar	15 g

The medium is sterilized in the usual manner. Before inoculation, separately sterilized fat (sunflower oil) (2 ml per 100 ml of medium) and 1% solution of Nile blue (0.2 ml per 100 ml of medium) are added to the melted medium.

The mixture is used for pour plates after prolonged shaking to emulsify the oil.

Rahn Medium

Distilled water	1000 ml
K_2HPO_4	5 g
$(NH_4)_3PO_4$	5 g
$MgSO_4 \cdot 7H_2O$	1 g
$CaCl_2 \cdot 6H_2O$	1 g
$FeCl_3 \cdot 6H_2O$	trace
NaCl	trace
Agar	15 g

Before inoculation either separately sterilized castor oil, beef fat, or pork lard is added to the hot, melted agar, which is then shaken vigorously for 10 minutes. The fat is thus emulsified with the agar. The top layer, rich in fat, is separated from the lower portion which contains a small amount of fat in the form of droplets. Only the lower layer of agar is used for pour plating. Sunflower oil, butyrin, and triolein may also be used as a source of fat.

Rahn liquid medium containing beef or pork fat may be used for the titer (extinction dilution) method. In this case, the following procedure is followed. Small Erlenmeyer flasks, closed with cotton stoppers, are sterilized by dry heat. The liquid medium is prepared and sterilized in an autoclave. At

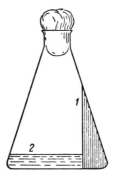

Fig. 108. Flask containing fat (1) and a nutrient medium (2) (Omelianski, 1940).

the same time, the fat is separately prepared and sterilized in an autoclave. The melted sterile fat is poured into the flasks, which are then tilted. The fat is allowed to solidify, after which a thin layer of the liquid medium is poured into the flask (Fig. 108). Sterile conditions are maintained at all times. After inoculation, fat-decomposing bacteria are detected when the fat turns white as fatty acids are formed.

All media for enumerating fat-decomposing bacteria in water masses are more suitable for growth of these microorganisms when yeast extract (0.5%) is added to the medium.

Cottonseed Oil Medium

20% Emulsion of cottonseed oil	25 ml
0.3% Solution of aniline blue	50 ml
Tryptone	10 g
Yeast extract	5 g
Agar	30 g
Distilled water	up to 1000 ml

The aniline solution is prepared by dissolving the dye in alcohol. The oil emulsion is prepared as follows: 10 g of finely ground gum arabic are triturated in a mortar in 100 ml of fresh cottonseed oil while 400 ml of warm distilled water are gradually added. After trituration, an opaque emulsion containing drops of fat less than 10 μ in diameter should be obtained. The agar, tryptone, and yeast extract are dissolved in 900 ml of distilled water by heating in an autoclave for several minutes. After these substances dissolve, the oil emulsion and the freshly filtered aniline solution are added. The volume of the mixture is brought to 1 liter with distilled water and it is sterilized in an autoclave. The medium should be kept in the cold. The colonies of fat-decomposing bacteria may be easily detected in the agar by the appearance of dark-blue coloring which surrounds them.

Anaerobic inducers of fat decomposition may be detected in the above media if the inoculated media are incubated under anaerobic conditions. The dilution (titer) method may be used, in which case inoculations are made in media poured into tall, thin tubes or bottles. After inoculation the bottles are filled with a sterile medium, closed with rubber stoppers, and sealed at the top with a hermetic sealer.

Anaerobic decomposition of fat is a very slow process which requires a significant amount of time.

To identify the species of lipolytic bacteria, the cultures are purified and studied by the usual methods.

DECOMPOSITION OF SALTS OF ORGANIC ACIDS

In the process of bacterial decomposition of proteins, fats, hemicelluloses, and other compounds, salts of fatty and other organic acids are formed. These compounds are almost constant components of the sediment deposits of various water masses. The organic acid salts in the water masses are decomposed by microorganisms to the point of formation of final gaseous end products, usually methane. Methane is formed in great quantities when organic substances decompose under anaerobic conditions in both fresh and sea water. The fermentation of organic acids salts is one of the sources of free methane.

Methane fermentation of various carbon compounds is brought about by different species of bacteria. The characteristics of species studied by various authors is given in Table 10.

The introduction of a certain carbon source into a nutrient medium enriches for a given species of bacteria, as well as the end products of fermentation.

Bacteria which decompose salts of fatty acids along with the liberation of methane may be detected in one of the following media containing the source of carbon which is under study.

Omelianski Medium

Water from the investigated water mass	1000 ml
Peptone	2 g
Salt of organic acid	20 g

Table 10. Methane-forming bacteria
(According to a summary by Nechaeva, 1953)

Species	Characteristics of Species	Utilizes with Formation of CH_4
Methanobacterium soehngenii Barker	rod-shaped cells, straight or slightly curved, non-motile, non-spore-forming, $5-7 \times 0.3-0.5\ \mu$, often combine in a filament of significant length, gram-negative, anaerobic	salts of the acids: formic, acetic, butyric, caproic, caprylic, capric; mixture of carbon dioxide and hydrogen
This species includes: (a) thermophilic methane bacteria, isolated by Coolhaas; (b) methane bacteria, isolated by Wiken	very thin rod-shaped bacteria, $3-6\ \mu$ in length, often form filaments, anaerobic	salts of the acids: formic, acetic, propionic, butyric, isobutyric, oxalic, lactic, gluconic; salts of acetic acid; acetone
M. omelianskii Barker	very thin, rather long, $9-10 \times 0.7\ \mu$, weakly curved, non-motile and non-spore-forming rods, filaments formed very seldom and are very short, gram-negative, anaerobic	alcohols: ethyl, propyl, isopropyl, butyl, isobutyl, amyl; mixture of CO_2 and hydrogen
This species includes methane bacteria isolated by Wiken, other than above		salts of formic acid, propyl alcohol
M. formicum Stephenson and Strickland	short, motile, non-spore-forming rods, gram-negative, develop in anaerobic and aerobic conditions; growth is poorer in mineral media than in beef-peptone media, especially when various carbohydrates are added; SO_2 is reduced to H_2S with H_2	in an atmosphere of hydrogen; formate, hexamethylene-tetramine, CO_2 carbon monoxide; methyl alcohol
M. formicum Schellen		formates; mixture of CO with H or H_2O

M. suboxydans Stadtman and Barker	thin, delicate, slightly curved rods of varied length; cells often contain granules of metachromatin; gram-positive; gram stain reaction changes with age of culture	salts of the acids: butyric, valeric, caproic, enanthic
M. propionicum Stadtman and Barker	short, relatively thick rods or cocci; motile; turbidity forms in liquid cultures; gram-positive	propionates (incomplete oxidation)
Methanococcus mazei Barker	fine, spherical, single cells, sometimes form groups surrounded by a mucous capsule; non-motile, non-spore-forming; gram-variable; aerobic	acetates and butyrates
This species includes: (a) coccus isolated by Gronevege, (b) coccus isolated by Wiken		(a) acetates, (b) butyrates, acetone
Methanococcus vannielli Stadtman and Barker	spherical cells with a diameter of .5 to 4 μ; cocci very motile, movement in a spiral; anaerobic	formates; mixture of CO_2 and hydrogen
Methanosarcina methanica (Smit) Kluyver and van Niel	large, spherical cells, non-motile and non-spore-forming, form packets, 8-10 μ size, tetrads, seldom in the form of diplococci; behavior to gram staining—variable	salts of the acids: acetic, butyric; methyl alcohol
This species includes sarcinae isolated by Wiken	large cocci, form typical cell packets; in liquid cultures the packets are large zoogloeic masses; anaerobic	salts of the acids: acetic, propionic, butyric
M. barkerii (Schnellen) Barker, Ruben and Kamen		acetates; methyl alcohol
This species includes sarcinae isolated by Schnellen		mixture of CO with H or H_2O

The addition of yeast extract in the amount of 0.5 per liter to Omelianski medium significantly improves the enrichment process. Introduction of a sterile solution of sodium sulfide (0.3 g per liter) into a sterile medium contributes to creation of anaerobic conditions in the medium. For the same purpose, a small piece of cotton soaked with an oxygen absorber is placed in the neck of the vessel.

Stadtman and Barker Medium

Distilled water	1000 ml
$CaCl_2 \cdot 6H_2O$	0.01 g
NH_4Cl	0.5 g
$MgCl_2 \cdot 4H_2O$	0.01 g
$FeCl_3 \cdot 6H_2O$	0.002 g
$MgSO_4 \cdot 7H_2O$	0.001 g
K_2HPO_4	3.48 g
KH_2PO_4	2.72 g
$NaMoO_4 \cdot 2H_2O$	0.001 g
Phenol red	0.003 g
Methylene blue	0.002 g

After sterilization, to each 100 ml of medium are added 20–40 mg of $Na_2S \cdot 9H_2O$; 200–250 mg of Na_2CO_3; and 100–500 mg of a sodium salt of a fatty acid. The pH of the medium is adjusted to 6.8–7.4 with a 10% solution of hydrochloric acid. To create anaerobic conditions, a small piece of cotton soaked with an oxygen absorber is placed in the neck of the vessel under the stopper.

Barker Medium

Distilled water	1000 ml
NH_4Cl	1 g
KH_2PO_4	0.4 g
$MgCl_2 \cdot 4H_2O$	0.1 g
Calcium salt of an organic acid	10 g
Yeast extract	0.3 g

Oxygen absorption in the medium is attained by introducing into it, before inoculation, 3 ml of the following sterile solution for every 100 ml of medium: $Na_2S \cdot 9H_2O$ (1%), Na_2CO_3 (5%). pH of the medium is 7.0.

Bychkovskaya Medium

Distilled water	1000 ml
Sodium or calcium salt of an organic acid	20 g
KH_2PO_4	0.3 g
$MgSO_4 \cdot 7H_2O$	0.1 g
NH_4NO_3 or $(NH_4)_2HPO_4$	1 g
Yeast extract	1 g

Wiken Medium No. 1

Tap water	1000 ml
$NaHCO_3$	3.5 g
$(NH_4)_2SO_4$	0.5 g
$MgSO_4 \cdot 7H_2O$	0.5 g
K_2HPO_4	1 g
Sodium salt of organic acid	10 g
Washed agar	15 g

pH 6.8

Wiken Medium No. 2

Tap water	1000 ml
C_2H_5OH (96%)	10 ml
$MgSO_4 \cdot 7H_2O$	0.1 g
$(NH_4)_2SO_4$	0.5 g
Na_2CO_3	3.5 g
K_2HPO_4	1 g
Agar	3 g

pH 6.8

The alcohol is added aseptically before inoculation. After inoculation, 0.8 mg of Na_2S is added per 100 ml of medium.

When using any of the above media, it is suggested that a small amount of sand, clay, or fine-grained asbestos be added to the test tubes before sterilization, in order to insure fermentation in both the upper and deep layers of the medium.

Gas produced on decomposition of the salts of organic acids may be collected in special vessels and subjected to analysis.

OXIDATION OF HYDROCARBONS AND HYDROGEN

Various hydrocarbons, in pure form and in mixture, are found in water masses. Because of their composition, hydrocarbons are not commonly utilized as nutrient sources by aquatic organisms with the exception of bacteria.

The bacteria that occur in hydrocarbon systems, their culture, growth, metabolism, etc., are discussed in the textbook, *Elementary Petroleum Microbiology*, by J. M. Sharpley, 1966, Gulf Publishing Co., Houston, Texas.

Oxidation of Liquid and Solid Hydrocarbons

To determine the number and activity of bacteria-decomposing petroleum and petroleum derivatives in water masses, the procedure and medium suggested by Tauson (1928), Voroshilova and Dianova medium, or another mineral medium may be used. Tauson medium is an aqueous mineral solution which includes all salts necessary for growth of bacteria except organic compounds. After inoculation, hydrocarbon or petroleum is added.

Tauson indicates that the medium can be used with both nitrate and ammonium nitrogen (some species grow better in ammonium nitrogen, others in nitrate).

In either case, two solutions are prepared.

Tauson Medium with Nitrate Nitrogen

1.	$Ca(NO_3)_2$	1 g
	KNO_3	0.25 g
	$MgSO_4 \cdot 7H_2O$	0.25 g
	$Fe_2(SO_4)_3 \cdot 9H_2O$	0.01 g
	Distilled water	800 ml
2.	$K_2HPO_4 + KH_2PO_4$ (1:1)	0.25 g
	Distilled water	200 ml

The solutions are sterilized separately and combined before inoculation in the proportion 4:1; pH of the medium is 6.6.

Tauson Medium with Ammonium Nitrogen

1.	$(NH_4)_2SO_4$	1 g
	$CaSO_4$	0.5 g
	$MgSO_4 \cdot 7H_2O$	0.3 g
	$FeSO_4 \cdot 7H_2O$	0.005 g
	Distilled water	800 ml
2.	$KH_2PO_4 + K_2HPO_4$ (1:1)	0.3 g
	Distilled water	200 ml

It is recommended that Fedorov trace elements solution (1 ml per liter) (p. 295) be added to the Tauson media.[2] A medium containing trace elements is more suitable for hydrocarbon-oxidizing bacteria.

The solutions are sterilized in an autoclave, combined before inoculation in the proportion 4:1, and poured out in a thin layer into presterilized Winogradsky flasks. Petroleum, jet fuel, or individual hydrocarbons (liquid or solid) are placed in each flask separately after inoculation. Both the petroleum and derivatives should be sterilized beforehand. Petroleum is added so that it forms a thin uniform layer (no thicker than 1.1 mm) on the surface of the mineral solution. Solid hydrocarbons are placed on the surface of the medium as separate little pieces or crystals, 0.1–0.6 g per 200 ml of medium.

When aerobic cultures are used, the hydrocarbons should not cover the entire surface of the medium. Inoculated media are incubated at 20–25 C.

Growth of petroleum-decomposing bacteria is detected after several days. First, turbidity of the liquid and a weak yellowish coloring is observed; then, on the boundary line of the mineral solution and petroleum, a bacterial film appears, after which certain changes in the petroleum itself may be noted. Turbidity and the formation of clear zones are seen where the mineral solution comes into direct contact with the air.

---

[2] Tauson media can be used for phenol-decomposing bacteria. In this case, phenol (0.1–0.5%) is added to the cited mineral solutions.

The process of bacterial oxidation of petroleum requires a significant amount of time (complete oxidation of the petroleum may take as long as 2 months).

The activity of the petroleum-oxidizing bacteria may be judged by comparing the weight of the petroleum placed in the medium with that which remains unoxidized after a specific period of time elapses (the petroleum is added by weight). The products of decomposition can be determined chemically.

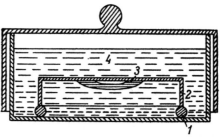

Fig. 109. Diagram of a culture of hydrocarbon-decomposing bacteria (Tauson, 1952). *1,* Glass triangle; *2,* petri dish; *3,* hydrocarbon; *4,* mineral solution.

Investigation of water-soluble hydrocarbons may be performed by the "diffusion inflow" method of Tauson. The hydrocarbon is placed in a nutrient medium, not on the surface of an aqueous solution, but under a layer of it. A triangular glass rod is placed on the bottom of a Koch dish under a layer of nutrient medium. The sterilized lid to a petri dish is placed in the Koch dish so that the former stands with its edges on the glass triangle and is covered by a layer of the nutrient medium. A curved pipette is used to place 1 or 2 ml of the liquid hydrocarbon under the top of the petri dish (Fig. 109). The inoculated dishes are placed in an incubator. Under these conditions, the hydrocarbon dissolves in the medium to saturation. Bacteria develop on the surface, utilizing the hydrocarbon dissolved at the interface between the hydrocarbon and the nutrient medium. The hydrocarbon is constantly replenished by new portions going into solution.

Voroshilova and Dianova Medium

Distilled water	1000 ml
NH_4NO_3	1 g

K_2HPO_4	1 g
KH_2PO_4	1 g
$MgSO_4 \cdot 7H_2O$	0.2 g
$CaCl_2 \cdot 6H_2O$	0.02 g
$FeCl_3 \cdot 6H_2O$ (saturated solution)	2 drops
Liquid hydrocarbon	1–2 ml
or solid hydrocarbon	0.5–1 g

Medium for Bacteria Decomposing Petroleum and Petroleum Derivatives

Tap water	1000 ml
K_2HPO_4	0.5 g
NH_4Cl	0.5 g
Chalk	trace amounts

The medium is poured into small flasks and sterilized, after which separately sterilized petroleum, jet fuel, gasoline, paraffin oil, or other hydrocarbons are added. Bacteria which oxidize liquid hydrocarbons (pentane, hexane, heptane) may be observed in Voroshilova and Dianova medium or in the following medium.

Medium for Bacteria Oxidizing Liquid Hydrocarbons

Tap water	1000 ml
KNO_3	1 g
$MgSO_4 \cdot 7H_2O$	0.2 g
NaCl	1 g
K_2HPO_4	1 g
pH 7.0–7.2	

Inoculated media are incubated in an atmosphere of hydrocarbon vapors.

Bacteria which oxidize liquid hydrocarbons demonstrate a characteristic fluorescence (Slavnina, 1947). *Bacterium aliphaticum* fluoresces an intensive green. *Serratia mogilewskii* in the early stages of development fluoresces grayish azure.

Petroleum-oxidizing bacteria can be isolated in the above solutions from which solid media are prepared by adding washed agar. Agar is added to 20 g per liter. Inocula in this case are transferred in the usual way to petri dishes. A thin layer of melted and cooled agar is poured over the inocula. The lid of the dish is opened slightly for several seconds and a very thin layer of a presterilized carbon source is poured over the surface of the dish so that a uniform layer results. Tauson has devised a series of complex methods for obtaining very thin layers of hydrocarbons.

Oxidation of Hydrocarbons—Methane and Propane

The formation of methane is a process which occurs throughout nature. Methane is liberated in large quantities from the bottom of water masses as a result of decomposition of organic substances by various bacteria. In these biotopes still another process is carried on—the oxidation of methane by specific microorganisms. Because of the activity of methane-oxidizing bacteria, a significant amount of the methane formed in benthic sediments is oxidized at the place of liberation and does not find its way up into the water mass itself.

A mineral medium is used for enriching for methane-, propane-, or ethane-oxidizing bacteria in water masses. Methane is used as the sole carbon source by methane-oxidizing bacteria, propane for propane-oxidizing bacteria, and so forth. The gas is introduced mixed with air, the presence of which should provide aerobic culture conditions. For methane oxidizers, the proportion of gases is as follows: one-third methane and two-thirds air; for propane oxidizers, one-fifth propane, four-fifths air.

Various media can be used to isolate and culture bacteria which oxidize methane and other gases.

Bokova, Kuznetsova, and Kuznetsov Medium

Tap water	1000 ml
KNO_3	1 g
$MgSO_4 \cdot 7H_2O$	0.2 g
K_2HPO_4	0.5 g
KH_2PO_4	0.5 g
NaCl	1 g

pH 7.2

Kaserer Medium

Distilled water	1000 ml
NH_4Cl	1 g
KNO_3	2 g
K_2HPO_4	0.5 g
$MgSO_4 \cdot 7H_2O$	0.2 g
$FeCl_3 \cdot 6H_2O$	trace

The medium is neutralized with a solution of $NaHCO_3$ to pH 7.0–7.2.

Omelianski Medium

Distilled water	1000 ml
NH_4MgPO_4	1 g
K_2HPO_4	0.5 g
$CaSO_4 \cdot 2H_2O$	0.1 g

Gromann Medium

Distilled water	1000 ml
Na_2CO_3	1 g
NH_4Cl	1 g
KH_2PO_4	0.7 g
$MgSO_4 \cdot 7H_2O$	0.2 g
NaCl	0.2 g
$FeCl_3 \cdot 6H_2O$	trace

One of the above media is poured in a thin layer into small Erlenmeyer flasks or sterile Koch dishes and placed in an anaerobic incubator or under a bell jar, the polished edge of which is fitted to the platform and smeared with vacuum grease. First the air is evacuated from the anaerobic incubator or bell jar to a residual pressure of 40 mm of mercury; the container is then filled with the corresponding gas mixture (Fig. 110). The anaerobic incubator or bell jar containing inoculated media is placed in an incubator for 14 days. It is recommended that sediments be placed in small sterile beakers into which a nutrient medium can then be poured.

The presence of the solid phase accelerates the develop-

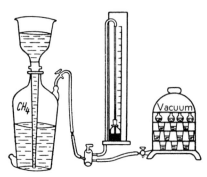

Fig. 110. Loading a bell jar, containing inoculated media, with a gas mixture (Mogilevskii, 1953).

ment of microorganisms. Thus sterile sand is often added when subsequent transfers are made.

Methane can be obtained by heating sodium acetate powder and barium hydroxide powder in a sand bath; the liberated gas is washed with bromine water and sodium hydroxide and collected under water into a collector (Omelianski, 1940).

Ethane can be obtained by electrolysis of sodium acetate in an acid medium, followed by purification of ethane from CO_2 and other impurities by passage through 20% potassium hydroxide and a solution of potassium permanganate.

In cultures in which a hydrocarbon utilizing microflora is present a bacterial layer or slime appears on the surface of the nutrient medium. Methane is oxidized by various bacteria (including the genera *Methanomonas*, *Pseudomonas*, and *Mycobacterium*). The development of pure cultures of methane-oxidizing bacteria may be possible only if pantothenic acid is added.

Often a film, at first thin and transparent and then thick, opaque, pigmented (brownish), and wrinkled, appears on the surface of the medium. Ethane-oxidizing bacteria yield a film which is white with a pinkish or reddish tint. Propane-oxidizing bacteria yield a mucous layer which is white or yellowish in color.

After purification by a series of transfers, the bacteria obtained in liquid culture may be isolated on an agar medium of the same composition. Well washed agar or Ion-agar is used.

Oxidation of Hydrogen

Hydrogen is a gas which is liberated in sediments of water masses during decomposition of organic substances rich in carbon. Hydrogen-oxidizing bacteria in a water mass are of great significance for the oxygen cycle in a body of water. Hydrogen-oxidizing bacteria are widely distributed in water masses; an especially large number are present in decaying sediments.

The hydrogen-oxidizing bacteria include autotrophs, those bacteria which are incapable of developing in organic media, *Pseudobacterium vitreum* (Niklewski) Krasil'nikov, and organisms capable of living both autotrophically, from energy liberated upon oxidation of gaseous hydrogen, and heterotrophically (*Bacillus pycnoticus* Ruhland and Gromann, *Hydrogenomonas pantotropha* (Kaserer) Orla-Jensen, *Pseudomonas facilis* (Schatz and Bovell) Davis et al.). None of the microorganisms which oxidize hydrogen is a strict autotroph. Under conditions of autotrophic growth, the oxygen concentration should not exceed 30%.

There are still no generally accepted methods for detecting hydrogen-oxidizing bacteria. An indispensable condition for enrichment is to incubate the inoculated media in a mixture of gases: hydrogen, oxygen, and CO_2 (6:2:1).

A number of media have been suggested for culturing these microorganisms.

Lebedeff Medium

Distilled water	1000 ml
KNO_3	2 g
KH_2PO_4	0.5 g
$MgSO_4 \cdot 7H_2O$	0.2 g
$FeCl_3 \cdot 6H_2O$	trace

Ruhland Medium

Distilled water	1 liter
$NaHCO_3$	1 g

NH$_4$Cl	1 g
KH$_2$PO$_4$	0.5 g
MgSO$_4$ · 7H$_2$O	0.1 g
NaCl	0.1 g

pH 7.1–7.2

After sterilization, a small amount of sterilized aqueous solution of FeCO$_3$ is added to the medium.

Cohen and Burris Medium for *Hydrogenomonas facilis*

Distilled water	1 liter
NaHCO$_3$	1 g
NH$_4$Cl	1 g
KH$_2$PO$_4$	0.5 g
MgSO$_4$ · 7H$_2$O	0.1 g
NaCl	0.1 g
CaCl$_2$ · 6H$_2$O	0.1 g
Fe(NH$_4$)$_2$ (SO$_4$)$_2$ · 6H$_2$O	0.008 g

The following mixture of trace elements is added to the medium:

H$_3$BO$_3$	228 μg
CaCl$_2$ · 6H$_2$O	80 μg
CuSO$_4$ · 5H$_2$O	8 μg
MnCl$_2$ · 4H$_2$O	8 μg
ZnSO$_4$ · 7H$_2$O	176 μg
Na$_2$MoO$_4$ · 2H$_2$O	50 μg

Cohen and Burris developed a special device in which a mixture of gas circulates through reservoirs containing gas into the culture vessel (Fig. 111). The pH of the medium, after equilibrium is established with the gas, is 6.8–7.2 and the pressure in the system is maintained approximately equal to atmospheric pressure. The gas is replaced with water, in proportion to its use in the reservoir. The parts of the system are sterilized before use. After sterilization and after the me-

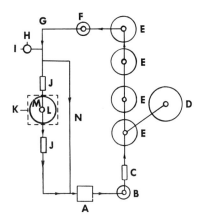

Fig. 111. Diagram of apparatus for culturing bacteria (view from above) (Cohen and Burris, 1955). *A*, Pump; *B*, bottle containing water for collecting oil; *C*, cotton filter; *D*, reservoir with substituting water; *E*, gas reservoirs; *F*, water interceptor; *G*, manometer; *H*, valve; *I*, air exit during evacuation; *J*, filters; *K*, temperature-controlled water bath thermostat; *L*, culture vessel; *M*, sampling tube; *N*, detour (arrows indicate the direction of the gas).

dium and the apparatus have cooled, the system is assembled and sealed with rubber stoppers. Bacterial growth is noted by the turbidity of the medium.

Hydrogen-oxidizing bacteria grow well on silicic acid gel plates saturated with a mineral medium and placed in an atmosphere of oxygen, hydrogen, and CO_2. They can be isolated easily into pure cultures. Kuznetsov and Romanenko (1963) suggest a simplified arrangement of inoculation whereby gas-oxidizing bacteria are included. A small vial is placed upside down in an ordinary test tube filled with a nutrient medium and is then filled with hydrogen or a gas mixture (Fig. 112). The medium is inoculated and the test tube is closed with a flame-sterilized rubber stopper.

When the vial is submerged, it is first closed with a small piece of thin sterilized paper and placed a few millimeters down into the test tube. The paper is removed and a capillary or needle from a hypodermic syringe is inserted into the medium under the small test tube and, with the syringe, the vial is filled with gas.

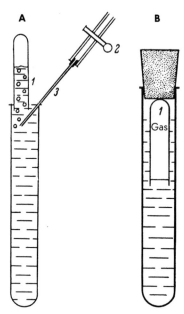

Fig. 112. Apparatus for culturing bacteria in an atmosphere of a mixture of gases (Kuznetsov and Romanenko, 1963). *A,* Filling the inverted test tube (*1*) with gas using a syringe (*2, 3*); *B,* closed test tube.

Oxidation of Carbon Monoxide

Autotrophic organisms which are capable of oxidizing CO to CO_2 may be found in water masses. Such organisms include *Carboxydomonas oligocarbophila* (Beijerinck and van Delden) Orla-Jensen, which develop well in media containing inorganic salts. A special medium is suggested for culturing these bacteria.

Medium for Carbon Monoxide
Oxidizing Bacteria

Distilled water	1000 ml
K_2HPO_4	0.1 g
$NaNO_3$	0.1 g
$FeCl_3 \cdot 6H_2O$	trace
$MgSO_4 \cdot 7H_2O$	trace

The atmosphere above the medium should contain no less than one-third CO.

Chapter 7

Methods of Studying Microorganisms
of the Nitrogen Cycle

PROTEIN MINERALIZATION

Decaying plants in a water mass and dead animals and organic material carried into a water mass undergo mineralization. The processes of decomposition of protein by microorganisms play an important role in the mineralization of organic material. Many bacteria decompose protein compounds to end products such as ammonia, hydrogen sulfide, and mercaptans.

Protein mineralization occurs in the water mass and in benthic sediments under aerobic and anaerobic conditions.

Quantitative Studies of Proteolytic Bacteria

A quantitative determination is always run when water samples are examined. The total number of proteolytic bacteria is determined by pour plate, generally with beef-peptone agar as the nutrient medium (see p. 56 for preparation). Fish-peptone agar is also used when working with natural water samples.

Recently, a number of media have been suggested which yield a larger number of species than ordinary beef-peptone agar. However, the use of beef-peptone agar provides a continuity of studies and permits comparison of the data with results of earlier investigations. Therefore, these media (formulas

are given below) can be recommended for growing a variety
of species of proteolytic bacteria and for determining the composition of heterotrophic microorganisms.

Colwell YE Medium

Yeast extract (Difco)	0.3%
Proteose peptone (Difco)	1.0%
NaCl	0.5%
Agar	2.0%

pH 7.2

Colwell SWYE Medium

Yeast extract (Difco)	0.3%
Proteose peptone (Difco)	1.0%
NaCl	2.4%
KCl	0.07%
$MgCl_2 \cdot 6H_2O$	0.53%
$MgSO_4 \cdot 7H_2O$	0.70%
Agar	2.0%

pH 7.2–7.4

Glycerol Beef-Peptone Agar

Tap water	900 ml
Beef water	100 ml
Peptone	2.5 g
Glycerol	20 g
K_2HPO_4	1 g
NaCl	3 g
$MgSO_4 \cdot 7H_2O$	0.25 g
$FeSO_4 \cdot 7H_2O$	0.01 g
$CaCO_3$	0.04 g
Agar	15 g

pH 6.9

Thornton Agar

Tap water	1000 ml
Asparagine	0.5 g
Mannitol	1 g

K_2HPO_4	1 g
$MgSO_4 \cdot 7H_2O$	0.2 g
$CaCl_2 \cdot 6H_2O$	0.1 g
NaCl	0.1 g
KNO_3	0.5 g
$FeCl_3 \cdot 6H_2O$	0.002 g
Agar	15 g

pH 7.4

To determine the number of proteolytic bacteria, the media are inoculated as soon as possible (not later than 3 hours after samples are taken).

The amount of water which should be collected is determined by the organic content of the water and, related to this, the number of proteolytic bacteria. For oligotrophic water, poor in organic matter, 0.1 to 1.0 ml water samples are usually inoculated onto the agar plates. Ponds and contaminated parts of rivers have a higher bacterial load, thus smaller quantities of water are used (0.1–0.0001 ml) and dilutions of the inoculum are also required. Sediments are usually inoculated from dilutions of 1:100, 1:1000, and 1:10,000 or higher (see p. 87 for a flow chart for dilutions). Three parallel inoculations are made from each dilution. However, inoculations can be done in duplicate but not less than this.

The appropriate quantity of inoculum is measured with a sterile graduated pipette and transferred to a petri dish. Cooled agar is then added. The inoculated dishes are incubated at 22–23 C. (Surface spread plate count procedure may be preferable to the pour plate in most instances.)

The colonies are counted after 48 hours. If slowly-growing bacteria are also to be enumerated the colonies are counted again after 10 days.

The diluents for preparing dilutions are extremely important. Distilled water is unsuitable since many bacteria suspended in distilled water rapidly die off, usually through lysis. Sterilized water from the same water mass from which the sample was taken is recommended. Weak solutions of amino acids (one or several) at a concentration of 0.1%, with pH of the solution adjusted to 7.0, yield good results. The solution is

poured into test tubes in 10-ml amounts, sterilized in an autoclave, and used to prepare dilutions.

If the bacterial content of this group in the water sample
is very small, the membrane filter method may be used. Appropriate amounts of water sample are filtered through bacteria-retaining membrane filters sterilized previously in distilled
water. When filtration is completed, the filters are placed back
side down on the surface of an agar nutrient medium in a petri
dish in order that cells which settle on the filter will have a
nutrient supply for growth. The dishes with the filters are incubated at a temperature of 23–25 C for 3–5 days. The number
of colonies growing on the filter is counted. Small colonies are
counted with a magnifying lens. When this method is used, however, some bacterial cells may settle on top of others. Thus, as
a consequence, the counts may be somewhat low.

If one of the purposes of a study is to determine the
types of proteolytic bacteria, the same inoculated plates (provided that growth is not too dense) may be used for isolating
pure cultures. If growth is heavy, fresh plates should be streaked
with smaller inocula, before procedures for isolating pure cultures are followed.

Beef-peptone gelatin is used to determine the number of
proteolytic bacteria in water samples. Either the pour plate or
spread plate method is employed and colonies are counted after
48 hours. Colonies surrounded by zones of liquefaction are
counted.

A gelatin medium containing peptone and yeast extract
may also be used.

Gelatin Peptone-Yeast Water Medium

Tap water	900 ml
Yeast water	100 ml
Peptone	5 g
Gelatin	120 g

pH 7.0–7.2

Frazier gelatin medium is rather convenient when working with water samples since it contains agar. With this medium
it is possible to work in hot weather without complete liquefication of the gelatin.

Frazier Gelatin Medium

NaCl	3 g
K$_2$HPO$_4$	1.5 g
KH$_2$PO$_4$	0.5 g
Gelatin	4 g
Dextrose	0.05 g
Peptone	0.1 g
Beef extract	5 ml
Agar	15 g
Distilled water	1000 ml

The medium is prepared as follows. The salts are dissolved in 100 ml of distilled water. The gelatin is separately dissolved in 400 ml of distilled water, and the dextrose, peptone, and beef extract are added. The solutions are poured together and boiled for several minutes. The agar is dissolved in 500 ml of water, the solutions are mixed, and pH is adjusted to 7.0. The mixture is then poured into flasks and sterilized. The following solution is prepared.

HgCl$_2$	15 g
HCl	20 ml
Distilled water	100 ml

After inoculation and at the end of the incubation period, the above solution is poured into the dishes containing Frazier gelatin and is allowed to stand for 5–10 minutes (15–30 minutes according to Frazier). The solution is carefully decanted and the colonies surrounded by transparent zones are counted. The medium not changed by bacteria becomes opaque, and the colonies surrounded by transparent zones, as a result of change in the gelatin, are very clearly visible (Fig. 113).

A saturated solution of ammonium sulfate or a 1% solution of picric acid may be used instead of the above solution.

When determining the quantity of proteolytic anaerobes, the same nutrient media and the same plating method is used. However, the conditions must be conducive to growth of anaerobes.

When enumerating proteolytic bacteria in water and

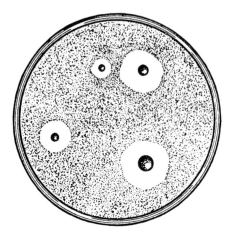

Fig. 113. Zones of gelatin liquefication in Frazier medium.

sediment, a mean is calculated from counts of duplicate plates. Results are expressed as numbers per milliliter of water and per gram of sediment. To avoid false accuracy, it is recommended that the numbers obtained be rounded off, converting the number of colonies to a per milliliter basis.

In the range of:	Round to the nearest:
	numbers obtained are
1–50	used without change
51–100	5
101–250	10
251–500	25
501–1000	50
1001–10,000	100
10,001–50,000	500
50,001–100,000	1,000
100,001–500,000	5,000
500,001–1,000,000	10,000

Plating methods, used for determining the number of heterotrophic bacteria in water samples, are based on the assumption that each colony develops from one cell. As a matter of fact, this is not always so. Some colonies, and in polluted water samples many colonies, develop from an aggregate or a chain of cells. Moreover, the nutrient medium used for plating is not suitable for every heterotrophic microorganism found in water. For some microorganisms, the culture conditions may prove unsuitable and suppress growth. For example, all anaerobes require special environmental conditions for growth. Microorganisms adapted to the low temperatures of cold-water conditions either may not withstand addition of agar at a temperature of 40–45 C or may not develop at an incubation temperature of 22–23 C. Thus, only an approximate number of heterotrophic bacteria found in a living state in the water sample at the time of study is obtained.

Enumerating Ammonifying Bacteria by the Dilution Method

This determination is performed by means of inoculations made with increasing quantities of water into test tubes containing beef-peptone broth or peptone water.

Peptone water is prepared as follows: 5 g of peptone are placed in 1000 ml of tap or distilled water (excess iron in tap water should be avoided) and boiled for 15 minutes. The following salts are added to the hot liquid:

K_2HPO_4	1 g
KH_2PO_4	1 g
$MgSO_4 \cdot 7H_2O$	0.5 g
NaCl	trace

After the salts dissolve, the solution is thoroughly shaken and filtered until completely transparent. The medium is poured into test tubes and sterilized for 20 minutes at 1 atm.

After inoculation, indicator papers are suspended over the medium to detect liberation of ammonia. Red litmus paper or paper reactive to NH_3 by Krup's method are used (see p. 143 for preparation). The test tubes are tightly closed with parchment paper or cellophane or rubber caps (Fig. 75). Ammonia liberated during protein decomposition causes blueing of the litmus paper and reddening of the strips of paper reactive to NH_3. Indicator papers prepared following Krup's method are much more sensitive.

Ammonia formed in culture may be determined quantitatively. For this purpose, cultures should be incubated in tightly closed flasks to prevent volatilization of the ammonia.

The amount of ammonia formed by bacteria in a medium depends not only on enzymatic activity of the bacteria but also on the nature of the proteins present. Pochon and Tardieux medium (1962) is used in some laboratories.

Pochon and Tardieux Medium

Winogradsky standard solution (see p. 197 for preparation)	50 ml
Asparagine	0.2 g
Solution of trace elements (p. 330)	1 ml
Distilled water	950 ml

pH 7.2

The appearance of ammonia is confirmed with Nessler reagent (see p. 274).

Ammonium in seawater can be determined by the phenol-hypochlorite method of Solorzano (1969) (*Limnol. Oceanogr.*, *14*:779–782).

Quantitative Enumeration of Spore-Forming Bacteria

Spore-forming bacteria are often used as an indicator for the presence of refractory organic compounds not utilized by non-spore-formers. To enumerate spore-formers, the inoculum is heated at an appropriate temperature for a specified time. However, it should be noted that counts obtained by this method may be minimized since all vegetative cells of spore-forming bacteria are killed by heating.

According to Mishustin (1948), the sediments must first be partially dried at room temperature to induce sporulation of vegetative cells of the spore-forming bacteria. An appropriate amount of sediment (1 g) is partially dried in sterile weighing bottles. A suspension (1:100 or 10:100) is prepared by dilution and 3–5 ml are transferred to small-diameter, sterile test tubes which are then placed in a special rack (Fig. 54) in a water bath set at 80 C for 10 minutes.

Inoculations from a pasteurized suspension are made in the usual way, i.e. either by the pour plate method or by surface spread plate inoculation. The culture medium used for spore-forming species is a mixture of beef-peptone and wort agar combined in a proportion of 1:1. The pH of the medium should be 7.0–7.2. The inoculated medium is incubated at 25–30 C for 2–3 days, after which colonies are counted.

Different species of spore-formers yield characteristic colonies on the medium cited above. Although it is a questionable practice, a rough estimate of the number of bacterial species is sometimes made on the basis of morphology of the colonies. For this kind of crude estimate, it is recommended that inoculated agar plates be retained for an additional 4–5 days at room temperature, after the total viable count is made.

Colonies of spore-formers most often observed (Fig. 114) belong to the following species: *Bacillus megaterium* Bary (smooth, white, butyrous, shiny colonies, consisting of typical rod-shaped cells); *B. cereus* subsp. *mycoides* (Flügge) Smith et al. (spreading over the surface, with curving filaments radiating from the colony); *B. mucosus* Zimmermann (mucoid, semitransparent, resembling drops of paste); *B. subtilis* Cohn (small, wrinkled, and coalescing with the agar); *B. agglomeratus* Migula (small, grayish, sometimes with a greenish tint); *B. brevis* Migula (butyrous, shiny colonies, smooth, convex); *B. cartilaginosus* Krasil'nikov (thick, round, compact colonies which can be lifted from the agar in their entirety); *B. filaris* Migula (semimucoid, wrinkled, wavy edged, slightly yellowish); *B. gracilis* Zimmermann (wrinkled, shiny, convex, yellowish); *B. cereus* Frankland and Frankland (flat colonies with flat center, weakly wavy edges, and a powdery surface); *B. idosus* Burchard (dry, lusterless colonies, laminated, finely wrinkled); *B. intricatus* Migula (widely spreading, whitish colonies, flat, mycelium-like, ingrowing into the agar, containing filaments with numerous septa); and *B. virgulus* (Duclaux)

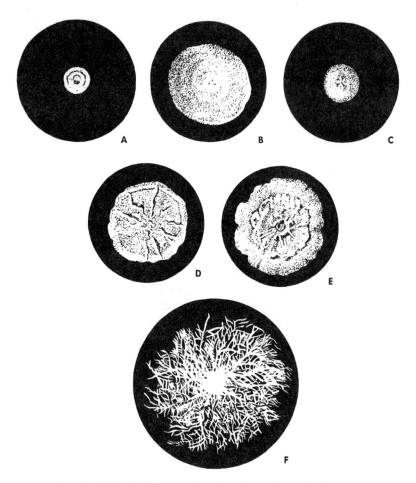

Fig. 114. Colonies of spore-forming bacteria on beef peptone agar-wort agar (Mishustin, 1948). *A, Bacillus agglomeratus* Migula; *B, B. cereus* Frankland and Frankland; *C, B. megaterium* Bary; *D, B. idosus* Burchard; *E, B. subtilis* Cohn; *F, B. cereus* subsp. *mycoides* (Flügge) (Smith et al.).

Macé (mucous, grayish colonies with fimbriate edges). *B. agglomeratus, B. cartilaginosus,* and *B. filaris* do not form gas from carbohydrates. *B. brevis* does form gas from carbohydrates.

A number of media have been suggested for spore-forming anaerobes.

Potato Medium

White potato	200 g
Glucose	5 g
$(NH_4)_2SO_4$	1 g
$CaCO_3$	3 g
Tap water	up to 1000 ml

The skin is removed from a potato. Water is added to the potato and it is steamed for 1 hour or boiled until it becomes soft. It is then passed through a fine sieve. The other ingredients are added and the volume is brought to 1 liter. After cooling, the medium is poured into test tubes and mixed constantly so that a uniform distribution of the potato granules is obtained.

Ethyl Alcohol Medium

Ethyl alcohol (96%)	8 ml
$CH_3COONa \cdot 3H_2O$	8 g
$KH_2PO_4–Na_2HPO_4$, buffer, 1 M, pH 7.0	25 ml
$(NH_4)_2SO_4$	0.5 g
Na_2CO_3	0.1 g
$MgSO_4 \cdot 7H_2O$	0.2 g
$CaSO_4 \cdot 2H_2O$	10 mg
$FeSO_4 \cdot 7H_2O$	5 mg
$MnSO_4 \cdot 4H_2O$	2.5 mg
$NaMoO_4 \cdot 2H_2O$	2.5 mg
Biotin	3 μg
Paraaminobenzoic acid	50 μg
Sodium thioglycolate	0.5 g
Distilled water	1000 ml

pH 7.0

The sodium thioglycolate may be replaced by addition of 0.2 g of $Na_2S \cdot 9H_2O$ to the medium after sterilization. Growth stimulants may be replaced by 0.5–1 g of yeast extract.

When isolating pure cultures, a medium of this composition with 2% agar is used; the inoculated plates are incubated under anaerobic conditions.

Methods of Culturing Protein-Degrading Spirilla

Spirilla (Fig. 115) are widely distributed in fresh water masses and in the oceans. These organisms are easily isolated in enrichment culture.

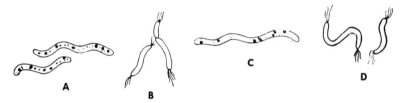

Fig. 115. Spirilla. *A, Spirillum volutans* Ehrenberg; *B, S.* tenue Ehrenberg; *C, S. proeclarum* Collin; *D, S. undula* (Müller) Ehrenberg.

The following procedure is used to obtain good growth of spirilla: wide-necked bottles are filled with water from the water mass under study. Algae or pieces of water plants from the same area are added, and the bottles are allowed to stand at room temperature. After 10–12 hours, large numbers of spirilla appear, growing most abundantly at the surface.

Other bacteria always grow in large numbers in association with spirilla. The following simple procedure is very useful for inducing numerical dominance of spirilla. Part of the enriched original culture is mixed with water from the original water sample in a 1:1 proportion and the medium is sterilized in an autoclave. This medium is used without addition of any other nutrient substances and is inoculated from the original culture. Spirilla develop well under these conditions. Enriched cultures are obtained by filling flasks with water sample and adding up to 0.5% beef extract, 0.3% peptone, and 0.1% yeast autolysate. Growth of spirilla in this medium usually begins after 2–3 days.

Soil extracts, wheat extracts, and yeast autolysates are components of many nutrient media for spirilla.

To prepare a soil extract, 100 ml of water are poured over 100 g of dry soil, heated to boiling, and cooled. After settling, the turbid liquid is clarified either by centrifugation or by filtration until clear. The filtrate is sterilized at 120 C.

Wheat extract is prepared by heating 10 g of whole

wheat grain in 100 ml of water. After cooling, the extract is filtered through filter paper and sterilized in an autoclave.

Yeast autolysate should be freshly prepared.

Other media used to culture the spirilla are as follows:

Soil Extract Medium

Soil extract	10 ml
Wheat extract	10 ml
Yeast autolysate	0.5 g
Tap water	80 ml

Casein Medium

Wheat extract	10 ml
Casein hydrolysate	2 g
Yeast autolysate	0.5 g
Tap water	90 ml

Calcium Lactate Medium

Water sample	1000 ml
Beef extract	5 g
Peptone	3 g
Calcium lactate	10 g
$MgSO_4 \cdot 7H_2O$	0.5 g
K_2HPO_4	0.5 g
NaCl	0.1 g
Yeast extract	0.1 g

Peptone-Glucose Medium

Tap water	1000 ml
K_2HPO_4	0.5 g
$MgCl_2 \cdot 6H_2O$	0.2 g
NaCl	0.2 g
Peptone	1 g
Glucose	1 g
Yeast extract	2 g

Beef Water Medium

Tap water	700 ml
Beef water	300 ml
Peptone	0.5 g

Beef Extract-Calcium Lactate Medium

Tap water	1000 ml
Peptone	5 g
Beef extract	3 g
Yeast autolysate	3 g
Calcium lactate	1 g
Agar	2 g

Pringsheim Soil Infusion Medium

A grain of wheat is placed in a large test tube, covered with 3–4 cm of garden soil and tap water, and allowed to stand for 48 hours, after which it is sterilized with flowing steam.

Rittenberg and Rittenberg Mineral Medium

Tap water	1000 ml
K_2HPO_4	0.5 g
$MgSO_4 \cdot 7H_2O$	0.25 g
Solution of trace elements	5 ml
pH adjusted to 7.0 with NaOH	

The solution of trace elements prepared in 0.1 N HCl contains per 5 ml:

$Fe_2(SO_4)_3 \cdot 9H_2O$	1.5 mg
$MnCl_2 \cdot 4H_2O$	0.9 mg
$Co(NO_3)_2 \cdot 6H_2O$	0.65 mg
$ZnSO_4 \cdot 7H_2O$	0.2 mg
H_2MoO_4	0.1 mg
$CuSO_4 \cdot 5H_2O$	0.005 mg
$CaCl_2 \cdot 6H_2O$	10 mg

The latter medium is used for *Spirillum volutans* Ehrenberg. To it are added:

| Wheat extract | 100 ml |
| Casein hydrolysate | 2 g |

The majority of the spirilla (though not all) grow on solid media. Pure cultures can be isolated after several transfers by streak inoculation on solid media. For this purpose, one of the above media is used with added agar. For isolation, an enrichment culture must be obtained (the best way to prepare an enrichment culture is to use a mixture of water and the original culture as a medium as described above).

Special methods of isolation are used for those species of *Spirillum* (namely, *S. volutans* Ehrenberg) which do not grow on solid media. Such methods have been developed by Rittenberg and Rittenberg (1962) for *S. volutans*. Isolation of pure cultures of spirilla is based on motility and on prolonged movement in one direction. These authors indicate that isolation of pure cultures is made difficult since it is suspected that associated bacteria produce certain growth factors required by *S. volutans*.

Flat capillaries are used for isolation of the bacteria. The flat capillaries are prepared by compressing with forceps a capillary softened in a burner flame. Capillary tubes 15 to 30 cm in length are sterilized and sealed (by drawing out the end).

For isolation of spirilla, the end of the tube is broken off with sterile pincers, a 10- to 20-cm layer of sterile medium is placed in the tube, and a 2- to 4-cm layer of mixed culture is introduced. The opening of the tube is sealed, leaving an air bubble space of several millimeters (Fig. 116). The tube is then placed on a microscope stage and examined under 100×. When there is a sufficient number of cells in the sterile liquid the tube is broken and the culture transferred. Special conditions are needed to isolate *Spirillum* species in pure culture, and to determine the metabolic products of associated bacteria contributory to their growth. Therefore, the spirilla are grown in a cellulose dialysis bag filled with sterile medium and placed in a mixed culture (Fig. 117). The pure culture is inoculated into the sterile medium after the bag has been in the mixed culture for 3 days.

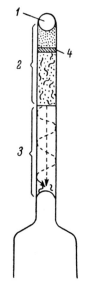

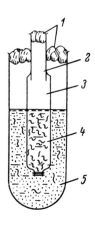

Fig. 116. Tube for isolating *Spirillum volutans* (Rittenberg and Rittenberg, 1962). *1,* Air bubble; *2,* mixed culture; *3,* sterile medium into which the spirilla migrate; *4,* layer of bacteria.

Fig. 117. Vessel for culturing *Spirillum volutans* in pure culture (Rittenberg and Rittenberg, 1962), *1,* Cotton stoppers; *2,* glass tube; *3,* small dialysis bag; *4,* pure culture; *5,* mixed culture.

DETECTING AMMONIA FORMATION FROM UREA

Urea-decomposing bacteria are widely distributed in nature. They are found in both the water and sediment in water masses. The decomposition of urea by the action of microbes is a very important stage in the recycling of vast amounts of nitrogen (Omelianski, 1940).

The total viable count of this group of bacteria in water masses is determined by inoculating solid media and counting those colonies identified by crystals of $CaCO_3$ and $CaHPO_4$, in the form of a thin iridescent film precipitated out around colonies of urea-decomposers. The solid media used contain urea and calcium salts. Adding an indicator to the medium also facilitates recognition of urea-decomposers since a color change occurs around the colonies as a result of alkalization of the medium by urease-positive bacteria.

The following media are used.

Beijerinck Medium

Yeast broth (20%)	1000 ml
Urea	20–30 g
Gelatin	100 g

Beef-Peptone Gelatin with Urea

Beef-peptone gelatin	100 ml
Urea	2 g
$(NH_4)_2CO_3$	0.3 g

The addition of ammonium carbonate greatly increases the selective properties of the medium.

Rubenchik Medium

Urea	50 g
K_3PO_4	1 g
NaCl	1 g
$MgSO_4 \cdot 7H_2O$	0.5 g
$FeCl_3 \cdot 6H_2O$	trace
Calcium salt of an organic acid (citric, lactic, tartaric, or malic)	10 g
Agar	15 g
Distilled water	1000 ml

Söhngen Medium

Urea	20 g
Calcium malate or calcium citrate	5 g
Ammonium citrate	0.5 g
Agar	20 g
Tap water	1000 ml

A separately sterilized 3% solution of K_2HPO_4 is added to the medium while it is still warm from sterilization. It is added dropwise until a weak opalescence appears. When the medium is prepared correctly, the plates should be almost transparent.

Allen Medium

Urea	20 g
K_2HPO_4	1 g
$CaCl_2 \cdot 6H_2O$	0.1 g
$MgSO_4 \cdot 7H_2O$	0.3 g
NaCl	0.1 g
$FeCl_3 \cdot 6H_2O$	0.01 g
Beef extract	5 g
Tap water	1000 ml

Soil Extract Agar with Urea

Urea	50 g
K_2HPO_4	0.5 g
Soil extract	100 ml
Tap water	900 ml
Agar	15 g

One rule must always be observed when preparing any medium with urea: The urea must be sterilized separately either with flowing steam or with dry heat (at 106 C for 30 minutes), since urea in liquid solutions partially decomposes with a rise in temperature. The following procedure is followed (Omelianski, 1940): the medium is first sterilized, without the urea, at 120 C and again, after the urea is added, with flowing steam. Filter-sterilized urea may also be used.

Tidwell, Heather, and Merkle (1955) suggest a medium which is sterilized in an autoclave together with the urea.

Tidwell, Heather, and Merkle Medium

Yeast extract	0.1 g
K_2HPO_4	9.1 g
NaH_2PO_4	9.5 g

Urea	20 g
Phenol red	0.01 g
Distilled water	975 ml

The medium is sterilized in an autoclave for 15 minutes. When first prepared the pH of the medium is low (about 4.8); after sterilization, it is 6.8 as a result of partial decomposition of the urea.

Urea and Glucose Agar

Beef-peptone broth	100 ml
Urea	1 g
Agar	1.5 g
Glucose	0.5 g
Bromothymol blue (1.5% alcohol solution)	1 ml

The urea, glucose, and bromothymol blue are added aseptically to the agar, which has been sterilized, melted, and cooled (to 50 C). pH is adjusted to 6.8–6.9 and the medium is poured into sterile test tubes. The medium is an olive green color. When urea-decomposing bacteria develop, the color of the medium changes to blue.

When liquid media are used, the number of urea-decomposing bacteria present is estimated by the dilution method.

The activity of urea-decomposing bacteria is determined in liquid media by the accumulation of ammonia, i.e. the ammonia is titrated with acid or measured colorimetrically with Nessler reagent.

DETECTING NITRIFICATION

The process of nitrification which leads to the formation of nitrogen compounds best assimilated by algae has tremendous significance in the overall biology of water masses. The recycling of nitrogen compounds within the nitrogen cycle after ammonification has taken place, i.e. the oxidation of ammonia into nitrites and then into nitrates, is an important step. The nitrates are further converted to organic nitrogen compounds in the microbial cells.

The number of nitrifying bacteria in the water and sediments of water masses is determined by dilution (transferring various quantities of inoculum into liquid media) and by counting colonies which appear on silicic acid plates treated with a selective medium.

An analysis is run separately for nitrifying bacteria of phase I, i.e. those forms which oxidize ammonium salts to nitrites, and for nitrifying bacteria of phase II, which oxidize nitrites to nitrates.

Nitrification, Phase I

Winogradsky's basic formula is used to detect the presence of nitrifying bacteria.

Winogradsky Medium

$(NH_4)_2SO_4$	2 g
K_2HPO_4	1 g
$MgSO_4 \cdot 7H_2O$	0.5 g
NaCl	2 g
$FeSO_4 \cdot 7H_2O$	0.4 g
Distilled water	1000 ml
Chalk	precipitated

A very small amount of chalk is added separately to each flask. The medium is sterilized in an autoclave for 10 minutes. When preparing a medium for nitrifying bacteria, great attention must be paid to the chemical purity of the reagents used. Before the medium is prepared, each of the necessary reagents must be checked for nitrite and nitrate impurities. Chalk especially often contains such impurities. If contaminants are discovered, the reagent or reagents must be discarded. If it is impossible to obtain a nitrite- or nitrate-free reagent, recrystallization is necessary. Chalk is freed of nitrite and nitrate impurities by repeated boiling with distilled water. The prepared medium is again checked, and only a medium which is perfectly free of nitrites and nitrates can be used for inoculations.

The medium must be poured in a thin layer (no more than 1.5–2 cm) into Winogradsky flasks or small Erlenmeyer flasks. All glassware must be treated with a dichromate cleaning solution and thoroughly flushed out with distilled water before use.

Ruban (1961) indicates that Winogradsky medium for nitrifying bacteria is improved by dilution of 1:10.

The presence and total number of nitrifying bacteria in a water mass under study is determined by adding 0.2–0.5% NH_4MgPO_4 to the sample.

Recently a number of media for growing nitrifying bacteria of phase I have been proposed. These media are, in essence, modifications of Winogradsky medium, except that they do not contain insoluble components, thus facilitating observation of growth of these microorganisms.

Medium Modification 1

Na_2HPO_4	13.5 g
KH_2PO_4	0.7 g
$MgSO_4 \cdot 7H_2O$	0.1 g
$NaHCO_3$	0.5 g
$(NH_4)_2SO_4$	2.5 g
$FeCl_3 \cdot 6H_2O$	14.4 mg
$CaCl_2 \cdot 6H_2O$	18.4 mg
Distilled water	1000 ml

pH 8.0

Medium Modification 2

$(NH_4)_2SO_4$	3 g
K_2HPO_4	0.5 g
$MgSO_4 \cdot 7H_2O$	50 mg
$CaCl_2 \cdot 6H_2O$	4 mg
Iron chelate	100 μg
Cresol red	50 μg
Doubly distilled water	1000 ml

After sterilization, the pH of the medium is adjusted to 8.0 with a sterile solution of K_2CO_3 and the culture is maintained at this pH.

Good growth of nitrosomonads may be obtained when a modification of Winogradsky medium is used.

Winogradsky Medium (Modified)

$(NH_4)_2SO_4$	2 g
K_2HPO_4	1 g
NaCl	0.5 g
$MgSO_4 \cdot 7H_2O$	0.5 g
$MnSO_4 \cdot 4H_2O$	trace
$Fe_2(SO_4)_3 \cdot 9H_2O$	trace
Distilled water	1000 ml
pH 8.5	

The medium is poured into flasks and sterilized in an autoclave. Flasks with fine pieces of washed marble are sterilized separately. The medium is poured into these flasks aseptically.

Stephenson Medium

$(NH_4)_2SO_4$	20 g
K_2HPO_4	7.5 g
KH_2PO_4	2.5 g
$FeSO_4 \cdot 7H_2O$	0.1 g
$MgSO_4 \cdot 7H_2O$	0.3 g
$CaCl_2 \cdot 6H_2O$	0.02 g
Distilled water	100 ml

The medium is dissolved with distilled water (1:100) for liquid media, or as silicic acid gel.

Meiklejohn Medium

$(NH_4)_2SO_4$	0.66 g
NaCl	0.3 g
KH_2PO_4	0.1 g
$MgSO_4 \cdot 7H_2O$	0.14 g

FeSO$_4$ · 7H$_2$O   0.03 g
Distilled water   1000 ml
Mixture of trace elements (trace amounts
    in 1000 ml of distilled water: Li$_2$SO$_4$,
    CuSO$_4$, ZnSO$_4$, H$_3$PO$_4$, Al$_2$(SO$_4$)$_3$, SnCl$_2$,
    MnCl$_2$, NiCl$_2$, CoSO$_4$, TiCl$_4$, KI, KBr)   1 ml

One gram of CaCO$_3$ is added per 100 ml of medium.

Inoculated flasks are placed in an incubator at 25 C. The following observations must be made on the cultures. Beginning with the fourth day after inoculation, chemical reactions for the presence of ammonia and nitrite are regularly conducted on alternate days and after a period of time, when positive reaction for nitrite is recorded, a survey of the developed microflora is made under a microscope.

A zinc-iodine-starch solution and Griess reagent serve as reagents for nitrite. Either may be used.

The zinc-iodine-starch solution is prepared as follows: 4 g of starch are ground in a porcelain mortar with a small amount of water and poured with constant mixing into a boiling solution of zinc chloride in distilled water (20 g of zinc chloride per 100 ml of distilled water). The mixture is boiled until the liquid is transparent. It is recommended that water be added to replace moisture lost in boiling to maintain the level of the solution. Two grams of dry zinc iodide are added to the transparent mixture; distilled water is added to obtain a volume of 1 liter, and the mixture is filtered. The reagent must be kept in the dark in a bottle with a ground glass stopper.

Nitrites may be detected in cultures with this reagent as follows: First, 3 drops of the reagent and 1 drop of dilute sulfuric acid are placed in the depression on a white porcelain plate. A drop of the sample under study is added. If nitrite is present in the medium, the liquid rapidly turns blue as a result of the formation of free iodine from the hydrogen iodide salt by the nitrite; the iodine reacts with the starch and yields a blue color. The intensity of color indicates the relative content of nitrite and, accordingly, the intensity of the process.

Griess reagent for detecting nitrite in cultures may be

obtained in dry form and is made up as a solution before use. The reagent possesses a very high sensitivity; unfortunately trace nitrite impurities with no relation to the nitrification process interfere in the reaction.

To prepare Griess dry reagent, 1 g of α-naphthylamine, 10 g of sulfanilic acid, and 89 g of tartaric acid are mixed thoroughly and ground in a mortar until a fine powder is obtained. The reagent is kept in a dark glass bottle with a ground glass stopper. Prepared Griess reagent can be obtained commercially.

Instead of tartaric acid, another acid (e.g. succinic or oxalic) is occasionally used. The proportion of separate components in this case is somewhat different: 1 g of α-naphthylamine, 10 g of sulfanilic acid, and 50 g of succinic (or oxalic) acid. When oxalic acid is used the onset of the reaction is somewhat delayed.

For the reaction, 7–10 mg of the dry reagent are taken with a scalpel and placed in the depression of the clean porcelain plate. About 4–5 drops of the culture (the number of drops must always be the same) are added and mixed with the end of a glass rod. If significant amounts of nitrites are present, coloring will appear immediately; if little nitrite is present, it will appear only gradually. The character of the color changes in the course of a few minutes. Observation of the change of the color over 5 minutes makes it possible to determine, of course very approximately, the energy of the process, i.e. the quantity of nitrites formed.

An accurate determination of the amount of nitrites formed in culture can be made with a colorimeter. This instrument is designed to determine concentrations of colored solutions. When it is used, the amount of nitrites formed by microorganisms is determined with great accuracy.

Nessler reagent for ammonia is prepared as follows: 50 g of potassium iodide are dissolved in 50 ml of hot distilled water; to this is added a hot concentrated solution of mercuric chloride until the red precipitate which forms stops disappearing. The liquid is filtered. A solution of 150 g of potassium hydroxide in 400 ml of distilled water and several additional milliliters of the mercuric chloride solution are then added. When the reagent cools, enough distilled water is added to bring the volume to 1 liter. The settled transparent liquid is taken up with a pipette to be tested.

The reagent is also prepared as follows: 22.5 g of iodine are dissolved in 20 ml of a solution of potassium iodide which contains 30 g of KI. After dissolution is complete, 30 g of metallic mercury are added and the mixture is shaken until the color disappears from the iodine solution. If the solution does not yield a reaction with starch for iodine, then an iodine in potassium iodide solution is added dropwise until the reaction occurs. The solution is diluted with distilled water to 200 ml and mixed thoroughly; to it are added 975 ml of a 10% solution of sodium hydroxide. Commercially prepared Nessler reagent is also available.

The reagent yields on reaction with ammonia a yellow-orange color of various tones, depending on the concentration of ammonia. When the concentration of ammonia is low, the color is yellow; when the concentration is high, the color is orange-yellow; when it is very high, the color is red-brown.

To obtain a reaction a few drops of the reagent are placed on a white porcelain plate, and a drop of the culture liquid is added from one edge to the reagent.

The disappearance of ammonia in cultures indicates its oxidation by bacteria only if the product of oxidation, nitrite, appears simultaneously and a microscopic investigation indicates the presence of nitrifying bacteria. The mere disappearance of ammonia does not indicate that the process of nitrification is taking place, because ammonia in an alkaline solution gradually volatilizes. Therefore, only the presence of nitrite serves as a reliable indication. Reactions for the presence of ammonia are, however, necessary, because, when it disappears completely, ammonia must be added to maintain growth of the bacteria. For this purpose a 10% solution of ammonium sulfate is prepared and decanted into a flask into which a graduated pipette has been inserted through a cotton stopper. The entire apparatus is sterilized. Two milliliters of the solution are added aseptically to each flask containing inocula in which all the ammonia has been oxidized. After several portions of ammonia are oxidized, transfers are made into fresh medium.

One or two control flasks must be incubated along with the culture, and the same reactions must be run as for the flasks containing inocula.

The number of nitrifying bacteria, both in water and in benthic sediments, is determined on gel plates (Winogradsky). The plates are prepared as follows: chemically pure hydrochloric acid is diluted with distilled water to specific gravity 1.1 (specific gravity is determined with a hydrometer), and sodium silicate solution (sodium silicate is best) is diluted with distilled water to specific gravity 1.1. The two dilutions are combined in a 1:1 proportion as follows: the sodium silicate is carefully added to the hydrochloric acid, with constant mixing. The mixture is blended well and poured into petri dishes. In a few hours the mixture solidifies into a solid jelly-like mass. To avoid the appearance of lenticular cavities in the gel mass, the hydrochloric acid and sodium silicate solutions should be heated to boiling and poured off, while hot, directly into petri dishes.

Gel plates prepared in this way contain chlorides which must be removed by washing in water. Moreover, the plates have an acid pH. The plates are washed over a period of 2–3 days. The petri dishes are opened and placed together in large deep crystallizers or pans. Water is poured over the plates and the water is changed several times a day. Alternatively, the plates are placed under a water faucet (the flow of the tap water should be such that the water does not beat against the plates). After the cold-water washing, the dishes with the gel plates are washed several additional times with hot distilled water. The washing is considered sufficient when a piece of gel taken from one of the dishes yields a green coloring, i.e. a neutral reaction, when placed in a drop of bromothymol blue solution.

Gel plates also are prepared from tetraethyl orthosilicate $[Si(OC_2H_5)_4]$ and ethyl alcohol combined in equal volumes (Ingellman and Laurell, 1947). Six volumes of water are added to the mixture. However, the water is added by aliquots with constant stirring (boiled water is used to avoid air bubbles). The solution is centrifuged until clear and poured into petri dishes. The gel plates are heated in an autoclave at 120 C for 30–40 minutes; to avoid cracks in the gel, the plates are cooled very slowly. The gel plates are thoroughly washed with water to remove traces of ether and alcohol. The gel plates are handled by standard procedures of inoculation. Gel plates prepared by either of these methods are sterilized either in boiling water or, when they must be autoclaved, in an autoclave under pressure. They are then soaked with either Winogradsky medium concentrated 10× (2 ml are placed in each dish) or a 0.5% solution of NH_4MgPO_4 (10 ml are placed in each dish). Both media should be hot when poured.

In the first case, the following solution is prepared:

Distilled water	200 ml
K_2HPO_4	0.5 g
$MgSO_4 \cdot 7H_2O$	0.3 g
NaCl	0.3 g
$FeSO_4 \cdot 7H_2O$	0.02 g
$MnSO_4 \cdot 5H_2O$	0.02 g
A salt of one of the following: zinc, titanium, molybdenum, or aluminum	trace

One milliliter of a sterile 5% solution of ammonium sulfate and 1 ml of chalk suspension are placed in petri dishes treated with the above solution. They are then partially dried in an incubator until the surfaces of the plates become lusterless. Over-drying the plates must be avoided since this results in formation of cracks. If they are over-dried, 1–2 ml of sterile water are added to each plate.

Inoculations are made onto the surface of gel plates with a tared [1] loop by introducing with it a specific amount of inoculum (for water, usually 1 ml). The inoculum is applied in straight lines equally spaced as drops of water or as small clumps of sediment (Fig. 118). To prevent drying and cracking of the gel, the inoculated petri dishes are placed in humidity-controlled incubators. Water liberated from the gel is removed from the lids of dishes with sterile strips of filter paper. The number of colonies counted includes only those show-

---

[1] A tared loop is a loop holding a known volume or weight of material (determined by appropriate measurement).

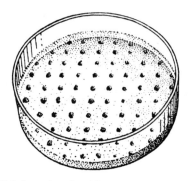

Fig. 118. Gel inoculated with small clumps of sediment.

ing zones of clearing which arise as a result of dissolving of the chalk in the first cited medium (see above) or the NH_4MgPO_4 in the second by action of the nitrite formed (Fig. 119). The colonies, colorless at first, become yellowish and finally dark brown. Examination of the colonies under a microscope reveals the presence of nitrifying bacteria. Gel taken from a zone of clearing yields a positive reaction for nitrite. Colonies with zones of clearing are counted with a magnifier lens.

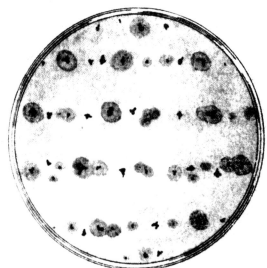

Fig. 119. Zones of dissolved chalk, i.e. nitrifying bacteria have developed around the clumps of soil.

When working with Winogradsky medium, it is very important to watch closely the disappearance of the ammonia. For this purpose, a small piece of gel is cut out with a platinum loop at the edge of the dish and placed in a drop containing an ammonia-detecting reagent. If the reaction for ammonia is negative, appropriate nutrient must be added. One or two drops of a 10% solution of ammonium sulfate are placed in the depression formed in the gel. It is best to inoculate four petri dishes containing gel (two treated with Winogradsky medium, two with the NH_4MgPO_4 solution) for each sample of water and soil since different species of nitrite-producing bacteria require a medium of different alkalinity.

A positive reaction for nitrite, obtained when cultures are tested three or four different times, serves as an indicator of the process of nitrification if microscopic examination reveals the presence of nitrifying bacteria.

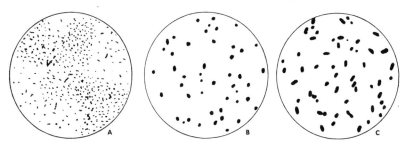

Fig. 120. Various strains (*A, B, C*) of *Nitrosomonas europea* (Winogradsky, 1952).

Various strains of *Nitrosomonas europea*, distinguished by cell size and shape (Fig. 120), are often encountered in water masses. Marine ammonia-oxidizing bacteria have been isolated. See Plates 4 and 5.

Pure cultures of nitrosomonads are difficult to obtain because of the presence of contaminating bacteria in cultures of this microorganism. Imshenetskii and Ruban (1953) developed two special isolation methods, the deep colony method and the drop method.

1. *Deep colony method.* One milliliter of an enrichment culture of a nitrosomonad is placed in 10 ml of a 75% sterile solution of NaCl, and a CO_2 stream, washed with sterile dis-

tilled water, is passed through the liquid for 20 minutes. In this way, the chalk, which is converted to calcium bicarbonate, is dissolved, and nitrosomonads are freed from solid particles to which they were adherent.

After clarification, 1 ml of the suspension of microorganisms in 0.75% NaCl is placed in a sterile petri dish with 1 ml of a sterile medium of the following composition: NaCl, 2%; $MgSO_4 \cdot 7H_2O$, 0.5%; Na_2HPO_4, 12%; KH_2PO_4, 4%; $FeSO_4 \cdot 7H_2O$, 0.01%; $MnSO_4 \cdot 5H_2O$, 0.01%; $NaHCO_3$, 1%; and 1 ml of a 5% solution of $(NH_4)_2SO_4$. Ten milliliters of a sterile solution of silica gel are then added. The plates harden after several minutes.

The silica gel is prepared beforehand, as usual, from a mixture of silicic and hydrochloric acids, immediately placed in a collodion or cellophane sack, and washed in distilled water for 1–1.5 days. The water is changed 6–8 times during this period. The temperature in the dialyzer should not exceed 10–12 C; otherwise the gel coagulates. After dialysis the solution is poured into flasks and sterilized in an autoclave at 120 C for 20 minutes. The mixture is kept in this form for use.

Microcolonies of nitrite-forming bacteria appear on the transparent, solid gel after 6–9 days. From these colonies, transfers are made either with a thin needle or with a capillary tube.

2. *Drop method.* The drop method is based on the idea that, if there is a very small number of cells in a culture medium, it is possible to place on a cover glass such small drops of the liquid, that a few of them will contain only one cell.

If the medium contains non-soluble components, immediately before isolation a stream of CO_2 washed with distilled water is passed through an enrichment culture for 20 minutes. Drops are taken from this now transparent culture with a Komarova (1949) micropipette and placed on a slide. The drops should be 0.2–0.3 mm in diameter. As soon as the drops are placed on the glass, they should immediately be taken up with small V-shaped pieces of filter paper. These filter paper pieces should be prepared, placed in a petri dish, and pre-sterilized. After the drop is collected, the filter paper is placed in a flask containing sterile medium. To obtain a pure culture, a large number of flasks must be inoculated.

The purity of the culture is first checked by microscopic examination of smears and then by transfer to a variety of media: beef-peptone agar, beef-peptone broth, potato agar, and starch-ammonia agar. The absence of growth in these media and the homogeneity of cells in culture indicate the purity of the culture.

Nitrification, Phase II

To detect the presence of nitrifying bacteria of phase II in a water mass, inoculations are made in Winogradsky medium:

Distilled water	1000 ml
$NaNO_2$	1 g
Na_2CO_3	1 g
NaCl	0.5 g
K_2HPO_4	0.5 g
$MgSO_4 \cdot 7H_2O$	0.3 g
$FeSO_4 \cdot 7H_2O$	0.4 g

When preparing the medium, great attention must be paid to the chemical purity of the reagents used. Sodium nitrite, especially, often contains nitrate impurities. Therefore, before the medium is prepared, all reagents should be checked. For this purpose, small quantities of distilled water solutions of the reagents in a concentration of 1 gram per liter should be prepared. Reactions for nitrates are performed with these solutions. If contamination of a salt by nitrates occurs, that salt must be recrystallized. Sodium nitrite is most often contaminated with nitrates; a method for recrystallizing this salt is provided by Karyakin (1936). To purify a commercial product (96–97%), 500 g of the salt are dissolved in 750 ml of water by heating. The salt solution is filtered; the filtrate is concentrated until crystals (a layer on the surface) begin to form, and is then cooled. Crystals of a pure preparation are suction filtered with paper filters in a Büchner funnel and the filtrates are in turn evaporated 1 to 2 times until crystallization begins. The yield of a pure (100%) preparation is approximately 400 g. Recrystallization must be repeated until an acceptable degree of purity is achieved. The purified liquid must be kept in a dry place and checked from time to time for presence of nitrate.

The prepared nutrient medium is poured in 50-ml amounts into Winogradsky flasks or small Erlenmeyer flasks previously washed with dichromate cleaning solution and rinsed with distilled water. The medium is sterilized in an autoclave at 120 C for 15 minutes.

The presence of phase II nitrification is noted by the disappearance of nitrite, the appearance of nitrate, and the presence of the causative agent or agents. The reagents listed above are used for detecting nitrites; dimethoxystrychnine or diphenylamine are usually used for detecting nitrates. The reaction with dimethoxystrychnine is carried out in the following way: a crystal of dimethyoxystrychnine and a drop of the culture are added to a drop of concentrated sulfuric acid placed in the well of a porcelain spot plate. If nitrates are present, a rose or cherry-red coloring appears, depending on the concentration of nitrate in the medium.

Dimethoxystrychnine is used as a 0.02% solution in sulfuric acid containing no nitrates.

Nitrates are also detected by reaction with pyrogallol: 0.2 g of pyrogallol is added to 10 ml of a culture and the two are mixed well. Two milliliters of concentrated sulfuric acid are carefully added with a pipette by placing the end of it in the liquid and then introducing the H_2SO_4. Following this step, 0.1 g of NaCl is added. "Boiling up" begins on the boundary of these layers and a purple ring appears. The nitrate content of the sample is judged according to the intensity of color and width of the ring.

This reaction is carried out with diphenylamine only after the unoxidized nitrites remaining in the medium are degraded, since this reagent yields the same reaction for nitrites. Therefore, when working with diphenylamine, often used to detect presence of nitrates in cultures, the unoxidized nitrites must first be broken down.

According to some workers, there is no single fully reliable qualitative reaction which permits detection of traces of nitrates, in water solutions, in the presence of large quantities of nitrites occurring in the initial stages of phase II nitrification in culture. Therefore, it is better that these reactions be run when nitrates have accumulated in the medium.

To break down nitrites in cultures, the reaction of nitrous acid with various compounds is used: sodium azide, amidosulfonic acid, urea, and methyl alcohol.

In the first method, an excess of sodium azide is added to a neutral solution which is acidified with acetic acid and after several minutes heated to boiling to break down excess of hydrazoic acid. The decomposition of nitrous acid occurs in the cold according to the equation:

$$2NaN_3 + 2HNO_2 \longrightarrow 3N_2 + 2NO + 2NaOH$$

By the second method, the sample under study is mixed with a 0.5% solution of amidosulfonic acid. The reaction proceeds so rapidly (even with concentrated solutions) that it is not necessary to worry about formation of nitric acid from oxidation of nitrous acid. The decomposition of nitrous acid proceeds according to the equation:

$$HNO_2 + NH_2SO_3H \longrightarrow H_2SO_4 + H_2O + N_2$$

Other methods of breaking down nitrites in cultures for subsequent detection of nitrates are examined in detail below.

1. Decomposition of nitrites with urea by a reaction run for 16–18 hours (Alekin). A series of test tubes are thoroughly washed with dichromate cleaning solution and dried. They are placed in an ordinary rack and on each one is recorded the number of a culture to be tested. Into each test tube is poured 0.1 g of urea; 7 ml of distilled water are added to each, followed by 1 ml of culture. Finally, 2 ml of sulfuric acid diluted with distilled water in the proportion 1:1 are added; this is best done in the cold. Test tubes containing this mixture are allowed to stand for 16–18 hours, usually until the following day. The nitrites break down completely during this time. Nitrate determinations can be carried out on the prepared mixtures the next day. An aliquot, 1 ml, of the mixture is taken (the basal medium is already diluted 10 times).

2. Decomposition of nitrites with urea in 15–20 minutes. Two test tubes washed with dichromate cleaning solution and rinsed well are prepared beforehand for each inoculation. To one test tube 2 ml of sulfuric acid diluted with distilled water (3:1) are added and to the other, 1 g of urea; 7 ml of distilled water are then added, and finally 1 ml of the culture is introduced. The resulting solution is mixed well with a clean glass rod until the urea dissolves completely. Next, the nitrite decomposition reaction is carried out in the cold as follows: the test tube containing sulfuric acid is placed in a beaker of ice water, after which the contents of the second tube are poured into the first in small portions, stopping each time gas begins to evolve and waiting until it is totally liberated. After the last portion is added and when strong liberation of gases has ceased, the decomposition of nitrites is considered complete. As the nitrite decomposition is carried out, the culture is diluted 10 times. The nitrate determinations are then performed with 1 ml of the solution.

3. Removing nitrites with methyl alcohol. Three milliliters of a culture are placed in a small volumetric flask (20–30 ml) calibrated to a volume of 4 ml. One milliliter of 2 N sulfuric acid and 2 ml of methyl alcohol are added. The solution is mixed thoroughly and evaporated to the mark, indicating total decomposition of nitrites. After this step, 8.75 ml of distilled water are added to 1 ml of the nitrite-free solution and the two are mixed well. One milliliter of the diluted solution is tested for nitrates.

Because urea often contains detectable amounts of nitrates, when breaking down nitrites with a new reagent of unknown purity, controls must be run as follows: After 1 g of urea is dissolved in 10 ml of distilled water, 2–3 drops of concentrated sulfuric acid are added and a nitrate determination is carried out. After the concentration of nitrates in the control reagent is calculated, it must be subtracted from the amount of nitrates in the sample of the water or sediment under study.

A solution of 0.017% diphenylamine in sulfuric acid is needed to detect nitrates after the nitrites in the medium have been decomposed. Pure concentrated sulfuric acid is selected for preparing the reagent (Alekin). Because commercial sulfuric acid often contains traces of nitric acid, before the reagent is prepared, the sulfuric acid must be freed of impurities by boiling for 15–20 minutes. Before the acid is boiled, KCl (5 g per liter) must be added so that the hydrochloric acid which forms as a result of boiling will reduce the nitric acid to gaseous oxides of nitrogen, which escape during boiling.

Once a pure solution is obtained, the reagent is prepared. Chemically pure crystal diphenylamine (170 mg) is weighed out on an analytical balance and placed in a 1-liter volumetric flask. Distilled water (150 ml) is added, followed by 50–100 ml of concentrated sulfuric acid. After the diphenylamine dissolves completely, the flask is brought to volume with concentrated sulfuric acid.

The prepared reagent must be perfectly transparent. It must be kept in a dark vessel with a ground glass stopper since diphenylamine decomposes under the action of light. Under these conditions, the reagent can be kept for a relatively long period of time. The sensitivity of the reagent when it is used to detect nitrates in Winogradsky medium is of the order of 0.001 mg of NO_3 per liter.

After the nitrites have been broken down and while the reaction with diphenylamine is carried out, by means of Table 11 approximate calculations may be made of the quantity of nitrates formed by the bacteria in the medium; thus the reaction rate may be followed with great accuracy. The color changes of the ring which forms at the border of the liquids as the nitrate content in the medium changes are given in Table 11.

Detection of presence of nitrates is as follows: 1 ml of the test culture in which nitrites have been degraded by one of the methods cited above is transferred to a test tube standing in a rack. One drop of a 20% solution of NaCl is added. Then 2 ml of a 0.017% solution of diphenylamine in sulfuric acid are collected with a pipette and carefully deposited down the wall of the test tube. The diphenylamine should not be dropped onto the surface of the liquid, and the test tube should not be shaken. If nitrates are present, a blue ring, the color intensity of which depends on the nitrate content, develops at the border of the two liquids. The character of the color change in the ring is observed on a white background in reflected light, and it is

Table 11. Approximate amounts of nitrates in cultures of nitrifying bacteria in Winogradsky medium

Detection with a 0.017% solution of diphenylamine in concentrated H_2SO_4

Dilution	Character of Ring Coloration				Concentration of Nitrate in Culture after Decomposition of Nitrites[1] (mg/liter)	
	Immediately	After 1 min	After 3 min	After 5 min	With Urea	With Methyl Alcohol
I (10⁻¹)	none	none	none	barely perceptible traces of ring	—	—
	barely noticeable trace of ring	weakly azure	azure	bright azure	2	3
	azure	bright azure	bright azure	blue	8	8.7
	bright azure	blue	bright blue	dark blue	23	24
	blue	bright blue	bright blue	dark blue	50	—
	bright blue	blue-black	blue-black	blue-black	75	—
II (10⁻²)	weakly azure	azure	bright azure	blue	100	—
	bright azure	blue	bright blue	dark blue	250	—
	blue	dark blue	blue-black	blue-black	500	—
	bright blue	blue-black	blue-black	blue-black	750	—
	dark blue	blue-black	blue-black	blue-black	1000	—

[1] The numbers cited have been converted to a concentration of NO_2 in milligrams per liter of medium.

compared with the data in the table on the first dilution. This makes it possible to determine approximately the amounts of nitrate in milligrams per liter of medium.

If a dark blue or blue-black ring appears immediately, the determination cannot be made with the first dilution. A second dilution must be prepared as follows: In a separate test tube, 9 ml of distilled water are added to 1 ml of the first dilution and mixed. The nitrate concentration determination is then repeated in 1 ml of the diluted solution. The results obtained are compared with the data in the table for dilution II.

After the nitrites have been degraded a qualitative reaction for nitrates may be performed with another method. A small crystal of diphenylamine is placed in the well of a porcelain spot plate. A drop of concentrated sulfuric acid is added. When the diphenylamine dissolves, a drop of the culture (in which the nitrites have been broken down) is carefully introduced at the side of the well. When the drops join (mixing is avoided) a blue spot will appear if nitrates are present.

Procedures for analysis of nitrite, nitrate, and organic nitrogen in seawater are given by Strickland and Parsons (1968) in *Manual of Seawater Analysis*, Bull. No. 167, Fisheries Research Board of Canada.

Chemical reactions alone are insufficient proof that the nitrification process is being effected. A microscopic investi-

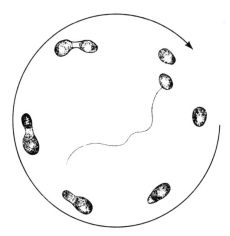

Fig. 121. Nitrobacter (Zavarzin).

gation to establish the presence of the causative agents is necessary. *Nitrobacter* Winogradsky—oval or slightly egg-shaped cells $1.1 \times 0.45 \mu$—are widely distributed in water masses.

Because cells in various growth stages are found in a culture (Zavarzin, 1958), this gives the impression of cell pleomorphism. The presence of pear-shaped cells (Fig. 121) is characteristic for a culture. A special feature of these cells is their poor staining reaction. Therefore, it is recommended that methods used for staining spores be applied for staining nitrobacters.

A quantitative determination of the numbers of nitrobacters may be made by dilution or by counting colonies which develop on gel plates. Two solutions are prepared:

1.	Distilled water	200 ml
	K_2HPO_4	0.5 g
	$MgSO_4 \cdot 7H_2O$	0.3 g
	NaCl	0.3 g
	$FeSO_4 \cdot 7H_2O$	0.02 g
	$MnSO_4 \cdot 5H_2O$	0.02 g
	A salt of one of the following: zinc, titanium, molybdenum, or aluminum	trace
2.	Distilled water	200 ml
	KNO_2	6 g

The solutions are sterilized. Well-ground kaolin (0.5 g) is sterilized separately. Two milliliters of the first solution, 1 ml of the second, and the 0.5 g of kaolin are combined in a small sterile flask, mixed well, heated, and placed, while hot, on the surface of a gel prepared as described above. The kaolin is added to provide a flat smooth surface on which the growth of colonies of nitrobacters is more easily observed. The mixture is distributed uniformly over the surface of the gel by rotating and swirling the dish in a figure eight. The petri dishes are then partially dried. Inoculations are made with a tared loop by depositing specific amounts of inoculum in rows. Small grayish, round colonies (slightly granular), which contain nitrobacters, are counted. The gel around these bacteria gives a positive reaction for nitric acid.

DETECTING MICROBIAL DENITRIFICATION

The process of denitrification, or reduction of nitrates to nitrites, ammonia, and molecular nitrogen, is brought about by a variety of bacteria widely distributed in water masses. Bacteria of this group vary in their morphological and biochemical features. When a mass development of denitrifying bacteria occurs in a body of water and when their activity is high, salts dissolved in water and vital to the development of phytoplankton may be totally destroyed, i.e. converted to molecular nitrogen and, therefore, inaccessible to the vast majority of algae.

In resolving the question of whether denitrifying bacteria are present in the water, the conditions present in the water at the time of sampling must be taken into consideration. One of the conditions which determine the vigorous activity of denitrificators is reduced oxygen content. Although denitrification proceeds when the medium is fully aerated, and although reduced oxygen content in the water is not mandatory for initiation of the process, nevertheless, if oxygen is present in the water in great amounts, conditions arise which reduce the rate of denitrification. An indispensable condition is the presence of nitrates. In addition, the presence of organic matter is a factor which determines the growth of denitrifying bacteria in a water mass. Denitrifying bacteria utilize the most diverse organic substances, among which are included products of cellulose decomposition. However, different species and even strains of one species possess a specific selective behavior with respect to certain organic compounds. Denitrifying bacteria are unable to carry out the reduction process when appropriate organic compounds are lacking, even though nitrates are present.

The presence of denitrifying bacteria in water is determined either by dilution, i.e. inoculation in a liquid enrichment medium, or by plate count when a solid medium is used. Giltay medium is often used for denitrifying bacteria.

Giltay Medium

Two solutions are prepared:

1. Distilled water	500 ml
Asparagine	0.5 g

Glucose	10 g
KNO$_3$	2 g
2. Distilled water	500 ml
Sodium citrate	2.5 g
KH$_2$PO$_4$	2 g
MgSO$_4 \cdot 7H_2O$	2 g
CaCl$_2 \cdot 6H_2O$	0.2 g
FeCl$_2 \cdot 4H_2O$	trace

The solutions are poured together and the pH is adjusted to 7.0 by adding a saturated solution of sodium bicarbonate. Several milliliters of the indicator bromothymol blue should be added to the medium so that the progress of the reaction may be observed by following changes in pH of the medium (when denitrifying bacteria are present, the pH shifts strongly to the alkaline side). A tall column of medium is prepared in test tubes and the tubes are sterilized with flowing steam for 30 minutes on 3 successive days. Inoculated test tubes are placed in an incubator at 25–30 C.

One of the following media is used.

Beef-Peptone Broth with Potassium Nitrate

Tap water	1000 ml
Beef extract	6 g
Peptone	2.5 g
KNO$_3$	1 g
pH 7.6	

Glucose Medium

Distilled water	1000 ml
KNO$_3$	1 g
K$_2$HPO$_4$	0.5 g
CaCl$_2 \cdot 6H_2O$	0.5 g
Glucose	10 g

Giltay medium, with the addition of 2% agar or a nitrate sucrose agar, is used as a solid medium.

Nitrate Sucrose Agar

Tap water	1000 ml
$NaNO_3$	2 g
K_2HPO_4	1 g
$MgSO_4 \cdot 7H_2O$	0.5 g
KCl	0.5 g
$FeSO_4 \cdot 7H_2O$	trace
Sucrose	30 g
Agar	15 g

pH 7.0

Denitrification is detected by the liberation of gas, which forms bubbles on the surface of the medium (Fig. 122) according to whether there is complete or partial disappearance of nitrates, and by the appearance of nitrites and ammonia in the medium. The same reagents used in nitrification are used

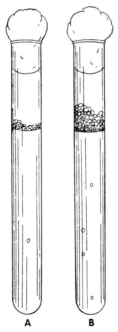

A   B

Fig. 122. Formation of gas bubble in a culture of denitrifying bacteria. A, Beginning of process; B, later development of process.

for nitrate, nitrite, and ammonia reactions. After 1 week of incubation, nitrate sucrose agar plates are treated with one of the nitrate reagents, and counts are made of colonies with no nitrate or reduced nitrate concentration. Parallel inoculations are tested with Griess reagent and colonies with zones of nitrite are counted.

A microscopic examination of the precipitate or colonies surrounded by zones reveals those nitrate-reducing bacteria which were present in the original inoculum. Isolation of pure cultures is necessary to determine the species of bacteria involved in denitrification. Giltay agar or nitrate sucrose agar are suitable for growth of cultures under conditions of reduced oxygen. The inoculated plates are placed in an anaerobic incubator, or the inoculum is overlaid with sterile aqueous agar. Colonies of denitrifying bacteria are characterized by the gases liberated and by the positive reaction of pieces of agar, taken near the colonies, with Griess reagent.

Suitable conditions for the growth of denitrifying bacteria are present in a variety of water masses. One source of energy is via oxidation of elementary sulfur or thiosulfate. The following media are used for isolating these bacteria.

Beijerinck Medium

Water sample	1000 ml
Sulfur (flowers of sulfur)	10 g
KNO_3	0.5 g
K_2HPO_4	0.2 g
$NaHCO_3$	0.3 g
$CaCO_3$	10 g
$MgCl_2 \cdot 6H_2O$	0.1 g

Trautwein Medium

Distilled water	1000 ml
KNO_3	1 g
$Na_2HPO_4 \cdot 2H_2O$	0.1 g
$Na_2S_2O_3 \cdot 5H_2O$	2 g
$NaHCO_3$	0.1 g
$MgCl_2 \cdot 6H_2O$	0.1 g

Thiosulfate–Iron Sulfate Medium

Distilled water	1000 ml
NH_4Cl	0.5 g
$MgCl_2 \cdot 6H_2O$	0.5 g
KH_2PO_4	2 g
$Na_2S_2O_3 \cdot 5H_2O$	5 g
KNO_3	2 g
$NaHCO_3$	1 g
$FeSO_4 \cdot 7H_2O$	10 mg
pH 7.0	

The solutions of the salts of iron, bicarbonate, and phosphate are sterilized separately.

Lieske Medium

Distilled water	1000 ml
$Na_2S_2O_3 \cdot 5H_2O$	5 g
KNO_3	5 g
$NaHCO_3$	1 g
K_2HPO_4	0.2 g
$MgCl_2 \cdot 6H_2O$	0.1 g
$CaCl_2 \cdot 6H_2O$	trace
$FeCl_3 \cdot 6H_2O$	trace

Tall cylinders or flasks are filled two-thirds full with medium. Sterilization is by flowing steam. After inoculations are made, an auxiliary nutrient medium sterilized in a separate vessel is poured in to the top. The cylinder is closed with a rubber stopper fitted with outlet tubes. If denitrifying bacteria are present (*Thiobacillus denitrificans* Beijerinck) the liberation of bubbles of nitrogen and nitrogen oxides begins after 4–5 days. Simultaneously, the disappearance of nitrates (reaction with dimethoxystrychnine) and appearance of sulfates (reaction with 1% solution of $BaCl_2$ in an acidified medium) is observed. An opalescent zone, containing *T. denitrificans*, appears in the medium at a distance from the surface.

The recent work of Watson and Remsen on marine ammonia-oxidizing bacteria should be noted. See ref. cited in Plates 4 and 5.

DETECTING MICROBIAL NITROGEN FIXATION

Nitrogen-fixing blue-green algae have been examined for nitrogen-fixing ability using the acetylene-ethylene reaction (Dilworth, *Biochim. Biophys. Acta, 127:*285–294, 1966; Harty et al., *Plant Physiol., 43:*1185–1207, 1968). The reactions involved in the binding of free nitrogen by bacteria are important in the nitrogen balance of a given body of water. Nitrogen-fixing bacteria play a great role in the replenishment of nitrogen compounds in natural bodies of water.

Nitrogen fixation in water is carried out by various species of bacteria belonging to several families and genera: nitrogen-fixing *Azotobacter;* aerobic rod-shaped bacteria; *Spirillaceae;* sulfur bacteria; anaerobic bacilli, i.e. *Clostridium pasteurianum;* and by yeast. Methods for detecting and culturing nitrogen-fixing *Azotobacter,* aerobic rod-shaped bacteria, *Spirillaceae,* and *C. pasteurianum* are given in this section.

The number of nitrogen-fixing bacteria is determined by dilution (inoculating liquid media with various amounts of water and sediment from the water mass) or by counting colonies on silica gel plates treated with suitable medium.

Aerobic Nitrogen Fixation

Nitrogen-fixing aerobes are widely distributed in water. All grow in the same media.

A number of media have been suggested for culturing azotobacters. Several of those found to be most suitable for selection and growth of azotobacters are listed below.

Fedorov Medium

Distilled water	1000 ml
Mannitol	20 g
K_2HPO_4	0.3 g
$CaHPO_4$	0.2 g
$MgSO_4 \cdot 7H_2O$	0.3 g
K_2SO_4	0.2 g
NaCl	0.5 g

FeCl$_3$ · 6H$_2$O   0.1 g
CaCO$_3$ 5 g
Trace elements   1 ml

Trace Elements Solution

Distilled water   1000 ml
H$_3$BO$_3$ 5 g
(NH$_4$)$_2$MoO$_4$ 5 g
KI   0.5 g
NaBr   0.5 g
ZnSO$_4$ · 7H$_2$O   0.2 g
Al$_2$(SO$_4$)$_3$ · 18H$_2$O   0.3 g

The mixture of trace elements is prepared separately. Concentrated solutions of trace elements are sterilized in an autoclave and used as needed; the necessary amount is withdrawn aseptically with a sterile pipette.

Mud Extract Medium

Tap water (water from the area
   under study is preferable)   900 ml
Sediment or soil extract   100 ml
K$_2$HPO$_4$ 0.3 g
Mannitol (or another carbon source)   10 g
Chalk   5 g

The addition of sediment or soil extract to the medium provides the trace elements for the azotobacters. To prepare the mud extract, 1 kg of sediment from the body of water under study or soil from another source is weighed out into a large Erlenmeyer flask. One liter of water is poured over it and the flask is then placed in an autoclave for 30 minutes at 115 C. The liquid is filtered through a thick, folded filter. Filtration usually goes very slowly, and the filtrate is clear, i.e. transparent. The filtrate is poured into bottles, sterilized in an autoclave, and used as needed for preparing media.

    In the above media, mannitol is replaced with starch, dextrin, sucrose, maltose, glucose, calcium salts of benzoic, acetic, or butyric acids, and other carbon sources. The advantage of using mannitol in media for azotobacters lies in the fact that mannitol is more slowly fermented by other bacteria, thus creating an advantage for the azotobacters. However, mannitol-negative strains are widely distributed in water and their presence may be detected only by using media containing other carbon sources. *Azotobacter insigne* Derx does not grow in media with mannitol. This species can be isolated only in media containing ethyl alcohol or calcium salts of organic acids (except formic acid).

When various carbon sources are used in media, both a higher number of azotobacters and more species are found than when a medium with only mannitol is used. A mixture of carbon sources, consisting of mannitol, glucose, sucrose, starch, dextrin, glycerin, and ethyl alcohol at a concentration of 1.9% and a salt of benzoic acid, 0.1–0.15%, are added to the medium. The following media modifications are recommended for selective isolation of azotobacters in water masses when used in parallel inoculations: (1) medium with mannitol (1%); (2) medium with mannitol (0.5%) and sodium benzoate (1%); (3) medium with ethyl alcohol (1%); and (4) medium with ethyl alcohol (1%) and sodium benzoate (1%).

Jensen Medium for *Azotobacter macrocytogenes*

Distilled water	1000 ml
Glucose	10 g (or 15 g of sucrose)
KH_2PO_4	0.2 g
K_2HPO_4	0.8 g
$MgSO_4 \cdot 7H_2O$	0.2 g
NaCl	0.2 g
$CaCl_2 \cdot 6H_2O$	0.05 g
$Fe_2(SO_4)_3 \cdot 9H_2O$	0.05 g
$Na_2MoO_4 \cdot 2H_2O$	0.05 g
Agar	20 g
	pH 5.5

Medium for *Azotobacter insigne*

Distilled water	1000 ml
Ethyl alcohol (96%)	10 g
K_2HPO_4	1 g

MgSO$_4$ · 7H$_2$O	0.2 g
CaCO$_3$	1 g
NaCl	0.2 g
FeSO$_4$ · 7H$_2$O	0.1 g
Na$_2$MoO$_4$ · 2H$_2$O	0.005 g
Agar	20 g

For other species of *Azotobacter*, another carbon source is substituted for ethyl alcohol. Fred, Baldwin, and McCoy suggest adding 100 ml of yeast extract (in 900 ml of distilled water) to a nutrient medium for azotobacters.

Medium for Nitrogen-Fixing Bacteria of the Genera *Pseudomonas* and *Achromobacter*

Distilled water	1000 ml
Sucrose	20 g
Na$_2$HPO$_4$	12.5 g
KH$_2$PO$_4$	1.5 g
CaCl$_2$ · 6H$_2$O	0.1 g
FeSO$_4$ · 7H$_2$O	0.003 g
Na$_2$MoO$_4$ · 2H$_2$O	0.00015 g

Liquid media for nitrogen-fixing aerobes are poured into small Erlenmeyer flasks to form a thin layer. The individual flasks are sterilized over a 3-day period with flowing steam.

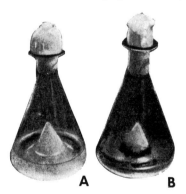

A   **B**

Fig. 123. Flasks containing liquid medium and filter for culturing azotobacters. *A,* Before inoculation; *B,* growth of an azotobacter on the filter at the border of the medium.

Azotobacters grow more profusely in a liquid medium into which is placed a filter folded in a cone (Fig. 123).

Flasks with inoculated media are either kept in an incubator at 25 C or allowed to stand at room temperature.

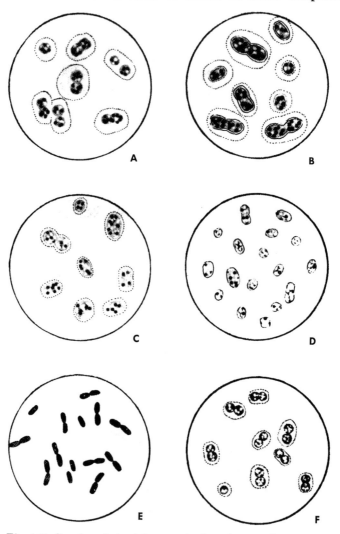

Fig. 124. Species of *Azotobacter. A, Azotobacter chroococcum* Beijerinck; *B, A. beijerinckii* Lipman; *C, A. agile* Beijerinck; *D, A. insigne* Derx; *E, A. vinelandii* Lipman; *F, A. macrocytogenes* Jensen.

Growth of azotobacters is confirmed by microscopic examination. Observations are made starting 3–4 days after inoculation. When azotobacters are present, a thin oily film develops on the surface of the medium. The presence of spirilla is suspected when the medium becomes very turbid. Typical large *Azotobacter* cells (Fig. 124) are easily recognized when examined under a microscope.

The total number of azotobacters in water and sediment is also determined on silica gel plates. The silica gel plates are prepared in the usual way (p. 276), treated with a 10× concentrated medium, partially dried, and inoculated. Inoculation from a water sample is made with a tared loop by placing drops in straight rows on the surface of the silica gel plate. A sediment inoculation is made as follows. One gram of sediment is aseptically weighed out in a small sterile weighing beaker; some sterile water is poured into another small sterile beaker. A glass rod with a curved hook at one end is wetted at the hooked end in sterile water and is used to select a small clump of sediment from the beaker and transfer it to the surface of the silica gel. The inoculum is distributed in rows, with each of the clumps of sediment a certain distance from the other.

Inoculated plates are placed in humidity controlled chambers or jars in an incubator at 25 C. If azotobacters are present, mucoid colonies grow within a few days. These colonies later usually turn brown and grow predominantly around the clumps of sediment (Fig. 125). Microscopic observation helps to distinguish colonies of azotobacters which are noted and counted.

Azotobacters in water are detected on submerged slides by pouring a thin layer of nutrient agar over the slides as soon as samples are taken and incubating the slides in sterile humidity controlled chambers.

Azotobacters in sediments are also observed on slides. For this purpose a drop of a mix consisting of a suspension of soil in sterile water is placed on a slide. One or two drops of a 1% agar medium are added and mixed with the inoculum. Colonies of azotobacters develop on the slides to rather large size.

Isolation of azotobacter in pure culture is a complicated task because the nutrient media used are insufficiently selective.

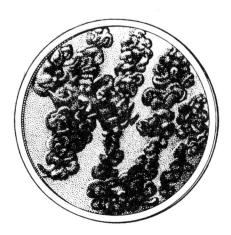

Fig. 125. Development of an azotobacter on gel (Winogradsky, 1952).

Associated bacteria, most often autotrophic microorganisms, develop along with the azotobacters. Moreover, some bacteria which utilize metabolic products of azotobacters are found in the mucoid capsules of the azotobacters. Other bacteria simply adhere to the mucoid capsules and are carried over to the fresh nutrient media.

One method of isolation is as follows. After a series of passages through liquid media containing less rich sources of carbon, namely, sodium benzoate or sodium acetate (if the azotobacter grows in these media), the material is inoculated onto silica gel. Often a microscopic examination is made and material is taken from a colony on the gel and transferred to a test tube containing sterile water and silica granules. The suspension is placed on a shaker where it is shaken for 10–15 minutes to separate the cells. A series of dishes containing gel or Fedorov medium with agar are inoculated. As soon as small colonies appear, they are examined under a microscope to assure that the azotobacters are growing. Inocula are taken from the colonies and the entire procedure is repeated. This is done until a pure culture is obtained.

In the presence of associated bacteria or contaminants not separated by standard procedures, the use of antibiotics is recommended. Species of *Azotobacter* are significantly more resistant towards chloramphenicol than many other bacteria.

A method of isolating azotobacters based on the use of antibiotics has been published (Depoux, 1953).

Depoux Medium

Distilled water	800 ml
Winogradsky standard solution (p. 197)	200 ml
Sodium pyruvate	15 g (or 5 g of mannitol)
Quinosol	0.15 g
Agar	20 g

Chloramphenicol is added to the medium to a concentration of 15 μg per milliliter. Other antibiotics are also used, generally in the same concentration.

According to Koleshko, isolation of azotobacters is easier when solid media containing an azotobacter culture killed by heating are used. Here a culture of azotobacter is grown on an agar slant. The culture is washed off with sterile water and the collected cells are heated to 70–80 C for 4 hours. When inoculations are made, 2–3 ml of such a preparation are added to each petri plate.

Prolonged growth of azotobacters is obtained in continuous culture. Malek (1956) and other investigators have grown azotobacters in a three-stage system of vessels (Fig. 41). Growth occurs in each stage. A nutrient solution enters the top vessel and filter-sterilized air is supplied.

The genus *Azotobacter* includes the group of free-living nitrogen-fixing bacteria. The characteristic morphological features of cells of azotobacters make identification to genus rather simple. However, identification of species is usually a more difficult task. Species identification is complicated by the fact that, in addition to the clearly defined species, a large number of varieties exist in nature.

To facilitate identification of *Azotobacter* species, special features, other than morphological and cultural characteristics, are used, e.g. difference in fluorescence of cultures under ultraviolet light, unequal sensitivity of various species to dyes, difference in utilization of carbon sources, and difference in resistance to certain chemical compounds. This information is included in the following descriptions of species.

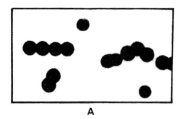 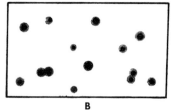

Fig. 126. Concentration of neutral red by colonies of *Azotobacter chroococcum* (A) and *A. beijerinckii* (B) (Callao and Hernandez, 1961).

Azotobacter chroococcum has motile cells which are 2.3–5.0 × 1.5–2.4 μ in size, rod-shaped in young cultures, and coccus-shaped in old cultures. They generally occur in pairs and more seldom in larger groups. The cells contain large amounts of granular intracellular reserve material; they form capsules, and are mucoid (Fig. 124, A). They form a dark brown pigment which is insoluble in water and does not penetrate the substratum (colorless strains are also encountered). A characteristic feature of this species is the ability to use starch as the sole source of carbon. Simple carbohydrates, alcohols, and fatty acids are the best carbon sources, but benzoates and complex carbon compounds are also used by this species. Mannitol, glucose, fructose, sucrose, maltose, galactose, raffinose, and ethyl alcohol are used; glycerin, arabinose, lactose, sodium citrate, mannose, dulcite, and xylose are not used (Jensen and Petersen, 1955). However, mannitol-negative strains of *A. chroococcum* are also found in water masses. This species concentrates the dyes neutral red, methylene blue, and methylene green (Callao and Hernandez, 1961) in the colonies when the dyes are added to the medium in concentrations of 1:50,000 (weight in grams to volume in milliliters) (Fig. 126, A). The bacterium grows on solid nutrient medium containing pyronine in a concentration of 1:25,000 and does not grow in a medium containing diamond fuchsin in a concentration of 1:50,000 (Callao and Montoya, 1960a). This species is cultured in a medium containing potassium tellurite when the maximum concentration of the latter is 1:5000–1:7000 (weight in grams to volume in milliliters) (Callao and Montoya, 1960b). Radially concentric cell distribution is observed in young cultures (Fig. 127, A) (Koleshko, 1960).

Fig. 127. Various distribution of cells in young azotobacter cultures (Koleshko, 1960). *A,* Radially concentric, *Azotobacter chroococcum; B,* stellate dispersed, *A. agile; C,* radial, *A. vinelandii.*

Azotobacter beijerinckii has slightly oval cells with rounded ends. They are larger than those of *A. chroococcum*, 3.3–6.6 × 2–2.8 μ, and occur more often in pairs. Tetrads and short chains are observed in old cultures (Fig. 124, *B*). Cells are non-motile, and are surrounded by capsules (Jensen, 1955).

A. beijerinckii develops at lower pH values (5.1 and even 4.75) than does *A. chroococcum*. Yellowish to light brown pigments which are insoluble in water and do not penetrate into the substratum are formed, or no pigment is formed. Starch is not the sole source of carbon; mannitol, ethyl alcohol, glucose, fructose, sucrose, and maltose are also utilized. Mannose, lactose, xylose, dulcitol, and sodium citrate are not utilized. When growing on solid nutrient media, this species concentrates, in the colonies, neutral red, methylene blue, and methylene green in concentrations of 1:50,000 (weight in grams to volume in milliliters) (Fig. 126, *B*). Growth is not observed on a solid medium containing diamond fuchsin, in a concentration of 1:50,000, or in a medium with pyronine, in a concentration of 1:25,000. The highest concentration of potassium tellurite at which growth is obtained is 1:15,000–1:25,000.

Azotobacter agile is typically a water species (Fig. 124, *C*). Cells are large (2.3 × 3.3 μ), spherical or slightly oval, single or in pairs, and highly motile (peritrichous). Cell movement of this species is steady, even, and in a straight line. Cells have an homogeneous cytoplasm with large granules of reserve materials. They are surrounded by a very thin capsule. This species yields abundant growth in nutrient media. Pigment formation is not always observed. Pigment is pinkish, yellowish, or green. This species grows well in media containing ethyl alcohol. Cultures possess a bluish white fluorescence under ultraviolet light. In colonies only neutral red is concentrated out of the medium; this occurs when dye concentration is 1:50,000 (weight in grams to volume in milliliters). Growth in solid medium with pyronine (1:25,000) is not observed but growth does occur in a medium with diamond fuchsin (1:50,000). The highest concentration of potassium tellurite at which growth is possible in the medium is 1:10,000–1:11,000. *A. agile* does not grow in media containing sodium benzoate. Young colonies are characterized by a stellate cell arrangement (Fig. 127, *B*).

Azotobacter insigne Derx is a typical water species, aerobic or microaerophilic (Derx, 1951). Cells are rod-shaped and oval, 2.5–3.8 × 1.6–2 μ in size, and often contain light refractive granules (Fig. 124, *D*). Cells are highly motile and have one or several polar flagella. The majority of strains are unable to use mannitol or glucose as a carbon source; however, there are strains which grow in glucose (Jensen, 1955). This species grows well in media containing ethyl alcohol and calcium salts of organic acids (with the exception of formic acid). Sodium benzoate is not utilized. In media with ethyl alcohol bluish gray and lilac-tinged water-soluble pigments which color the medium are formed. Growth is poor on solid media; colonies seldom reach 1 mm in diameter.

Azotobacter vinelandii Lipman cells, in young culture, are rod-shaped, 2–3 × 0.8–1.0 μ in size, and motile, and have peritrichous flagella. The cells contract or "shrink" with growth (Fig. 124, *E*). This species forms water-soluble pigments, greenish brown to dark red, and demonstrates green fluorescence under ultraviolet light. According to Jensen (1961), this species is easily distinguished from others by its rapid growth (3–5 days) in liquid medium containing rhamnose as the sole carbon source. *A. vinelandii* concentrates neutral red, methylene blue, and methylene green in the colonies. It grows in a solid medium containing pyronine (1:25,000) and diamond fuchsin (1:50,000). It is very sensitive to sodium tellurite: the greatest concentration at which growth is possible is 1:175,000–1:200,000. In young colonies, cells are distributed radially (Fig. 127, *C*).

Azotobacter indicus Starkey and De cells are small (1.7–2.7 × 0.5–1.2 μ) with large granules of fat and large slimy capsules. Acid is formed in nitrogen-free glucose medium. Mannitol, glucose, mannose, fructose, galactose, sucrose, and lactate are utilized, but lactose, maltose, sodium benzoate, and dextrin are not utilized. Films are not formed in liquid media and growth is not observed in protein media. The addition of small amounts of nitrogenous substances (ammonia salts, peptone, casein hydrolysate) to the nutrient medium stimulates growth of this microorganism. Rust-red and brownish red pigments are formed.

Azotobacter macrocytogenes Jensen is easily distinguished from other species. Acid-resistant, it grows at pH 4.5–5.5 and in a nitrogen-free glucose medium. In the latter, cells are spherical, 2–2.5 μ in diameter, and have a homogeneous cytoplasm. The cells are grouped and surrounded by a capsule (Fig. 124, *F*). At this stage of growth, the cells are non-motile. After 3–6 days, the capsules dissolve and the non-encapsulated cells are weakly motile, with slow movement. Strongly light-refracting inclusions (granules of reserve material) are observed. As a source of carbon, glucose, maltose, fructose, galactose, sucrose, ethyl alcohol, mannitol, sorbitol, acetates, lactate, butyrate, and succinate are utilized. Xylose, arabinose, rhamnose, mannose, lactose, dextrin, starch, inulin, glycerin, amyl alcohol, dulcite, oxalates, propionates, tartrates, citrates, and benzoates (the latter in a concentration of 0.2 to 0.5%) are not utilized. Very large cells (up to 8 μ in diameter) are formed in ethyl alcohol agar. A streak on a solid glucose medium is colorless or light yellow. In a liquid glucose medium, a rose-colored water-soluble pigment is formed. Acid is produced in media with glucose and sucrose.

The last two species—*Azotobacter insigne* and *A. macrocytogenes*—are included by some authors in the genus *Beijerinckia* Derx, the characteristic features of which are considered to be great acid resistance, formation of acid in media with glucose, and the absence of cysts.

In addition to azotobacters, other nitrogen fixers grow in nutrient media inoculated with water samples: various rod-shaped microorganisms, namely, representatives of the genus *Azotomonas* (Fig. 128), small motile rod-shaped gram-negative cells; species of the genus *Pseudomonas;* and spirilla, including *Spirillum azotocolligens* (Fig. 129).

The mechanism and rate of nitrogen fixation by azotobacters is detected by placing cultures in an atmosphere containing N^{15} (35%), and by subsequent determination of the

 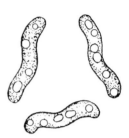

Fig. 128. Azotomonas fluorescens   *Fig. 129. Spirillum azotocolligens*
Krasil'nikov (Krasil'nikov, 1949).   Rodina.

amount of fixed nitrogen with the mass spectrometer. It has
been shown that there is no isotopic effect in the biological
fixation of nitrogen. Thus, the method is particularly useful.
The amount of nitrogen fixation in the aquatic sphere, i.e. the
water masses or natural bodies of water, has been established by
this method (Dugdale et al., 1959). A description of the method
is given by Rittenberg et al. (1939) and Neess et al. (1962).

   The amount of nitrogen fixed by bacteria is determined
chemically by one of the modifications of the Kjeldahl micromethod. Liquid cultures and cultures on silica gel are used to
determine fixed nitrogen. Liquid cultures permit a more accurate expression of results per gram of consumed sugar. Cultures are grown in media with glucose because quantitative determination of mannitol content in a medium is complicated.

   To determine the total nitrogen in broth cultures, an
accurately measured volume of the contents of the flasks is
taken. When determining the amount of nitrogen fixed by bacteria on silica gel, the test plates (with colonies) and control
plates are acidified with weak sulfuric acid and carefully dried
at 60 C. The products of acid hydrolysis in each plate are converted to a powder. A portion is weighed out on analytical
scales in a tall test tube, from which material is then transferred to a round-bottomed digestion flask with a long neck.
When transferring dry material from the test tube to the flask,
care must be taken not to contaminate the neck of the flask.
After the material is poured out, the test tube is again weighed,
and the weight of the sample is calculated.

A graduated pipette is used to transfer material from liquid culture to a flask for digestion. When a film is present, the culture is poured out of the flask onto a porcelain mortar of the appropriate size and the film is ground with the liquid until it is evenly distributed in the medium. A pipette is first washed with the liquid being analyzed, and then the liquid is drawn into it. This measured portion of the liquid is transferred to the flask for digestion. In this step, the pipette is placed as deeply as possible into the flask, without touching the sides of the wall, and the contents are deposited on the bottom of the flask. Before the pipette is removed from the flask, the tip is washed with several drops of water from a wash bottle. To remove the solution from the pipette completely, the pipette is rinsed twice with distilled water; the pipette wash water is also transferred to the digestion flask.

The total nitrogen is determined by digestion of the organic material, with concentrated sulfuric acid, over a flame. Organic matter is oxidized to carbon dioxide and water; the nitrogen is released in the form of ammonia, which reacts with the sulfuric acid to form ammonium sulfate. The ammonia obtained during digestion is released with alkali and is distilled into an accurately measured amount of 0.01 N solution of sulfuric acid.

To speed the process, a catalyst (salts of copper or mercury) and potassium sulfate are added during digestion.

There are many modifications of the Kjeldahl micromethod. The following is an accepted modification (Niderl' and Niderl', 1949; Belozerskii and Proskuriakov, 1951).

Reagents:

1. Concentrated sulfuric acid (specific gravity, 1.84)
2. Catalysts: powdered nitrogen-free mixture of 1 part potassium sulfate and 3 parts copper sulfate, powdered selenium acetate, or mercury sulfite
3. Perhydrol (nitrogen-free hydrogen peroxide, 30% solution)
4. 30% Solution of sodium hydroxide; this solution is prepared from 240 g of chemically pure sodium hydroxide and 560 ml of distilled water.
5. 0.01 N Solution of sodium hydroxide

6. 0.01 N Solution of sulfuric acid
7. 0.3% Solution of an indicator (methyl red)
8. Mixture of 0.1% alcohol solutions of methyl red and methylene blue (4:1); mixture is diluted with distilled water (1:2).

The indicator solution is prepared by treating 0.15 g of methyl red with 40 ml of a 0.01 N sodium hydroxide solution. The mixture is filtered and the filtrate is stored in a dropping bottle with a glass stopper. The capillary of the glass stopper should be drawn out in the form of a glass rod about 1 mm thick. Because only traces of the indicator are applied, the glass rod is simply immersed in the indicator solution. The inside of the vessel is then touched with the rod and the indicator is thus transferred to the vessel for titration. The flask is rotated and the indicator and the solution being titrated are mixed.

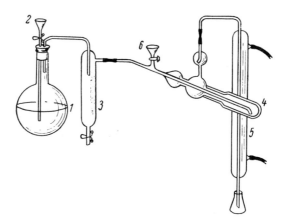

Fig. 130. Apparatus for determining total nitrogen by the Kjeldahl micromethod (see text for explanation).

The micro-Kjeldahl apparatus shown in Fig. 130 [1] is convenient for carrying out analyses. It consists of a steam evaporator (*1*), condenser for the flow of sample residue (*3*), funnel (*2*), distilling flask (*4*), and cooler (*5*).

---

[1] Micro-Kjeldahl apparatus are commercially available from scientific equipment supply houses in the U.S.A.

A round-bottomed Pyrex liter flask two-thirds filled with distilled water serves as a steam generator. Small pieces of silicon carbide, boiling stones, or traces of zinc dust are placed in the water for uniform boiling. The boiling flask is connected with the side-arm tube of the cylindrical condenser, which is provided with an outlet tube at the bottom. At the upper part of the condenser is a glass tube bent at a right angle and connected to a junction with the sealed-in inlet tube of the distilling flask.

This tube goes to the bottom of the flask and has at the top a side outlet, to which a small funnel (6) of about 10-ml capacity is joined by a short rubber tube with a clamp. The funnel serves to introduce the sample to be analyzed into the flask. The entire distilling flask is 300–350 mm long and consists of a distilling or reaction part, which passes up into two spherical foam interceptors, and a third such sphere, which is provided with a bent outlet tube and which is connected to the second sphere. The lower portion of the distilling flask consists of a double-walled vessel; the space between the walls is evacuated to provide insulation. Thus there is no need to heat this part of the apparatus. The distilling flask is secured in an inclined position such that the end of the inlet tube is situated vertically. To reduce danger of contamination of the distillate where it comes in contact with the rubber, the outlet tube of the upper sphere is connected by a junction to the short vertical elbow of the tube of the cooler (with a quartz tube, or one made of hard glass). To obtain the tightest possible connection of the glass tube with the quartz tube, the outer diameters must be equal.

Digestion of the material is carried out according to the following method. An accurately weighed portion of the sample or an accurately measured amount of liquid is placed in a large test tube or a 50-ml Kjeldahl flask made of Pyrex. Concentrated sulfuric acid (1 ml) and a mixture of copper sulfate and potassium sulfate (about 0.3–0.5 g) are added in a proportion of 1:3. The liquid is heated, at first over a low flame and then vigorously until it becomes completely transparent and straw-yellow or light green in color. The duration of the boiling depends on the sample and usually is limited to 15–20 minutes. If the digestion is still incomplete, 2–3 drops of

Perhydrol are added to the digesting sample and heating is continued for 5–10 minutes. If the substance digests with very great difficulty, the last operation is repeated.

Before distillation, the apparatus is usually steamed for 15–20 minutes. The conical flask serving as collector must also be steamed.

After digestion the liquid is carefully poured through the funnel (Fig. 130, 6) into the distilling flask (Fig. 130, 4). The funnel is rinsed several times with small amounts of distilled water. About 3–7 ml of a 0.01 N solution of sulfuric acid are accurately measured out from a buret into the collecting vessel. A drop of an indicator (methyl red) is added. The nitrogen concentration in samples varies. Thus, the amount of acid may be insufficient or in too great an excess. The first distillation will serve as indicator. Subsequent distillations will require more or less acid, according to the initial results. The excess should in no case be less than 1–2 ml. The sample cooled in the digestion flask is then diluted with 2 ml of distilled water and again cooled under a faucet. The flask is raised until the end of the cooler is submerged deeply in the liquid.

A pipette is used to introduce 3–7 ml of a 30% solution of sodium hydroxide through the funnel into the distilling flask. This amount is sufficient to create the necessary alkalinity.

Before distillation is begun, the clamps at the funnel and condenser are closed and the delivery of vapor through the condenser into the distilling flask is increased. When distilling ammonia, vapor coming from the boiler, in which the water is first brought to a boil, is passed through the vessel (Fig. 130, 3). The alkaline liquid heats up quickly and boils.

The duration of distillation is 5–10 minutes, beginning from the time the condensate reaches the outlet of the second spherical foam interceptor located on the distilling flask. The Erlenmeyer flask is then lowered until the end of the tube of the cooler is 10 mm higher than the liquid; distillation is continued for another minute. The Erlenmeyer flask is lowered and the end of the tube of the cooler is washed above the flask with several drops of distilled water. If the rose color of the liquid in the Erlenmeyer flask becomes too pale, another drop of indicator (methyl red) must be added. The liquid in the Erlenmeyer flask is back titrated from a microburet with an alkali of the same strength as the acid selected, and in the presence of an indicator, to the end point. The appearance of a canary-yellow coloring is considered to be the end point.

If during distillation the liquid becomes yellow, which indicates a shift in pH from acidic to alkaline, the analysis must be considered unsuccessful, since in this case a sufficient amount of 0.01 N solution of acid is lacking. The required amount is such that quantitative absorption of the ammonia being distilled occurs. When the distillation is completed, the residue automatically passes from the distilling flask to the condenser, where a sufficient vacuum is formed after the burner is removed from the boiler. Further cleaning of the distilling flask is not necessary, and the apparatus may be considered ready for subsequent determination. The residue is removed from the condenser by opening the clamp underneath.

Microdeterminations must be accompanied by control digestion runs with subsequent ammonia distillations in order to check reagents for purity and to establish the necessary correction factors.

When the titrated solutions have the same normality, the nitrogen calculation is performed by subtracting from the number of milliliters of alkali needed for titration of the given acid that amount of alkali necessary for the final back titration. A mixture of methylene blue and methyl red is a convenient indicator.

Calculation: 1 ml of 0.1 N sulfuric acid corresponds to 1.4 mg of nitrogen.

Example: 7 ml of 0.1 N sulfuric acid are taken; this corresponds to 7 ml of 0.1 N sodium hydroxide. After distillation, 2 ml of 0.1 N NaOH are used for titration; accordingly, $7 - 2 = 5$ ml of sulfuric acid are neutralized by ammonia. From this, one calculates $1.4 \text{ mg} \times 5 = 7$ mg of N_2 in the original sample.

Calculation must be made of the amount of glucose required by the bacteria for the amount of nitrogen fixed. Analyses of glucose content in sterile solutions and in the sample are run before the nitrogen analysis. This method is based on the ability of reducing sugars, possessing free carboxyl, to reduce cupric copper to cuprous copper in an alkaline solution.

The reaction proceeds quantitatively. The precipitate of cuprous oxide formed strictly corresponds to the amount of sugar in the solution. For the determination, the precipitate is dissolved with ferric sulfate in the presence of sulfuric acid. The cuprous oxide is quantitatively oxidized by the ferric iron and reduces it to ferrous iron. The latter is in turn also quantitatively oxidized by potassium permanganate.

Required reagents are as follows:

1. Solution of copper sulfate. Recrystallized $CuSO_4 \cdot 5H_2O$ (40 g) is dissolved in 1 liter of distilled water; if the solution obtained is turbid, it is filtered through filter paper.

2. Solution of ferric sulfate. $Fe_2(SO_4)_3$ (50 g) is dissolved in distilled water; 200 g (108 ml) of concentrated sulfuric acid (specific gravity 1.84) are added and carefully mixed. The volume is increased to 1 liter with distilled water. Oxidation must be thorough. In order that the solution will possess no reducing properties, permanganate is added dropwise until a persistent, though very weak, rose tint appears.

3. Alkali solution. Sodium hydroxide (150 g) is dissolved in 1 liter of distilled water. Potassium sodium tartrate $(C_4H_4O_6KNa \cdot 4H_2O)$ (200 g) is added.

4. Titrated solution of potassium permanganate. This may be prepared from fixanal [2] or as follows: 5 g of potassium permanganate are dissolved; distilled water is added to 1 liter. This is heated to boiling and boiled for 30–40 minutes. The flask is removed from the heat and closed with a stopper fitted with a tube filled with dry soda lime. It is then allowed to cool. The solution is filtered through an ordinary paper filter directly into a storage bottle. The vessel is closed with a stopper and allowed to stand for several days. The bottle should be made of dark glass; if this is not available, an ordinary colorless glass bottle wrapped with black paper may be used.

After 4–5 days, the titration standard is established and it is determined with sodium oxalate or oxalic acid. Several aliquots of 0.20–0.25 g of sodium oxalate predried in an in-

---

[2] A pH standard solution (*fixanal* refers to a Russian standard pH solution).

cubator are measured directly into flasks for titration. To each are added 50 ml of distilled water and 2–2.5 ml of concentrated sulfuric acid. This is heated to 70 C and the hot liquid is titrated with the permanganate solution until a persistent rose color is obtained.

Dilution calculation is performed in the following manner. Oxidation is according to the equation:

$$5Na_2C_2O_4 + 2KMnO_4 + 8H_2SO_4 =$$
$$10CO_2 + K_2SO_4 + 2MnSO_4 + 5Na_2SO_4 + 8H_2O$$

Two molecules of $KMnO_4$ in an acidified medium yield 5 atoms of oxygen. Thus, 5 atoms of oxygen are required for the oxidation of 5 molecules of sodium oxalate, 1 oxygen atom for 1 molecule of sodium oxalate. From this $134:16 = a:(bT)$, where a is the weighed portion of $Na_2C_2O_4$, and b the number of milliliters of $KMnO_4$ required for titration. The titer of $KMnO_4$, expressed in oxygen, is

$$T = \frac{16a}{134b} \times 2O_2.$$

The mean is derived from the experimental results and is accepted as the dilution. Deviation between dilutions from different individual titrations is admissible only in the fifth place.

The dilution obtained is recalculated for copper. It is absolutely necessary to express the dilution, obtained for oxygen, for copper because weight values of copper and corresponding weight equivalents of sugar are given in Table 12, below. To convert the dilution, one proceeds from the fact that 1 oxygen atom is required for 2 atoms of copper:

$$Cu_2O + Fe_2(SO_4)_3 + H_2SO_4 = 2CuSO_4 + 2FeSO_4 + H_2O$$
$$10FeSO_4 + 2KMnO_4 + 8H_2SO_4 =$$
$$5Fe_2(SO_4)_3 + K_2SO_4 + 2MnSO_4 + 8H_2O,$$

that is $(63.6 \times 2):16 = x:T$.

If the dilution of $KMnO_4$ for oxygen was 0.001155, then

$$x = \frac{127.2 \times 0.001155}{16} = 0.00918, \text{ or } 9.18 \text{ mg of Cu.}$$

Filtration of a liquid containing cuprous oxide obtained

Table 12. Milligrams of glucose corresponding
to milligrams of copper

Glucose	Copper	Glucose	Copper	Glucose	Copper
10	20.4	41	79.3	71	131.4
11	22.4	42	81.1	72	133.1
12	24.3	43	82.9	73	134.7
13	26.3	44	84.7	74	136.3
14	28.3	45	86.4	75	137.9
15	30.2	46	88.2	76	139.6
16	32.2	47	90.0	77	141.2
17	34.2	48	91.8	78	142.8
18	36.2	49	93.6	79	144.5
19	38.1	50	95.4	80	146.1
20	40.1	51	97.1	81	147.7
21	42.0	52	98.9	82	149.3
22	43.9	53	100.6	83	150.9
23	45.8	54	102.3	84	152.5
24	47.7	55	104.1	85	154.0
25	49.6	56	105.8	86	155.6
26	51.5	57	107.6	87	157.2
27	53.4	58	109.3	88	158.8
28	55.3	59	111.1	89	160.4
29	57.2	60	112.8	90	162.0
30	59.1	61	114.5	91	163.6
31	60.9	62	116.2	92	165.2
32	62.8	63	117.9	93	166.7
33	64.6	64	119.6	94	168.3
34	66.5	65	121.3	95	169.9
35	68.3	66	123.0	96	171.5
36	70.1	67	124.7	97	173.1
37	72.0	68	126.4	98	174.6
38	73.8	69	128.1	99	176.2
39	75.7	70	129.8	100	177.8
40	77.5				

in the sugar determination is done through a special filter made
from glass wool. It is prepared as follows: a cylindrical glass
tube, constricted at one end (Fig. 131), or a glass funnel
is used. Several layers of coarsely cut glass wool are laid cross-
wise on the bottom of the tube or funnel. The glass wool
should be distributed so that it covers the entire bottom. Glass

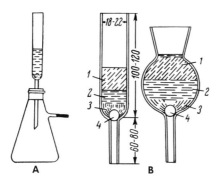

Fig. 131. Tubes for filtering cuprous oxide sugar determinations. A, General view of tube and flask; B, views of tubes and the distribution of glass wool: 1, layer of finely ground glass wool; 2, layer with coarser particles; 3, coarsely cut wool; 4, glass sphere (dimension in millimeters).

wool, finely cut with scissors, is placed in the tube. Finally, the top portion of the filter is filled with glass wool ground in a porcelain mortar.

The final layer is the main part of the filter. It is compressed with repeated washing of the filter with hot water. When such a filter is used, filtration proceeds easily, without application of suction or with a weak aspirator. A strong aspirator or pump must not be used, since the filter might be destroyed and part of the precipitate lost.

Sugar determinations are conducted as follows. A pipette is used to transfer 5–20 ml (depending on expected sugar content) of the sample solution to a 150- to 200-ml Erlenmeyer flask. The amount of sugar in the unknown should not be less than 10 or more than 100 mg (40–60 mg are recommended). Determination requires preservation of definite ratios of solutions of reacting substances; therefore, if a greater sugar content in the sample is expected, the sample should be diluted as follows: 5–10 ml of the solution are taken, and 5–10 ml of distilled water are added. A graduated cylinder is used to add to the flask 20 ml each of a copper sulfate solution and an alkaline solution containing potassium sodium tartate (these reagents are added in excess and the amount is not taken into account). Careful mixing is achieved by slightly rotating the flask on a hotplate. The mixture must be boiled for exactly 3 minutes from the moment it starts to boil, after which the red cuprous oxide precipitate obtained is allowed to stand for 1–2 minutes. If the sample contains significantly more sugar than expected, or, on the other hand, less than expected, then after the 3-minute boiling the result may be that no cuprous oxide precipitate was formed (little sugar in the solution) or that all the copper was reduced to the cuprous form and the cupric form of it did not remain at all (apparent by the disappearance of the blue coloring and by the abundant precipitate); in the latter case there was more sugar than expected in the given conditions. In both cases (more or less sugar) the analysis must be repeated with a different amount of sample tested.

If no precipitate forms when 10 ml of the solution are used, then 20 ml are used. Water need not be added, but the solutions of potassium sodium tartrate and copper sulfate are added in the same amounts, i.e. 20 ml of each. If all the copper sulfate is used for oxidation of sugar (apparent by the disappearance of the blue color), the amount of the solution is decreased, e.g. instead of 10 ml, 5 are used, or instead of 20, 10. If this amount is still too much, 10 ml of the solution can be diluted in a volumetric flask with distilled water to 100 ml, and 10 or 20 ml may be taken from this solution.

The subsequent stages of analysis are filtration and dissolution of the cuprous oxide precipitate. The liquid is filtered through a filter made of glass wool with an aspirator. As far as possible, the solution should be decanted to avoid transferring the precipitate to the filter. The precipitate which remains in the flask is washed by decantation (the greater portion of the precipitate is not transferred to the filter) with hot distilled water, which is poured off onto the filter.

The walls of the flask and filter are carefully washed down with a stream of water. The rinsing step is continued until an alkaline reaction with litmus disappears. The funnel with the filter is then transferred to another boiling flask. The rinsed cuprous oxide precipitate is dissolved in the same Erlenmeyer flask with 10–15 ml of a solution of ferric sulfate and is added to the filter. Suction is not applied immediately, so that the precipitate on the filter dissolves completely; usually it filters through without the application of suction. The flask is then washed twice with warm distilled water, and all the wash water is poured through the filter.

When all the dissolved cuprous oxide is collected in the suction flask, the ferrous oxide formed is immediately titrated with a permanganate solution. Titration is continued until a rose color appears. The change in the color of the liquid from green to rose is very distinct. Samples are titrated in duplicate.

On the basis of the amount of milliliters of permanganate required for titration, calculation of the corresponding amount of copper (in milligrams) and the corresponding content of sugar (in milligrams) in the sample is given in Table 12.

Spectrophotometric methods for sugar determination are now more commonly used. Standard biochemistry texts can be consulted for these methods.

Anaerobic Nitrogen Fixation

To detect and make total counts of the anaerobic nitrogen-fixing bacterium, *Clostridium pasteurianum* Winogradsky, inoculations are made using dilutions into a liquid nutrient medium. The classical medium for *C. pasteurianum* is that of Winogradsky.

Winogradsky Medium

Distilled water	1000 ml
Glucose	20 g
K_2HPO_4	1 g
$MgSO_4 \cdot 7H_2O$	0.5 g
NaCl	trace
$MnSO_4 \cdot 5H_2O$	trace
$FeSO_4 \cdot 7H_2O$	trace
$CaCO_3$	40 g

For the aquatic forms of *C. pasteurianum* this medium is optimal when the following are added: a mixture of trace elements (1 ml), ascorbic acid (1 g), and yeast autolysate (1 ml). These ingredients are added aseptically to the prepared medium.

Winogradsky medium for growth of *C. pasteurianum* is used with the following composition:

Winogradsky basic medium	1000 ml
Solution of trace elements	1 ml
Yeast autolysate	1 ml
Trilon B (sodium ethylenediaminetetra-acetate, EDTA)	0.1 g

With this composition, conditions are created in the medium which are favorable for the growth of *C. pasteurianum* and unfavorable for contaminating microflora.

Augier medium is used for isolation and culture of *C. pasteurianum*.

Augier Medium

1% Solution of KH_2PO_4	75 ml
$NaHCO_3$ (0.1 N solution)	33 ml
Soil extract	10 ml
Glucose	10 g
Solution of trace elements	1 ml
Winogradsky standard solution (p. 197)	50 ml
$CaCO_3$	0.05 g
Phenosafranin (0.2% solution)	8 ml
Distilled water	up to 1000 ml

Phenosafranin is introduced into the medium as an oxidation-reduction indicator. The medium is poured into test tubes in 10-ml amounts, and sterilized for 20 minutes at 100 C on each of 3 days. Some authors recommend adding potato extract to Winogradsky medium.

Winogradsky medium is poured into test tubes in 15-ml amounts and sterilized in an autoclave at 100 C for 20 minutes. After inoculation, the test tubes are placed in an incubator at 25–28 C. The beginning of the nitrogen fixation process is measured by the liberation of gas bubbles at the bottom of the test tube. Gas formation increases gradually and a luxurious foam appears on the surface of the medium. The precipitate is examined under a microscope for the presence of *C. pasteurianum*. The cells of this species have a characteristic spindle-

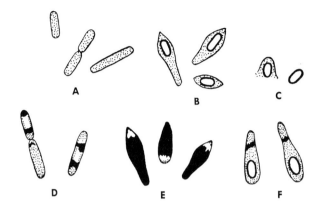

Fig. 132. Clostridium pasteurianum Winogradsky. *A,* Young cells; *B,* cells with spores; *C,* spores; *D,* granulose inclusions in young cells; *E,* granulose inclusions in mature cells; *F,* granulose inclusions in cells with spores.

shaped form and contain oblong spores (Fig. 132). A granulose reaction with iodine is used in identifying the species. A drop of the culture taken from the precipitate is placed on a slide and covered with a cover glass. A drop of a solution of iodine in potassium iodide is placed under the cover glass, and the surplus iodine is removed with a small piece of filter paper. The cytoplasm of *C. pasteurianum* turns blue from the iodine; the spores do not stain and, therefore, are visible against the blue background.

The medium is poured into test tubes in 10-ml amounts, and growth is obtained as follows. After inoculation the test tubes are placed in an incubator for several days. When abundant liberation of gas bubbles is noted, the test tubes are removed from the incubator and placed in a slanted position for 7–9 days. On the walls of the test tubes a precipitate of *C. pasteurianum* cells forms. The precipitate or cell sediment is examined under the microscope.

Growth of an anaerobic nitrogen fixer is also obtained on solid media.

To isolate *C. pasteurianum* in pure culture, precipitate is taken from an old enrichment culture in which sporogenesis has already taken place. The precipitate is transferred to a test tube containing 1 ml of sterile water, and the tube is heated for 10 minutes at 80 C. Dilutions are prepared and inoculations are made onto one of the following solid media.

Winogradsky Agar Medium (with trace elements, yeast autolysate, and ascorbic acid or EDTA)

Potato-Dextrose Agar

Potato broth	1000 ml
Dextrose	20 g
K_2HPO_4	0.2 g
$MgSO_4 \cdot 7H_2O$	0.2 g
Ascorbic acid	1 g
Yeast autolysate	1 ml
Agar	20 g
$CaCO_3$	3 g

Water (1000 ml) is poured over 20 g of cleaned, sliced potato. The mixture is boiled for 1 hour and filtered. Water is added to 1 liter total volume; shredded agar is added and the mixture is boiled until the agar dissolves. The salts and dextrose are added to the hot agar and allowed to dissolve; the solution is mixed well, and poured into test tubes, which are sterilized at 100 C for 20 minutes. According to some investigators, addition of raw potato juice (0.3 ml per 9 ml of water) as a diluent increases the number of colonies on the agar. The potato juice should also be added to the agar.

Soil Extract Agar

Distilled water	500 ml
Aqueous soil extract	500 ml
Glucose	5 g
K_2HPO_4	1 g
$MgSO_4 \cdot 7H_2O$	0.5 g
Biotin	0.001 mg
Paraaminobenzoic acid	0.1 mg
$CaCO_3$	5 g
Agar	20 g

Emtsev Potato-Carrot Agar

Potato broth	500 ml
Carrot broth	500 ml
Glucose	20 g
K_2HPO_4	1 g
$MgSO_4 \cdot 7H_2O$	0.5 g
NaCl	trace
$FeSO_4 \cdot 7H_2O$	trace
$MnSO_4 \cdot 5H_2O$	trace
Peptone	5 g
Yeast autolysate (p. 59)	0.02 ml
Chalk	40 g
Agar	15 g

Potato and carrot broths are prepared as follows: 0.5 kg of thoroughly washed vegetable is covered with double the amount of distilled water and boiled for 10 minutes, after which the liquid is decanted and filtered. The filtrate is increased to the original volume with distilled water.

Inoculated media must be placed under anaerobic conditions, i.e. in anaerobic incubators or dryers in which air is replaced with a mixture of nitrogen and CO_2. Incubation of the plates is usually for 2–4 days at 25 C. Inoculation into capillaries is successful. The individual colonies are transferred to a liquid medium. When growth is observed, the isolation is repeated. Emtsev recommends that the isolation be repeated three to five times.

The purity of an isolated culture is followed by inoculations into beef-peptone agar and beef-peptone broth (in short or tall columns of agar) and into media specifically designed for anaerobes.

The number of nitrogen-fixing anaerobes is also determined by counting colonies on silica gel plates. The silica gel plates are prepared as cited above. The plates are saturated with Winogradsky medium at 10× concentration and, after a specific amount of material is inoculated, the plates are incubated in anaerobic incubators under an inert gas atmosphere with CO_2. Fermentation begins after a few days; the chalk around the colonies dissolves (as a result of acid formation) and a transparent aureole is observed. On silica gel plates and around colonies of *C. pasteurianum*, small cracks are caused by the gas which forms. Smears from the colonies examined under the microscope reveal the cells typical of the species.

The amount of fixed nitrogen is determined in liquid medium by the Kjeldahl micromethod.

Chapter 8

Methods of Studying the Sulfur Cycle

FORMATION OF HYDROGEN SULFIDE
FROM PROTEINS

Bacteria which decompose proteins with the liberation of hydrogen sulfide are widely distributed in water masses. They may be detected and enumerated either by the plate count method or by the extinction dilution, or titer, method. In the first case, beef-peptone or nutrient agar with lead acetate or carbonate added is used (p. 147). When inoculation is to be made, lead acetate or carbonate is added to prepared, sterile melted agar and the two are mixed well to obtain a uniformly turbid medium. This is poured over the inoculum in petri plates. Inoculated plates are incubated at 22–23 C. As the H_2S forming colonies of bacteria develop, brown or almost black zones are formed around the colonies as a result of PbS formation. The colonies are usually counted on the sixth day of incubation. To prevent volatilization of the hydrogen sulfide, a thin layer of aqueous agar may be added after the colonies have appeared.

When the extinction dilution method is used, inoculations are made from a series of dilutions to test tubes, each of which contains 10 ml of beef-peptone broth and either suspended indicator papers or stoppers soaked in a solution of lead acetate (p. 146).

DETECTING SULFATE REDUCTION

The reduction of sulfates, sulfites, and thiosulfates to hydrogen sulfide is a biogenous process, as a result of which

323

significant, and sometimes very large amounts of H_2S accumulate in water masses. Especially large amounts of H_2S resulting from the activity of sulfate-reducing bacteria accumulate in water masses with a high concentration of sulfates in the water and with a high production of organic nutrient, since the presence of sulfates and organic nutrient is required by sulfate-reducing bacteria for growth. Significant amounts of H_2S produced during growth of sulfate-reducing bacteria may be found in reservoirs in the first few years of use, owing to decomposition of sulfates of soils rich in organic substance. H_2S production may also be very high in pond water, which is rich in sulfates, particularly after addition of plant fertilizers. However, even in water masses low in organic nutrient but rich in sulfates, sulfate reduction may proceed vigorously in the sediment deposits because the utilization of organic nutrient in the deposits leads to the accumulation of great quantities of H_2S in the bottom layers of water and the water-sediment interface.

Agents which initiate the process of sulfate reduction are motile vibrios—*Vibrio desulfuricans* (Beijerinck) Holland (Fig. 133), which are $2–4 \times 0.7–0.94 \mu$ in size.

All varieties of this species reduce sulfates under anaerobic conditions; associated with this is the formation of a black ferrous sulfide precipitate from which the substratum acquires a black color. Growth of sulfate-reducing bacteria in cultures is easily detectable by formation of a black precipitate around the colonies (Fig. 134), blackening of the medium, and precipitation of the hydrate of ferrous sulfide on the bottom and walls of the flasks containing liquid cultures.

Various media are used for detecting sulfate-reducing bacteria.

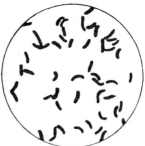

Fig. 133. Vibrio desulfuricans (Beijerinck) Holland.

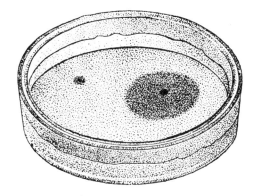

Fig. 134. Colonies of sulfate-reducing bacteria in agar.

Sturm Modification of Tauson Medium

Water sample	1000 ml
Calcium lactate	3.5 g
$(NH_4)_2SO_4$	4 g
K_2HPO_4	0.5 g
$MgSO_4 \cdot 7H_2O$	1 g
$CaSO_4 \cdot 2H_2O$	0.5 g
Mohr salt—$(NH_4)_2Fe(SO_4)_2 \cdot 6H_2O$	0.5 g

The medium is neutralized with a solution of potassium hydroxide to pH 7.0. Mohr salt is sterilized separately and added to the medium before inoculation.

Imshenetskii (1949) indicates that this medium is optimal for growth of sulfate-reducing bacteria when 5% yeast water (final concentration of 1%) is added to it.

Van Delden Medium

Water sample	1000 ml
Sodium lactate	5 g
Asparagine	1 g
$MgSO_4 \cdot 7H_2O$	1 g
K_2HPO_4	0.5 g
$FeSO_4 \cdot 7H_2O$	trace

The medium is made alkaline with KOH to pH 7.0. The

medium is prepared without $FeSO_4 \cdot 7H_2O$. The latter is added at the time inoculations are made by placing a small crystal in each flask. Each crystal must be first sterilized in a burner flame.

Starkey Modified Medium

Water sample or tap water	1000 ml
Peptone	5 g
Beef extract	3 g
Yeast extract	0.2 g
$MgSO_4 \cdot 7H_2O$	1.5 g
Na_2SO_4	1.5 g
$Fe(NH_4)_3(SO_4)_3$	0.1 g
Glucose	5 g

pH 7.4

Medium with Calcium Lactate

Distilled water	650 ml
Beef water	350 ml
Peptone	2 g
Yeast extract	2 g
K_2HPO_4	0.3 g
$MgSO_4 \cdot 7H_2O$	0.3 g
$(NH_4)_2SO_4$	1 g
$Fe(NH_4)_3(SO_4)_3$	0.15 g
Na_2SO_3	0.6 g
Calcium lactate	3 g
Ascorbic acid	0.15 g

pH 7.6

Cultures of sulfate-reducing bacteria may be grown in 100-ml bottles with ground glass stoppers. The bottles are filled two-thirds full with nutrient medium and closed with cotton stoppers. The glass stopper to each bottle is wrapped well in paper and tied to the mouth of each bottle. Bottles are sterilized in an autoclave at 112 C. After inoculation the bottles are filled to the stopper with additional sterile medium of the same composition, closed with the ground glass stoppers, and sealed with either paraffin, Mendeleev putty (a hermetic sealer), plasticine, or modeling clay. Care must be taken that

no air bubbles remain in the inoculated bottles after stopper-
ing. If bottles with ground glass stoppers are not available,
inoculations can be made in bottles or test tubes closed tightly
with fitted rubber stoppers. The bottles must be filled so that
no air bubbles remain under the stoppers. Long (150 mm) test
tubes are sometimes used. Before inoculation the test tubes
with the medium are heated at 100 C for 10 minutes, cooled
quickly, and inoculated. It is not necessary to fill the test tubes
if a cotton stopper is pushed in as close to the liquid as possible
(1 cm from it). Two pieces (0.5 mm) of pyrogallic acid and
8–10 drops of a concentrated solution of NaOH are then placed
on the stopper. The test tubes are closed at the top with rubber
stoppers and incubated.

Many typical vibrios (Fig. 133) are observed when the
precipitate from these cultures is examined under a microscope.

Sulfate-reducing bacteria may also be enumerated by the
pour plate method. Instead of liquid media, agar of corresponding composition (Tauson or Starkey medium) is prepared. Inoculations in plates are either made in the usual way and are
incubated anaerobically (in vacuum dryers or in Aristovskij
apparatus (p. 77), or they are made in plates placed together
as in the Sturm method (Fig. 50). The black colonies which
develop are easily observed (Fig. 134).

Final investigation of sulfate reduction in water masses
may be accomplished by determining the titer or enumerating
the sulfate-reducing bacteria present in the sample. However,
when sulfate reduction has proceeded extensively, the energy
involved, i.e. the amount of H_2S formed by bacteria, should be
determined. It is better to do this after pure, or at least well
characterized, cultures of the sulfate-reducing bacteria have
been obtained. To isolate these bacteria the pour plate method
in petri dishes (plates incubated in an atmosphere of hydrogen)
or the Sturm pour plate method may be used. Isolation of the
bacteria may also be accomplished by the Isachenko method,
which involves thorough mixing of the inoculum with the nutrient agar and drawing up the agar-inoculum mixture into
sterile pipettes (0.3 cm in diameter). When the agar hardens
in the pipettes the ends are closed off with Mendeleev putty or
sealed with a torch.

Colonies in plates and tubes usually appear after 4–5

Fig. 135. Colonies of sulfate-reducing bacteria in a tube.

days of incubation. The black colonies which appear (Fig. 135) are removed from the tubes as follows: a cut or nick is made with a glass file on the tube opposite the colony; the tube is flamed at this place and broken by bending. The end of a needle is used to remove a small piece of the colony and to transfer it to a liquid medium. It is usually necessary to repeat the isolation procedure several times.

To determine the amount of H_2S formed by the bacteria, inoculations are made into 0.5-liter bottles. After 20–30 days, cultures are tested for production of H_2S.

OXIDATION OF HYDROGEN SULFIDE, SULFUR, AND THIOSULFATES

Bacteria which oxidize H_2S, sulfur, and sulfur compounds are widely distributed in various water masses. In those water masses in which an accumulation of hydrogen sulfide is noted in the bottom layers, either sulfur bacteria or thiobacteria (more seldom, both occur together) develop in mass quantities at the boundaries of layers containing H_2S.

The growth mat of those microorganisms which oxidize hydrogen sulfide is a powerful bacterial filter which prevents penetration of H_2S into the higher layers of the water.

Large accumulations of purple sulfur bacteria give the water a noticeable pink color; when green sulfur bacteria develop en mass, the water has a green color. More often the sulfur bacteria develop in water masses in small quantities and may be detected by the following special methods through examination of (1) membrane filters through which a water sample has been filtered, (2) specimens prepared from benthic sediments or overgrown areas, and (3) submerged slides. Sulfur bacteria are very clearly observed when fluorescent microscopy is used because these bacteria fluoresce well with specific fluorochromes (e.g. coriphosphine and acridine orange).

Methods of Culturing Purple Sulfur Bacteria

Many media and methods have been suggested for obtaining purple sulfur bacteria in culture. Several of them are listed below.

1. Mud from the water mass is placed in a tall cylinder, into which water from the same water mass is poured. A small crystal of asparagin or a small piece of a boiled chicken egg is added and the cylinder is closed with a cork stopper. Paraffin is poured on the top and the cylinder is placed in an illuminated incubator at a temperature of 20–30 C. If an illuminated incubator is not available, the cylinder should be placed near a window but not in direct sunlight. If sulfur bacteria are present in the sediment of the water mass, growth appears after 3–5 weeks. The sulfur bacteria obtain the necessary H_2S from the growth of sulfate-reducing bacteria, which obtain the compounds they require from the growth of saprophytic decomposers.

The growth of purple sulfur bacteria in a medium is discernible by the appearance of a pink color of varied intensity in the water and rose-red spots on the surface of the sediment and on the walls of the cylinder. To identify the sulfur bacteria which have developed, water or a part of a colony is taken from the sediment and examined under a microscope. Both the shape and size of the cells are noted.

2. The following culture apparatus is very favorable for good growth of sulfur bacteria. Tall test tubes (about 30 cm high) are filled three-fourths full with a solid medium containing sulfates. After sterilization, the medium is inoculated with an active culture of sulfur-reducing bacteria and incubated until a black precipitate of ferrous sulfide appears. The precipitate indicates the accumulation of hydrogen sulfide in the medium. The water sample (if sulfur bacteria are present) or water and sediment samples (if growth of sulfur bacteria from sediment is desired) are then poured into test tubes. After inoculation the test tubes are kept in diffuse light. Sulfur bacteria develop well under these conditions.

3. Rapid growth of purple sulfur bacteria may be obtained in mineral media containing Na_2S.

van Niel Medium

Distilled water	1000 ml
NH_4Cl	1.0 g
K_2HPO_4	0.5 g
$MgCl_2 \cdot 6H_2O$	0.2 g
$NaHCO_3$	0.5 g
$Na_2S \cdot 9H_2O$	1 g
$Na_2S_2O_3 \cdot 5H_2O$	0.2 g
Trace elements	
solution	1 ml

The solutions of $NaHCO_3$, $Na_2S \cdot 9H_2O$, and $Na_2S_2O_3 \cdot 5H_2O$ are sterilized separately by the cold method (i.e. filtration) and are added to the medium immediately before use.

Trace Elements Solution (sterilized separately)

Distilled water	1000 ml
$FeCl_3 \cdot 6H_2O$	1.5 g
H_3BO_3	0.1 g
$ZnSO_4 \cdot 7H_2O$	0.01 g
$Co(NO_3)_2 \cdot 6H_2O$	0.05 g
$CuSO_4 \cdot 5H_2O$	0.005 g
$MnCl_2 \cdot 4H_2O$	0.005 g

To obtain growth of purple sulfur bacteria the pH of the medium is adjusted with a solution of H_3PO_4 at 8.0–8.4.

Inocula are placed in bottles with ground glass stoppers, after which the medium is added. Paraffin is poured on the stoppers and the bottles are then placed in an illuminated incubator at 25–30 C. According to Kondratieva (1963), continuous illumination with a 50- to 100-watt lamp with the inoculated media at a distance of 30–50 cm from the lamp is sufficient for rapid growth of purple sulfur bacteria.

According to Postgate (1959), somewhat larger concentrations of individual salts are necessary: K_2HPO_4, 1 g; $MgCl_2 \cdot 6H_2O$, 0.5 g; $NaHCO_3$, 2 g per liter. van Niel (1944) indicates that the introduction of inoculum in the amount of 1–5% (by volume) leads to a more rapid development of purple sulfur bacteria.

When working with material from brackish ponds or sea-water, an amount of NaCl corresponding to the salinity of the sample source is added to the medium.

Modification of van Niel Medium

Water from mass under study	93 ml
KH_2PO_4	0.1 g
NH_4Cl	0.1 g
$CaCl_2 \cdot 6H_2O$	0.01 g
NaCl	2.5 g
$MgCl_2 \cdot 6H_2O$	0.05 g

One milliliter of a solution of trace elements of the composition given above is added to the medium. Before it is used, the medium is brought to a volume of 100 ml by addition of these previously sterilized solutions:

10% Solution of $Na_2S \cdot 9H_2O$	1 ml
5% Solution of $NaHCO_3$	4 ml
10% Solution of $Na_2S_2O_3 \cdot 5H_2O$	1 ml

Glycerin, mannitol, asparagine, and peptone in a final concentration of 0.15–0.20% and yeast extract in the amount of 0.5% may be added to van Niel medium and its modifications until the pH is adjusted. Addition of 0.1% peptone and 0.5% yeast extract is recommended for more rapid growth of the sulfur bacteria.

EDTA Medium

Distilled water	1000 ml
NaCl	5 g
NH_4Cl	1 g
KH_2PO_4	0.5 g
Na_2CO_3	6.3 g
$Na_2S_2O_3 \cdot 5H_2O$	0.3 g
EDTA (Trilon B)	0.1g
pH 8.2	

Schlegel and Pfennig (1961) indicate that the following

media are optimal for purple sulfur bacteria (*Chromatium okenii* (Ehrenberg) Perty, *C. warmingii* (Cohn) Migula, *Thiospirillum jenense* (Ehrenberg) Winogradsky, and *Amoebobacter* sp.).

Schlegel and Pfennig Medium 1

Distilled water	3000 ml
$CaSO_4 \cdot 2H_2O$	2 g
KH_2PO_4	1 g
$(NH_4)_2SO_4$	1 g
$MgSO_4 \cdot 7H_2O$	0.1 g
KCl	0.5 g
$CaCO_3$	0.15 g
$NaHCO_3$	4.5 g
$Na_2S \cdot 9H_2O$	2.25 g
EDTA (ethylenediaminetetra-acetic acid)	75 mg
$FeSO_4 \cdot 7H_2O$	10 mg
$ZnSO_4 \cdot 7H_2O$	5 mg
$MnCl_2 \cdot 4H_2O$	1 mg
Vitamin B_{12}	30 μg
Trace elements solution (p. 330)	1.5 ml

By bubbling CO_2 through the medium, the pH is adjusted to 6.6–6.8 for *Chromatium okenii* and *Thiospirillum jenense* and 6.8–7.2 for *Chromatium warmingii*. Growth temperature is 18–23 C.

Schlegel and Pfennig Medium 2

Distilled water	1000 ml
Solution of vitamins	10 ml
Solution of trace elements (p. 330)	1.5 ml
$FeSO_4 \cdot 7H_2O$	10 mg
$ZnSO_4 \cdot 7H_2O$	5 mg
$(NH_4)_2SO_4$	1 g
KCl	0.5 g
KH_2PO_4	1 g
$MgSO_4 \cdot 7H_2O$	1 g

NaHCO$_3$	2 g
CaSO$_4 \cdot$ 2H$_2$O	0.5 g
CaCO$_3$	0.05 g
Na$_2$S $\cdot$ 9H$_2$O	0.5 g

pH 6.5

Solution of Vitamins

Distilled water	100 ml
Biotin	0.1 mg
Calcium pantothenate	0.1 mg
Paraaminobenzoic acid	1 mg
Nicotinic acid	2 mg
Vitamin B$_1$	2 mg

To obtain an enriched culture, 35 parts of medium and 65 parts of water (or sediment extract) are used. Sediment extract is prepared as follows. Fresh sediment, soil, and CaSO$_4$ in the proportion 1:1:0.5 are placed in a 10-liter flask with a fivefold volume of rain water or tap water and left to stand in the dark for several weeks. The transparent, hydrogen sulfide-saturated water is removed and added to the medium. Each time some liquid is withdrawn, an equivalent volume of water is added to the flask.

To obtain an homogeneous enriched culture, several successive re-inoculations are made. When such a culture is obtained, isolations may be made of single strains. For this, inoculations are made in a medium of the same composition, with the addition of 1.5–2.0% washed agar. Isolation is as follows: dilutions are prepared from the culture; several drops of each dilution are placed in each sterile test tube. Agar is melted and cooled to 45 C, solutions of NaHCO$_3$ and Na$_2$S are added; pH is adjusted as necessary; and the agar is quickly poured into the test tubes containing the inoculum. The medium is mixed and allowed to solidify. When the columns of agar solidify, melted sterile paraffin is poured in at the top so that oxygen does not penetrate into the medium.

When continuous illumination is applied, colonies of sulfur bacteria appear after 5–7 days. The test tube is sawed with a glass file and broken at a point at which colonies are well

isolated from each other, and the column of agar is transferred
to a sterile petri dish. The colonies are cut out with a sterile
needle or spatula and re-inoculated into an agar medium.
Glass capillaries may be used. After growth appears, one of the

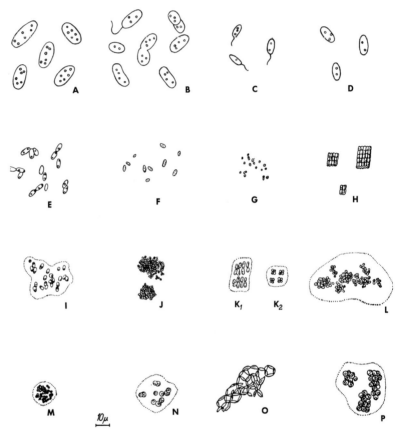

Fig. 136. Purple sulfur bacteria. *A, Chromatium gobii*
Isachenko; *B, C. warmingii* (Cohn) Migula; *C, C. weissei*
Perty; *D, C. okenii* (Ehrenberg) Perty; *E, C. minus* Winogradsky; *F, C. vinosum* (Ehrenberg) Winogradsky; *G, C.
minutissimum* Winogradsky; *H, Thiosarcina rosea* (Schroeter) Winogradsky; *I, Thiocapsa roseopersicina* Winogradsky;
J, Thiopolycoccus ruber Winogradsky; *K₁, Thiopedia elon-
gata* Rodina; *K₂, T. rosea* Winogradsky; *L, Lamprocystis
roseopersicina* (Kützing) Schroeter; *M, Amoebobacter roseus*
Winogradsky; *N, Thiothece gelatinosa* Winogradsky; *O, Thio-
dictyon elegans* Winogradsky; *P, Thiocystis violacea* Winogradsky.

colonies which develop is transferred to a liquid medium. The purity of the obtained cultures of sulfur bacteria is verified in the usual way (with inoculations into beef-peptone agar, beef-peptone broth, yeast extract agar, and a medium for sulfate-reducing bacteria).

In a culture of purple sulfur bacteria containing no heterotrophic bacterial contaminants, although the cells seem uniform, a microscopic investigation may reveal that more than one species is present. To avoid this, Kondratieva (1963) recommends inoculating the cultures into media which have a different pH and different concentrations of Na_2S, and which contain different organic compounds.

There are many species of purple sulfur bacteria. The species are identified according to cell shape, capsule formation (common for the group), presence of swarming stages, and color of pigments (Fig. 136). A more or less elongated cell shape is characteristic of purple sulfur bacteria. Some species have a spherical shape. The dimensions of the cells and the shapes of the aggregates which they form may change significantly according to conditions of the medium.

Purple sulfur bacteria are motile and have flagella at polar locations. They are gram-negative, and their cells contain bacteriochlorophyll and carotinoids. H_2S and other sulfur and hydrogen compounds, as well as organic compounds serve as donors of hydrogen; CO_2 and organic compounds serve as sources of carbon.

Methods of Culturing Green Sulfur Bacteria

van Niel medium, described above, with trace elements but a different pH, and Larsen medium are the basic media for culturing green sulfur bacteria. pH 7.0–7.3 is more favorable for green sulfur bacteria. Postgate indicates that different concentrations of sodium sulfide or sodium thiosulfate are favorable for different species of green sulfur bacteria. Thus, for *Chlorobium thiosulfatophilum*, $Na_2S \cdot 9H_2O$ (0.2 g per liter) is added to the medium; for *C. limicola*, 1 g per liter; and for *C. thiosulfatophilum*, $Na_2S_2O_3 \cdot 5H_2O$ is added in the amount of 1 g per liter.

Larsen Medium

Distilled water	1000 ml
NH_4Cl	1 g
KH_2PO_4	1 g
$MgCl_2 \cdot 6H_2O$	0.5 g
NaCl	1 g
$NaHCO_3$	2 g
$Na_2S_2O_3 \cdot 5H_2O$	1 g
$Na_2S \cdot 9H_2O$	1 g
$CaCl_2 \cdot 6H_2O$	0.1 g
Solution of trace elements	1 ml

pH 7.0–7.5

Solution of Trace Elements

Sodium ethylenediaminetetra-acetate	50 g
$ZnSO_4 \cdot 7H_2O$	22 g
$CaCl_2 \cdot 6H_2O$	5.54 g
$MnCl_2 \cdot 4H_2O$	5.06 g
$FeSO_4 \cdot 7H_2O$	4.99 g
$(NH_4)_6Mo_7O_2 \cdot 4H_2O$	1.1 g
$CaSO_4 \cdot 2H_2O$	1.57 g
$CoCl_2 \cdot 6H_2O$	1.61 g
Distilled water	1000 ml

Na_2S and $Na_2S_2O_3$ are sterilized separately as 10% solutions and added to the basic medium before it is inoculated. Cultures are grown under continuous (24-hour) illumination with 75- to 100-watt lamps.

However, very often the growth of green sulfur bacteria is suppressed by purple sulfur bacteria. In these cases, enrichment cultures should be set up as described by Winogradsky. Shredded filter paper, calcium sulfate, and sediment from the water mass are placed in a narrow glass cylinder which is filled with water from the water sample, closed tightly, and set out in the light. Complex bacterial processes occur within the cylinder. As a result of aerobic bacterial growth, all of the oxygen in the cylinder is utilized, and conditions for anaerobic growth are created. Decomposition of the cellulose begins, and carbon dioxide and hydrogen are liberated. Development of sulfate-reducing bacteria causes liberation of hydrogen sulfide, thus creating conditions for the development of purple and green sulfur bacteria. Colonies of green sulfur bacteria appear on the walls of the cylinder as separate little spots or flecks and are isolated from the purple sulfur bacteria.

Chlorobium limicola grows rapidly and *C. thiosulfato-philum* grows slowly. The colonies are removed with a thin pipette and are transferred to van Niel or Larsen medium and grown under anaerobic conditions in the light. Green sulfur bacteria are strict anaerobes; therefore, even the smallest traces of oxygen depress their growth.

Further purification is achieved by inoculation onto an agar medium which is incubated under anaerobic conditions according to the method given for purple sulfur bacteria.

For isolation, one of the media given above may be used with 1.5–2.0% agar at pH 7.0–7.3 (attained by adding several drops of $NaHCO_3$), or Larsen medium may be used.

Larsen Medium (Basic Solution)

Distilled water	1000 ml
NH_4Cl	1 g
$MgCl_2 \cdot 6H_2O$	0.05 g
NaCl	0–30 g
Agar	20 g

The medium is sterilized in an autoclave. The following solutions are sterilized separately: $NaHCO_3$ (2% solution), $Na_2S \cdot 9H_2O$ (2%), and $FeCl_3$ (1.5%). They are added aseptically before inoculations are made so that the concentrations in the medium are as follows: $NaHCO_3$, 0.2%; $Na_2S \cdot 9H_2O$, 0.02–0.12%; and $FeCl_3$, 50 μg. pH is adjusted to 7.3 by adding several drops of a dilute solution of H_3PO_4.

The purity of a culture of green sulfur bacteria is established by the usual methods. It is absolutely necessary to eliminate purple sulfur bacteria, for which inoculations are made in suitable media with higher pH values.

To maintain cultures, Kondratieva (1963) recommends maintaining them under weak illumination or in the dark at a temperature no greater than 19–20 C.

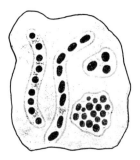

Fig. 137. Chlorobium limicola Nadson (according to Krasil-'nikov, 1949).

The green sulfur bacteria contain free-living species. The cells of various species, usually small in size, are very short rods, often connected in short chains. They do not form spores and reproduce by transverse fission of the cells; they are gram-negative and contain bacterioviridine and specific carotinoids. The majority of species are non-motile, although motile forms have been isolated. Some species do not form colonies (*Chlorobium*, Fig. 137); others are united into characteristic aggregates (*Pelodictyon, Clathrochloris*). Representatives of the genus *Pelodictyon* Lauterborn (Fig. 138) do not contain droplets of sulfur in the cells, whereas representatives of the genus *Clathrochloris* form sulfur deposits within the cells. Green sulfur bacteria also contain species which form sym-

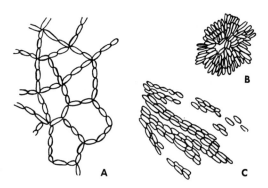

Fig. 138. Representatives of different species of *Pelodictyon* Lauterborn (Krasil'nikov, 1949). A, *P. aggregatum* Perfilyev; B, *P. clathratiforme* (Szafer) Lauterborn; C, *P. parallelum* (Szafer) Perfilyev.

biotic aggregates with other organisms of the genera *Chloro-bacterium*, *Chlorochromatium*, and *Cylindrogloea*. More complete descriptions and a key for determination are given by Kondratieva (1963). Also, see Skerman, *A Guide to the Identification of the Genera of Bacteria*, 2nd ed., The Williams and Wilkins Co., Baltimore, 1967.

Methods of Culturing Colorless Sulfur Bacteria

The following procedure of Winogradsky is used to culture colorless sulfur bacteria: a piece of root stalk of rush or another aquatic plant is placed at the bottom of a tall glass cylinder. A small amount of gypsum or sediment is added; the cylinder is then filled with water from the same water mass. Under these conditions colorless filamentous sulfur bacteria which form a film will generally develop. Omelianski medium yields good results.

Omelianski Medium

5% Solution of malt	1000 ml
Glucose	5 g
Peptone	1 g
Flowers of sulfur	10 g
Agar	20 g

Melted agar is inoculated with *Oidium lactis* and poured into petri dishes. After a few days, when the liberation of H_2S begins, the agar is cut into pieces. The agar pieces are placed at the bottom of a tall glass cylinder and covered with the water sample. After 10–15 days, a thick film consisting of colorless sulfur bacteria and various other microorganisms appears on the surface of the liquid.

Good growth of colorless sulfur bacteria is obtained with a special culture apparatus. Agar of the following composition is poured onto the bottom of a cylinder:

Distilled water	1000 ml
K_2S	3 g

K$_2$HPO$_4$	0.2 g
NH$_4$Cl	0.1 g
MgCl$_2 \cdot$ 6H$_2$O	0.1 g
Agar	30 g

Sediment from the water mass under study is placed over the agar (once it solidifies); the cylinder is then filled completely with water from the same source.

A simpler method for obtaining a culture of colorless sulfur bacteria is the method proposed by Shaposhnikov. Sediment is placed on the bottom of a cylinder and covered with water. Small ampoules filled with hydrogen sulfide solution serve as a source for H$_2$S. The hydrogen sulfide enters the water gradually, thus permitting growth of the sulfur bacteria.

To obtain enrichment cultures of *Beggiatoa*, Cataldi (1940) recommends a medium consisting of dried hay and 0.05% Na$_2$S. The medium is prepared as follows. Dry hay is covered with water and extracted at 100 C. The water is changed three times. The hay is finally drained and dried at 37 C and is used to prepare the medium. Water (70 ml) and hay (0.8%) are placed in 125-ml Erlenmeyer flasks, which are then sterilized in an autoclave. A large amount of sediment is inoculated into the flasks. Beggiatoae develop as a film on the surface.

To obtain pure cultures, a part of the film is rinsed several times with sterile tap water and the bacterial filaments are placed on the surface of a sterile agar (0.2% beef extract, 1% agar). Beggiatoae which develop on the surface of the agar migrate to the periphery of the dishes. Agar blocks containing a single filament are cut out and transferred to other dishes of fresh sterile agar of the same composition.

Pure cultures are grown in Scotten and Stokes (1962) medium.

Scotten and Stokes Medium

| Tap water [1] | 1000 ml |
| CaCl$_2 \cdot$ 6H$_2$O | 0.2 g |

---

[1] If distilled water is used for the medium instead of tap water, trace elements are needed.

K_2HPO_4	0.5 g
$MgSO_4 \cdot 7H_2O$	0.5 g
Yeast autolysate	1 g
Beef extract	1 g
Peptone	1 g

The authors indicate that it is desirable to add vitamins in the amount of 0.4 ml per 100 ml of medium. The concentration of vitamins in 1 ml is as follows:

Pantothenic acid	25 μg
Nicotinic acid	100 μg
Biotin	0.5 μg
Paraaminobenzoic acid	10 μg
Riboflavin	100 μg
Thiamine	100 μg
Pyridoxine	100 μg
Inositol	500 μg
B_{12}	100 μg
Folic acid	1 μg

pH 7.2–8.2

Cultures are grown at 25–30 C.

Scotten and Stokes indicate that a number of strains of *Beggiatoa* are aerobes and growth is obtained only when the medium is aerated. In this situation beggiatoae produce a white surface film on the medium. Other strains are microaerophiles and grow as spots on the bottom and walls of the culture vessel.

Faust and Wolfe (1961) point out the effectiveness of Cataldi's method but state that the media used are somewhat modified. To obtain enrichment cultures, dry roadside weeds are selected. Weeds which have been subjected to the weather for several months are cut into fine pieces and boiled in tap water (100 g of grass per liter of water). The water, which contains dissolved grass components, is boiled for about 10 minutes and then decanted. Fresh water is poured in and the boiling is repeated. After the water has been changed five times, the extracted hay is left in the water overnight, and on the following day the procedure is repeated three times. The hay is then removed and dried. The extracted hay (0.5 g) and river water (60 ml) are added to 125-ml Erlenmeyer flasks. Each flask is inoculated with a small clump of silt taken from a polluted river and is incubated at about 28 C.

A second method involves using a corncob both as a source of trace elements, and as a site of attachment for the organisms. River water is poured into wide-necked, orange-glass bottles. Several milliliters of sediment and a few decomposing leaves from the polluted river are added. The weathered corncob is placed in a bottle so that one end touches the bottom and the other remains at the top of the bottle. Water is poured in, leaving about 50 ml of air space, and the bottle is stoppered. It is incubated at room temperature.

The process of isolation of pure cultures is a modification of Cataldi's method. A cluster of filaments is transferred with a capillary pipette from an enrichment culture into sterile water to remove the accompanying bacteria. The clump of fila-

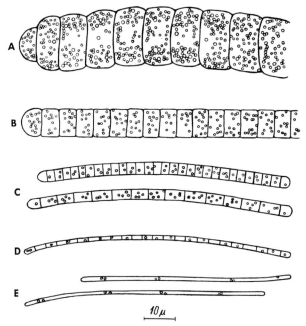

Fig. 139. Colorless sulfur bacteria of the genus *Beggiatoa* Trevisan. *A, B. mirabilis* Cohn; *B, B. arachnoidea* (Agardh) Rabenhorst; *C, B. alba* (Vaucher) Trevisan; *D, B. leptomitiformis* Trevisan; *E, B. minima* Winogradsky.

ments is taken from the sterile water and placed on the dry
surface of a solid medium of the following composition:

Distilled water	90 ml
Tap water	10 ml
Yeast extract	0.2 g
Agar	1.5 g

pH 7.0

The medium is sterilized in the usual manner, poured

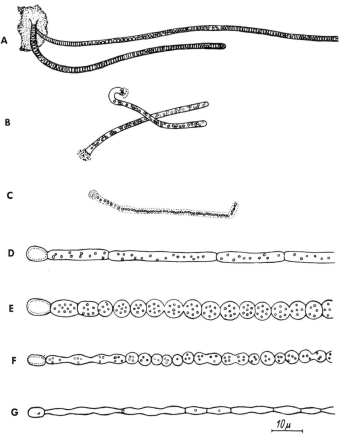

Fig. 140. Colorless sulfur bacteria of the genus *Thiothrix*
Winogradsky. *A, T. annulata* Molisch; *B, T. nivea* (Rabenhorst) Winogradsky; *C, T. tenuis* Winogradsky; *D, T. longi-
articulata* Klas; *E, T. torquis* Rodina; *F, T. torquis* forma
minor Rodina; *G, T. undulata* Rodina.

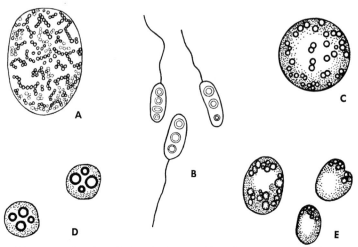

Fig. 141. Single-celled colorless sulfur bacteria. *A, Achromatium oxaliferum* Scheviakoff; *B, A. mobile* Lauterborn; *C, Thiophysa macrophysa* Nadson; *D, Thiosphaerella amylifera; E, Thiovolum majus* Hinze.

into petri dishes, and partially dried in an incubator at 30 C for 2 hours with the dishes just slightly opened. Each dish is examined between the 4th and 6th hours after inoculation under a microscope at 30×. Small agar blocks which contain well-isolated filaments are cut out and transferred aseptically to 2 ml of the medium given above (without the agar). Aggregations of filaments become noticeable in the original cultures after 1–3 weeks. Such a colony consists of actively motile filaments.

Species of the genus *Beggiatoa* Trevisan are widely distributed in water masses (Fig. 139). They are long motile filaments which do not adhere to the substratum; they consist of cells of uniform thickness in the entire length of the filament; and they are covered with a thin mucous film.

No less widespread are the various species of the genus *Thiothrix* Winogradsky (Fig. 140). Colorless single-celled organisms (Fig. 141) are less often encountered.

Methods of Culturing Thiobacteria

Thiobacteria actively oxidize hydrogen sulfide in water masses. They are widely distributed in nature and are of great

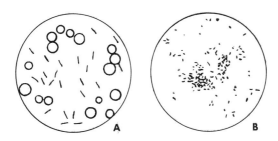

Fig. 142. Thiobacteria. *A, Thiobacillus thioparus* Beijerinck;
B, Thiobacillus thiooxidans Waksman and Joffe.

significance in the processes of the self-purification of water
masses when contamination from domestic or industrial sewage
is present. Thiobacteria are involved in the destruction of concrete pipes and structures in both fresh water and sea water.

The thiobacteria group includes species identified by
their physiological properties: they possess characteristic metabolic capabilities with respect to utilization of organic substrates (some species are capable of developing in the presence
of organic substances, others are not) and energy requirements
(some oxidize H_2S and elemental sulfur, others only sulfur, a
third type sulfur and thiosulfates, a fourth group only thiosulfates, a fifth group sulfides of heavy metals, changing Fe^{++}
to Fe^{+++}). Species which oxidize thiosulfates deposit free sulfur
outside of their cells; others, which oxidize elemental sulfur,
produce only sulfuric acid during active growth. The demands
of the various representatives of this group with respect to pH
of the medium are very diverse: *Thiobacillus thioparus* Beijerinck (Fig. 142, *A*) grows in an alkaline medium (pH
9.0–9.6); *Thiobacillus thiooxidans* Waksman and Joffe (Fig.
142, *B*) develops only in a very acidic medium (pH 2.0–4.0).
Some forms, e.g. species from Leningrad water masses described
by Salimovskaya (Rodina) and species from Japanese sources
require a weakly acidic medium for growth (pH 5.4–6.4). Therefore, to detect thiobacteria in water masses, specific features
of the water mass under study must be considered so that the
correct nutrient medium is selected. In many cases several kinds
of media must be inoculated.

The following media are used with considerable suc-
cess in growing bacteria such as *Thiobacillus thioparus*.

Beijerinck Medium

Distilled water	1000 ml
$Na_2S_2O_3 \cdot 5H_2O$	5 g
$NaHCO_3$	1 g
$Na_2HPO_4 \cdot 2H_2O$	0.2 g
NH_4Cl	0.1 g
$MgCl_2 \cdot 6H_2O$	0.1 g

Sulfur Powder Medium

Distilled water	1000 ml
K_2HPO_4	0.5 g
NH_4Cl	0.5 g
$MgCl_2 \cdot 6H_2O$	0.2 g
$CaCO_3$	20 g
Sulfur (powder)	10 g

Starkey Medium

Distilled water	1000 ml
$(NH_4)_2SO_4$	0.2–0.4 g
KH_2PO_4	3–4 g
$MgSO_4 \cdot 7H_2O$	0.5 g
$CaCl_2 \cdot 6H_2O$	0.25 g
$Fe_2(SO_4)_3 \cdot 9H_2O$	0.01 g
Sulfur (powder) (or 5 g of $Na_2S_2O_3 \cdot 5H_2O$)	10 g

pH 7.5

Rodina Medium
(for bacteria which require a more acidic medium)

Distilled water	1000 ml
K_2HPO_4	3 g
NH_4Cl	2 g
$MgCl_2 \cdot 6H_2O$	0.5 g
$CaCl_2 \cdot 6H_2O$	0.2 g
$Na_2S_2O_3 \cdot 5H_2O$	10 g

pH 6.0–6.2

The medium is poured into small conical flasks in a thin layer. If Jakobson medium is used, a small amount of flowers of sulfur is poured on the surface of the medium, with care that the sulfur lies on the surface of the medium in an extremely thin layer. The medium is sterilized at 110 C for 30 minutes. It is recommended that a solution of trace elements (p. 330) be added to 1% for culturing *Thiobacillus thioparus*.

T. *thioparus* (Fig. 142, *A*) is an autotroph. Other species of this group (*T. trautweinii* Bergey et al. and *Thiobacterium beijerinckii* Isachenko and Salimovskaya) grow well in the media given above and in organic media.

Starkey medium, adjusted to pH 3.5, and Waksman media are used to grow bacteria of the species *Thiobacillus thiooxidans* (Fig. 142, *B*), a very short, motile, rod-shaped, non-spore-forming, gram-negative autotroph.

Waksman Medium 1

Distilled water	1000 ml
$(NH_4)_2SO_4$	0.2 g
$MgSO_4 \cdot 7H_2O$	0.5 g
KH_2PO_4	3 g
$CaCl_2 \cdot 6H_2O$	0.25 g
Sulphur (powder)	10 g

Waksman Medium 2

Distilled water	1000 ml
$MgSO_4 \cdot 7H_2O$	0.5 g
$(NH_4)_2SO_4$	0.2 g
KH_2PO_4	1 g
$Ca_3(PO_4)_2$	2.5 g
Sulfur (powder)	10 g

In both media pH 3.0 is attained with a molar solution of phosphoric acid.

The number of bacteria of the species *Thiobacillus thiooxidans* in a water mass can be estimated with silica plates treated with Waksman medium. Separately sterilized dry sulfur powder is placed on the surface of the plates in a very fine layer. Inoculation is made with drops of water or small clumps of sediment. Colonies of thiobacteria are clearly visible on the plates owing to the formation of transparent zones around them as a result of oxidation of sulfur to sulfates.

A quantitative determination of sulfates, sulfites, and polythionates formed in liquid cultures of thiobacteria can be made to establish the degree of activity of these bacteria.

Growth in media containing hyposulfite can be detected by the appearance of a very thin film, consisting of droplets of sulfur and bacteria, and a fine, uniform turbidity. The film contracts, or wrinkles, with time and acquires a yellowish color from the liberated sulfur. A yellow precipitate appears on the walls and bottom of the culture vessel. The medium becomes more clear as the released sulfur precipitates. The greatest amount of sodium thiosulfate is oxidized in the first days after inoculation. Therefore, in the first days of growth, oxidation of $Na_2S_2O_3$ may be measured by a chemical test: a small amount of the medium is placed in a test tube and acidified with several drops of diluted nitric or hydrochloric acid (the acid must be examined beforehand for absence of contamination with sulfuric acid); when a 1–5% solution of $BaCl_2$ is added to the medium, an abundant precipitate of $BaSO_4$ settles out. One of the media listed above, with agar added to 1.5%, is used to isolate thiobacteria.

A solid agar medium may also be used for a quantitative determination of the occurrence of thiobacteria in water and sediment. Thiobacteria of the appropriate subgroup grow in suitable mineral media and form small, round, yellowish (owing to the liberated sulfur) colonies. Under a microscope they appear blue-green, and visible droplets of sulfur may be seen along the edges of the colonies. Fine motile (peritrichous) rod-shaped bacteria develop with deposits of sulfur outside the cells.

Thiobacillus ferrooxidans Temple and Colmer, a short rod 0.8–1 × 0.4 μ in size, is gram-negative, non-spore-forming, and motile. This species is isolated from acidic mine waters. The acidity of the medium and the presence of sulfides and compounds of ferrous iron are necessary conditions for growth of this microorganism. *T. ferrooxidans* is capable of oxidizing ferrous iron compounds and reduced compounds of sulfur. Growth may be obtained in one of the following media.

Colmer Medium

Distilled water	1000 ml
$FeSO_4 \cdot 7H_2O$	2 g

$(NH_4)_2SO_4$	0.1 g
KH_2PO_4	0.1 g
$MgSO_4 \cdot 7H_2O$	0.1 g

pH is adjusted to 2.5 with sulfuric acid.

Leathen, McIntyre, and Braley Medium

A. Basic Medium

Distilled water	1000 ml
$(NH_4)_2SO_4$	0.15 g
KCl	0.05 g
$MgSO_4 \cdot 7H_2O$	0.5 g
K_2HPO_4	0.05 g
$Ca(NO_3)_2$	0.01 g

B. 10% Solution of $FeSO_4 \cdot 7H_2O$

Ten milliliters of $FeSO_4$ solution are added aseptically to 1000 ml of basic medium. The pH should be 3.5.

T. ferrooxidans is isolated on silica plates with Leathen, McIntyre, and Braley medium. Young colonies of this species have irregular edges and are cream-colored to reddish brown. With time, i.e. as the culture ages, the colonies acquire a dark brown coloring.

Thiobacillus novellus may be grown both autotrophically and heterotrophically in a liquid medium or in a synthetic medium with citric acid. The synthetic medium with the addition of citric acid or glutamic acid is optimal for this species.

Chapter 9

Methods of Studying Transformations of Phosphorus Compounds by Microorganisms

The phosphorus cycle has great significance in the productivity of water masses. Phosphorus is present in water masses in the form of organic compounds (as a part of phosphorylated compounds, nucleoproteins, phosphatides, and other compounds of plant and animal tissues) and inorganic compounds.

Phosphorus is found in fresh water masses in the form of the phosphate ion in negligible amounts (usually hundredths and thousandths of a milligram per liter of water) because the soluble phosphorus compounds in water masses quickly form weakly soluble or insoluble complexes which settle out as a precipitate.

The conversion from insoluble inorganic to soluble phosphorus compounds proceeds through the activity of a special group of microorganisms which are distributed in water masses.[1]

The breakdown of organic phosphorus-containing compounds and the accompanying liberation of soluble phosphates takes place under the influence of another group of microorganisms and constitutes a second source of increase of soluble mineral compounds of phosphorus in water masses.

---

[1] The conversion of insoluble phosphates to the soluble state is caused not only by the direct action of specific microorganisms, but also occurs as a result of the metabolic activity of various groups of bacteria which act on these compounds.

Despite its importance, the phosphorus cycle in water masses has not been thoroughly elucidated and methods of investigating and enumerating the bacteria which bring it about need further study.

SOLUBILIZATION OF INSOLUBLE PHOSPHORUS COMPOUNDS BY MICROORGANISMS

The number of microorganisms capable of converting insoluble phosphates to the soluble state may be determined by the pour plate method when a solid medium with glucose and insoluble phosphates, namely $Ca_3(PO_4)_2$, is used.

Soil Extract Medium

Tap water	990 ml
Glucose	10 g
$(NH_4)_2SO_4$	0.5 g
K_2HPO_4	0.8 g
KH_2PO_4	0.2 g
NaCl	0.2 g
KCl	0.1 g
$MgSO_4 \cdot 7H_2O$	0.3 g
Yeast extract	2.5 ml
Soil extract	10 ml
Trace elements Solution (p. 295)	1 ml
Agar	20 g
$Ca_3(PO_4)_2$	2.5 g

pH 7.0

Soil Extract Medium

Tap water	750 ml
$(NH_4)_2SO_4$	0.5 g
$MgSO_4 \cdot 7H_2O$	0.05 g
K_2HPO_4	0.4 g
$MgCl_2 \cdot 6H_2O$	0.1 g
$FeCl_3 \cdot 6H_2O$	0.01 g

$CaCl_2 \cdot 6H_2O$	0.1 g
Peptone	1 g
Yeast extract	1 g
Soil extract	250 ml
Glucose	5 g
$Ca_3(PO_4)_2$	25 g
$CaCO_3$	15 g

Novogrudskii Medium

Distilled water	1000 ml
$(NH_4)_2SO_4$	1 g
$MgSO_4 \cdot 7H_2O$	0.5 g
KCl	0.5 g
$CaCl_2 \cdot 6H_2O$	0.5 g
NaCl	0.1 g
Phosphate fertilizer	10 g
Yeast autolysate	1 ml

Sediment Extract Medium

Tap water	500 ml
Sediment extract	500 ml
Glucose	10 g
$Ca_3(PO_4)_2$	2.5 g
Yeast extract	0.5 ml
Agar	20 g

pH 7.0

Carrot Broth Medium

Carrot broth (1 kg of carrot in 1 liter of water)	1000 ml
Asparagine	1 g
Glucose	10 g
Yeast extract	0.5 g
Agar	20 g
Insoluble phosphate	until uniform turbidity is obtained

$Ca_3(PO_4)_2$ is prepared separately. The phosphate salt (2.5 g) is ground to a very fine powder in a mortar. Gum arabic (0.5 g) is ground separately in a porcelain mortar. Both powders are mixed thoroughly, transferred to a flask containing 100 ml of water, and then sterilized. When inoculations are made, a pipette is used to add aseptically 1 ml of the shaken suspension to the dish with the inoculum. Nutrient medium is added and mixed well. The addition of gum arabic prevents the formation of clumps of insoluble phosphate and yields a medium of uniform and equal turbidity.

Tricalcium phosphate may be substituted for apatite, phosphorite, or phosphate fertilizer.

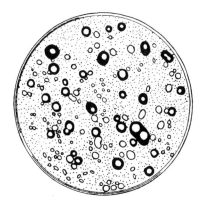

Fig. 143. Zones of tricalcium phosphate dissolved by phosphoroclastic bacteria.

In all media, colonies surrounded by transparent zones are counted (Fig. 143). Among microorganisms which form zones of dissolved phosphate in media containing tricalcium phosphate are species which liberate soluble phosphates into the medium, and others which use completely the dissolved phosphate in cellular metabolism.

Asporogenous rods are found in both groups of microorganisms, including pigmented types (*Pseudomonas rubigenosa* (Catiano) Krasil'nikov, *Chromobacterium denitrificans* Topley and Wilson) and many aquatic strains of yeast belonging, in general, to the *Torulopsis* and *Rhodotorula* groups.

When pure cultures of microorganisms which convert

insoluble phosphates to soluble forms are isolated, their biochemical activity may be examined in a synthetic liquid medium. The medium is poured into small flasks in 100-ml amounts
and inoculated with pure cultures. After 2–3 weeks a portion
of the culture is filtered through a No. 3 membrane filter. The
phosphate ion is detected in the filtrate with the aid of a colorimeter. Sterile medium serves as control.

LIBERATION OF PHOSPHATES FROM
ORGANIC COMPOUNDS

The presence of bacteria which bring about the process
of liberation of phosphorus compounds from organic compounds
may be detected on silica gel plates treated with one of the
following media.

Menkina Medium

Distilled water	1000 ml
$(NH_4)_2SO_4$	0.5 g
$MgSO_4 \cdot 7H_2O$	0.3 g
KCl	0.3 g
NaCl	0.3 g
$MnSO_4 \cdot 5H_2O$	trace
$FeSO_4 \cdot 7H_2O$	trace
Glucose	10 g
Chalk	5 g
Lecithin	See text

Three to five milliliters of this medium are placed on
each plate. Then lecithin or nucleic acid is added to a final
concentration of 3–5 mg of P_2O_5 per plate. The lecithin must
be dissolved beforehand in 96% alcohol. The plates are sterilized in an autoclave at ½–¾ atm for 20 minutes. Inoculation is performed by aseptically introducing small clumps of
sediment or droplets of water into the medium. The same medium may be used with agar.

Modified Version of Menkina Medium

Tap water	1000 ml
$(NH_4)_2SO_4$	0.5 g
$MgSO_4 \cdot 7H_2O$	0.25 g
K_2HPO_4	0.8 g
KH_2PO_4	0.2 g
NaCl	0.1 g
Soil extract	10 ml
Yeast extract	2.5 ml
Trace elements solution (p. 295)	1 ml
Chalk	5 g
Lecithin, or a sodium salt of nucleic acid, or nucleic acid	See text

The sources of organic phosphorus are added separately to each flask to a final concentration of 3–5 mg of total phosphorus per 10 ml of medium. Lecithin is prepared as follows: 1 g of lecithin is dissolved in 2 ml of ethyl alcohol. Distilled water (18 ml) is then added. The solution is neutralized to pH 7.0 with a saturated solution of sodium bicarbonate. It is then added to a small flask by means of a graduated pipette and sterilized with flowing steam for 40 minutes. When inoculations are made, 0.1 ml of the solution is placed in each dish after the inoculum is added. Melted and cooled agar is added and mixed well. Nucleic acid is added to 30–50 mg per plate (content of total phosphorus, 3–5 mg).

Odoevskaia (1961) has devised media which do not require nucleic acid compounds and lecithin.

Potato Agar

Raw cleaned potato (500 g) is boiled for 2 hours in 1 liter of water; the amount of liquid after boiling should be 0.5 liter. Ten milliliters of the sterilized broth are increased to 100 ml with the addition of water. Agar (2%) and a 1% solution of phenol (4 ml) are added.

Potato-Mineral Salts Agar

Tap water	900 ml
Potato broth (described above)	100 ml

Glucose	3 g
$(NH_4)_2SO_4$	0.2 g
KCl	0.1 g
NaCl	0.1 g
$MgSO_4 \cdot 7H_2O$	0.1 g
1% Solution of phenol	4 ml
Chalk	precipitated
Agar	20 g

According to Odoevskaia, colonies of phosphorus bacteria are easily distinguished in this medium owing to their characteristic morphology. They are large and mucoid and appear to be surrounded by clearly visible zones. The author mentions the significant decrease in the number of phosphoroclastic bacteria as a result of addition of phenol.

The hydrolysis of organic compounds containing phosphorus is effected by many microorganisms which possess phosphatases. There is probable involvement of 15 to 20 enzymes, each of which is distinguished by the chemical nature of the substratum hydrolyzed. Phosphatase acts on nucleic acid and lecithin together with other enzymes. The result is a separation of phosphates through a series of intermediate compounds.

The following method (Krasil'nikov and Kotelev, 1957; Kotelev, 1958) may be used to detect the presence of microorganisms which demonstrate phosphatase. Inoculations are made in a solid medium (beef-peptone agar, Czapek medium, or a starch-ammonia medium) which is diluted so that colonies develop separately from each other. After the colonies germinate, the petri dish containing the colonies is covered with a layer of agarized buffer medium containing sodium phenolphthalein phosphate and is placed in a refrigerator for 24–48 hours. The phosphatase is detected by pouring 2–3 drops of ammonia on the inside surface of the lid of the inverted dish. Under the action of the phosphatase, the reagent breaks down into free phenolphthalein and phosphorus. Ammonia diffuses into the medium and produces, in addition to the liberated phenolphthalein, a red coloring in both the colonies which possess phosphatase and the agar around them.

The medium containing sodium phenolphthalein phosphate is prepared as follows: 0.1% $MgSO_4 \cdot 7H_2O$, 0.05%

sodium acetate, and 0.7% washed or purified agar are added to a 0.1 N solution of NH_4Cl. The agar is dissolved in a hot water bath and the solution is cooled to 45–50 C. Sodium phenolphthalein phosphate (0.1%), dissolved beforehand in a small amount of distilled water (2–5 ml), is then added.

The above medium with sodium phenolphthalein phosphate is used to determine phosphatase activity of pure cultures. Pure cultures are grown in Menkina nutrient agar in a uniform lawn. A freshly prepared medium with sodium phenolphthalein phosphate is poured into large (12 mm in diameter) petri dishes in a 2- to 3-mm layer. After the agar hardens, blocks of the agar medium on which the culture under study has been grown are placed on fresh agar. The blocks are cut out with a cork borer (about 8 mm in diameter) and placed on the agar with the side containing the bacterial inoculum down. To preserve sterility, several drops of toluene are placed in the dishes with the agar blocks. The dishes are allowed to stand for 2 days to permit the bacterial phosphatase to diffuse into the agar containing the sodium phenolphthalein phosphate. After 2 days the dishes are inverted so that the lids are at the bottom, and several drops of concentrated ammonia are placed on the inside of the lids. Gaseous ammonia penetrates into the agar with the sodium phenolphthalein phosphate, and a red coloring is observed in the places in which free phenolphthalein appeared as a result of phosphatase activity.

The following procedure may also be used (Naumova, 1961). A culture is grown in beef-peptone agar or in Menkina medium in petri dishes. The bacterial deposit is washed off, and the suspension obtained is diluted 10 times with distilled water and poured into test tubes or flasks. One or two milliliters of buffered substratum (0.1% sodium phenolphthalein phosphate in 1 N ammonium buffer mixture at pH 9.8) and 3 drops of chloroform, as a bactericide, are poured into each test tube. The test tubes are closed with stoppers and placed in an incubator at 37–38 C. After various periods of time, the color of the liquid in the test tubes is noted. A red coloring indicates the presence of phosphatase.

Chapter 10

Methods of Culturing Iron Bacteria

Iron bacteria are widely distributed in various water masses. Often they achieve massive growth, stain water a yellow color, and cover the bottom with a layer of rust. Iron bacteria play a fundamental role in the formation of iron-manganese nodules in lakes and in the ocean as well.

The presence of iron compounds is the condition which determines the growth of iron bacteria in water masses. Iron in natural waters is often encountered in the most diverse inorganic compounds in dynamic equilibrium. In addition to the iron compounds, organic compounds of iron, complicated complexes containing iron, and suspensions of minerals containing iron may also be present in natural waters.

The iron bacteria include species which behave differently towards organic substrates. Some are autotrophs and live in water masses with clean water containing dissolved ferrous bicarbonate. Others are mixotrophs, and a third group is heterotrophic and dwells on the surface of aquatic plants and algae in waters containing significant amounts of organic substances.

Some iron bacteria require a lower pH for growth. Others, which use the organic part of iron humates, are encountered in a water mass containing a more alkaline water.

There are no general methods for culturing iron bacteria, and a number of forms have not yet been obtained in pure culture. The taxonomy of iron bacteria demands further accumulation of data, special studies of individual species, and more exact descriptions of them. Therefore a variety of nutrient media must be used to study the iron bacteria in a water mass.

The simplest method of determining the presence of iron bacteria in water masses is microscopic examination of submerged slides and membrane filters through which a specific volume of water has been filtered. The specimens are treated with a 2–5% solution of potassium ferrocyanide for 2–3 minutes, and then with 5% hydrochloric acid for 1–2 minutes. They are then stained with erythrosin (5% solution in 5% carbolic acid water). Thus the cells of ferrobacteria, which stain red, are distinguished from their iron sheaths and small stalks, which stain blue. If these solutions are too concentrated, more dilute concentrations (2%) may be used. When submerged slides are examined under a microscope, the iron bacteria can be readily identified.

A modification of this staining method for iron bacteria has been suggested by Meyers (1958). According to this method, 10 ml of a suspension of iron bacteria are centrifuged and a drop of the pellet is placed on a clean slide, dried in the air, and fixed by submersion for 15 minutes in methyl alcohol. Equal volumes of a 2% solution of potassium ferrocyanide and a 5% solution of acetic acid, both prepared in distilled water, are mixed and heated to boiling. The specimen is submerged for 2 minutes in the hot solution, rinsed with water, and dyed with 2% aqueous safranin. It is again rinsed and dried. The cells stain red; deposits of iron are blue.

Iron bacteria may be detected in the following ways.

1. Water taken with a sampler from the water mass under study is placed in wide-necked bottles. If iron bacteria are present, by the next day flakes resembling cotton wool will appear in the jar. After the samples are treated as indicated above, iron bacteria may be detected in the flakes.

2. Water is placed in an aquarium jar together with sediment from the same water mass. When the sediment precipitates, cotton-like accumulations followed by rusty spots appear above the precipitate in a layer of the water. Cholodny (1953) recommends inserting several cover glasses into the lower side of a flat cork (Fig. 144) and then placing it on the surface of the water in an aquarium. The cork is allowed to float freely in the aquarium for about 24 hours. During this time many iron bacteria attach themselves to the surface of the cover glass. The cork is removed, and filter paper is used to carefully remove

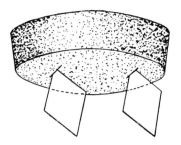

Fig. 144. Cork float with cover glasses for attachment of iron
bacteria (Cholodny, 1953).

the water from the cover glasses, which are then air-dried,
fixed, and stained. Slides are placed on the bottom of the
aquarium in a vertical position.

There are a number of media for culturing iron bacteria.

Winogradsky Medium

Distilled water	1000 ml
NH_4NO_3	0.5 g
$NaNO_3$	0.5 g
K_2HPO_4	0.5 g
$MgSO_4 \cdot 7H_2O$	0.5 g
$CaCl_2 \cdot 6H_2O$	0.2 g
Ferric ammonium citrate	10 g

This medium may be used as a solid medium if 1.5%
agar is added.

Lieske Medium

A small amount of sediment from the water mass under
study, iron filings, and some extract from fallen leaves (enough
so that the liquid does not turn yellow) are added to the water
sample. Several re-inoculations are required to obtain enriched
cultures.

Iron Carbonate Medium

Distilled water	1000 ml
$(NH_4)_2SO_4$	1 g

$MgSO_4 \cdot 7H_2O$	0.05 g
K_2HPO_4	0.1 g
$Ca(NO_3)_2 \cdot 4H_2O$	0.02 g
$FeCO_3$	0.05 g

Prévot Medium

Distilled water	1000 ml
Agar	0.25 g
$FeSO_4 \cdot 7H_2O$	0.005 g
$(NH_4)_2Fe(SO_4)_2 \cdot 6H_2O$	0.005 g
$NaHCO_3$	0.03 g
KCl	0.03 g
$(NH_4)_2SO_4$	0.03 g
$(NH_4)_2CO_3$	0.03 g
Glucose	0.2 g
Asparagine	0.1 g
K_2HPO_4	trace

pH 6.0

The medium is incubated at 13–18 C.

Iron Bacteria Medium

Distilled water	1000 ml
$(NH_4)_2SO_4$	1.5 g
KCl	0.05 g
$MgSO_4 \cdot 7H_2O$	0.5 g
K_2HPO_4	0.05 g
$Ca(NO_3)_2 \cdot 4H_2O$	0.01 g
$NaHCO_3$	0.03 g
Solution of trace elements (p. 336)	1 ml

The medium is poured in 100-ml amounts into 250-ml conical flasks and sterilized in an autoclave. Iron solution (10 ml) is poured aseptically into each of the flasks. The iron solution is prepared as follows:

$FeSO_4 \cdot 7H_2O$	10 g
Distilled water	100 ml

The solution is sterilized by filtration through an ultra-filter (membrane filter No. 1 or No. 2 or a Berkefeld filter may be used). The solution is slightly opalescent and has a pH of about 6.5.

Lieske Agar

Distilled water	1000 ml
$Mn(CH_3COO)_2$	0.1 g
Agar	10 g

Growth of heterotrophic iron bacteria may be obtained as follows:

1. In peptone water with organic iron:

Tap water	1000 ml
Iron citrate	1 g
KH_2PO_4	0.05 g
Peptone	5 g

2. In solutions of calcium salts of organic acids with nitrates and traces of K_2HPO_4.

3. In water of the mass under study when concentrated hay infusion (20 ml per liter) with iron silicate (20–40 mg per liter) is added.

Kalinenko indicates that to obtain growth of heterotrophic iron bacteria, use should be made of a solid medium onto which inoculations are made with material freshly taken from the water mass. Microscopic amounts of the sample are smeared on the surface of the medium.

Special media and culture methods have been devised for individual species of iron bacteria; selection of particular media and methods is determined by physiological features of the bacteria.

Iron bacteria of the genus *Leptothrix* (Fig. 145, *A–D*) are easily obtained in culture by the Winogradsky method. A small amount of boiled hay is placed on the bottom of a glass cylinder about 50 cm high. Some freshly precipitated ferric hydroxide is added and the cylinder is filled with the water sample. Anaerobic decomposition of plant residues rapidly begins in the

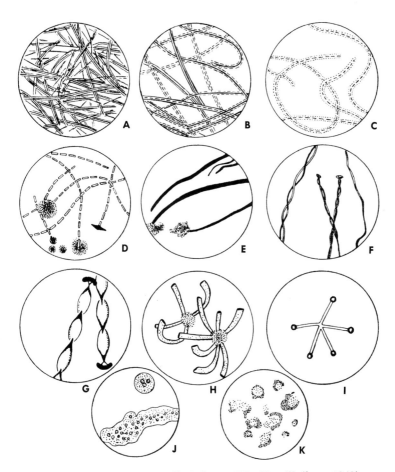

Fig. 145. Iron bacteria (Cholodny, 1953; Krasil'nikov, 1949).
A–D, Genus *Leptothrix; F–K,* cultures of the autotrophic
species *Gallionella ferruginea.*

vessel. The ferric hydroxide is reduced and the end product is
water-soluble ferrous bicarbonate which diffuses upward. Flocs
of iron bacteria appear on the diffusion boundary of oxygen in
the vessel.

Luxurious growth of species of the genus *Leptothrix* may
be obtained in a peptone-manganese medium.

Leptothrix Medium

The following solutions are prepared and sterilized: (1)

0.025% peptone solution; (2) 0.005% solution of $Mn(CH_3COO)_2$ (manganese acetate); and (3) willow leaves broth.

The willow leaves broth is prepared as follows: 12–15 g of leaves are covered with 1 liter of water, steamed for 15 minutes in Koch apparatus, filtered, poured into test tubes, and sterilized with flowing steam. The medium must contain:

Willow leaves broth	1 ml
Solution of $Mn(CH_3COO)_2$	1 ml
Peptone water	1 ml
Distilled water	8 ml

This medium, with pH 5.8–6.8, yields optimal conditions for growth of species of the genus *Leptothrix*. Pure cultures are obtained by serial transfers in the described medium with the addition of 1% agar.

Growth of organisms belonging to the genus *Leptothrix* can be obtained in other solid media, particularly in a medium containing ferric ammonium citrate.

Ferric Ammonium Citrate Medium

Water sample	1000 ml
$Mn(CH_3COO)_2$	0.75 g
Ferric ammonium citrate	0.75 g
Peptone	0.25 g
Agar	10 g

Colonies of *Leptothrix* are easily recognized by their brown color.

Special media are used to obtain cultures of the autotrophic species *Gallionella ferruginea* (Fig. 145, *F*).

Lieske Medium

Distilled water	1000 ml
$(NH_4)_2SO_4$	1.5 g
KCl	0.05 g
$MgSO_4 \cdot 7H_2O$	0.05 g
K_2HPO_4	0.05 g
$Ca(NO_3)_2 \cdot 4H_2O$	0.01 g

To prepare this medium, salts of a high degree of purity

must be used. When culturing autotrophs, glassware should be washed with an acid-dichromate cleaning solution.

The medium is poured in 30-ml amounts into tall straight bottles. After sterilization, the bottles are left to stand until a gaseous equilibrium with the atmosphere develops. Before inoculation, iron in the form of iron filings or flat pieces (3% according to weight), or freshly precipitated ferrous sulfide (FeS) is added to each vessel. FeS is precipitated from a solution of $FeSO_4$ with Na_2S. The precipitate is centrifuged with glass distilled water, freshly boiled to remove the air.

Iterson indicates that it is best to grow cultures of gallionellae in an atmosphere enriched with carbon dioxide (up to 1% CO_2), although this is not mandatory. It is easier to obtain growth of gallionellae at low temperatures (6 C). Gallionellae in an inorganic medium rich in iron ions usually develop as flakes on the bottom of the vessel.

Lieske medium may be used for other species of iron bacteria (Iterson, 1958) if leaf extract is added. Leaf extract is prepared by heating, in an autoclave, fresh leaves of table lettuce (*Lactuca sativa*) in tap water for 20 minutes. The liquid contains Fe^{++} ions and has a pH of about 6.6. pH is adjusted to 7.4, and the liquid changes to a reddish color. The liquid is poured into small bottles and again sterilized.

van Niel Medium

NH_4Cl	0.05 g
K_2HPO_4	0.03 g
$MgSO_4 \cdot 7H_2O$	0.02 g
Na_2CO_3	0.01 g
Distilled water	1000 ml

Freshly precipitated FeS is used as a source of iron ions. A mixture of gases, composed of nitrogen with 7–15% O_2 and 5% CO_2 added, is passed through the prepared medium. This causes a change in the pH of the medium (original pH 8–9, final 6–7). Inoculated media are incubated at room temperature in containers under an atmosphere of 5% CO_2.

Solutions of vitamins and trace elements should be added to the above media to create the most optimal conditions for iron bacteria (Iterson).

Solution of Vitamins per Liter of Medium

Vitamin B_1	0.2 μg
Riboflavin (B_2)	0.2 μg
Calcium pantothenate	0.2 μg
Pyridoxine (B_6)	2.0 μg
Paraaminobenzoic acid	0.6 μg
Biotin	0.002 μg
Nicotinic acid	0.04 μg
Folic acid	0.2 μg
Vitamin B_{12}	trace

Solution of Trace Elements (added to the medium in an amount of 1%)

$ZnSO_4 \cdot 7H_2O$	4.4 mg
$CaCl_2 \cdot 6H_2O$	1 mg
MoO_3	3 mg
$BaCO_3$	15 mg
$Co(NO_3)_2 \cdot 6H_2O$	2.5 mg
H_3BO_3	56 mg
KI	1.3 mg
$Al_2(SO_4)_3 \cdot 6H_2O$	250 mg
Silicic acid	trace
Distilled water	1000 ml

Iron Bacteria Medium

Distilled water	1000 ml
NH_4Cl	1 g
K_2HPO_4	0.5 g
$MgSO_4 \cdot 7H_2O$	0.2 g
$CaCl_2 \cdot 6H_2O$	0.1 g

Sterile flasks are filled to one-third to one-half volume with sterile nutrient medium.

CO_2 is passed through the medium for 1–2 minutes and a sterile suspension of excess FeS is carefully added, without mixing. The suspension is prepared by mixing equal volumes of molar solutions—heated to boiling—of ferric ammonium alum and Na_2S. After precipitating, the iron sulfide is washed, decanted, and sterilized. For convenience of observation, slides may be submerged in the inoculated flasks. Gallionellae will attach themselves to the slides.

To maintain cultures, transfers must be made no less than once a week.

Growth of the autotrophic species *Ferrobacillus ferrooxidans* may be obtained in Silverman and Lundgren medium.

Silverman and Lundgren Medium

Distilled water	700 ml
$(NH_4)_2SO_4$	3 g
KCl	0.1 g
K_2HPO_4	0.5 g
$MgSO_4 \cdot 7H_2O$	0.5 g
$Ca(NO_3)_2 \cdot 4H_2O$	0.01 g
0.1 N Solution of H_2SO_4	1 ml
14.74% Solution of $FeSO_4 \cdot 7H_2O$	300 ml

The authors indicate that *Ferrobacillus ferrooxidans* grows better in shaken culture.

Methods similar to those mentioned for growing iron bacteria are used for bacteria which oxidize manganese.

Medium for Bacteria which Oxidize Manganese

Tap water	900 ml
Saturated solution of manganese bicarbonate	100 ml
$NaHCO_3$	0.1 g
$(NH_4)_2SO_4$	0.1 g
K_2HPO_4	trace
$MgSO_4 \cdot 7H_2O$	trace

Manganese bicarbonate is obtained by passing a stream of CO_2 through a suspension of $MnCO_3$. The medium is sterilized with flowing steam.

Zavarzin Medium for *Metallogenium symbioticum*

Lieske medium without iron (p. 360)	1000 ml
$Mn(CH_3COO)_2$	0.1 g
Washed agar	15 g

Culturing Pigmented and Luminous Microorganisms and Streptomycetes; Methods of Isolating Streptomycetes

CULTURING PIGMENTED MICROORGANISMS

Water masses provide a habitat for many pigmented microorganisms. Bacterial pigments are diverse in chemical composition. Their formation is often connected with the composition of the medium. Media containing magnesium ion are recommended for isolating and culturing pigmented bacteria. There are several genera of bacteria with pigmented species: *Pseudomonas* and *Chromobacterium* spp. are shown in Plate 6.

Medium for Red-Pigmented Bacteria

Distilled water	1000 ml
Asparagine	10 g
Glucose	10 g
Na_2CO_3	0.25 g
K_2HPO_4	3 g
$MgSO_4 \cdot 7H_2O$	0.4 g
Agar	15 g

Omelianski Medium

Distilled water	1000 ml
Beef extract	1 g
Glycerin	10 g
$(NH_4)_2HPO_4$	2 g

$MgSO_4 \cdot 7H_2O$	0.5 g
KH_2PO_4	1 g
Agar	20 g

Salle Medium

Distilled water	1000 ml
Yeast extract	5 g
Peptone	5 g
Glucose	10 g
Agar	20 g

Kossovich Medium

Distilled water	1000 ml
Saccharose	30 g
KH_2PO_4	2.5 g
$(NH_4)_2HPO_4$	2.5 g
$MgSO_4 \cdot 7H_2O$	2.5 g
NH_4Cl	2 g
$Ca_3(PO_4)_2$	0.05 g

Chromobacterium violaceum Medium

Distilled water	1000 ml
Fibrin	10–20 g
KCl	0.02 g
Agar	20 g

Pseudomonas aeruginosa Medium

Distilled water	900 ml
Beef-peptone broth	100 ml
Peptone	10 g
Glucose	5 g
K_2HPO_4	1 g
$MgSO_4 \cdot 7H_2O$	2 g

Serratia marcescens Medium 1

Distilled water	1000 ml
$(NH_4)_2HPO_4$	5 g

K₂HPO₄	5 g
Na₂SO₄	0.5 g
MgSO₄ · 7H₂O	4 g
NaCl	4 g
FeSO₄ · 7H₂O	4 g
MnSO₄ · 5H₂O	4 g

$$K_2HPO_4 \quad 5\ g$$
$$Na_2SO_4 \quad 0.5\ g$$
$$MgSO_4 \cdot 7H_2O \quad 4\ g$$
$$NaCl \quad 4\ g$$
$$FeSO_4 \cdot 7H_2O \quad 4\ g$$
$$MnSO_4 \cdot 5H_2O \quad 4\ g$$

pH 7.2

Serratia marcescens Medium 2

Distilled water	1000 ml
Peptone	10 g
Asparagine	2 g
$CH_3COONa \cdot 3H_2O$	2 g
K_2HPO_4	1 g
$MgSO_4 \cdot 7H_2O$	0.5 g
Agar	20 g

The medium is acidified with 2 drops of lactic acid.

Omelianski Medium for *Bacillus subtilis*
subsp. *aterrimus* and varieties

Distilled water	1000 ml
Peptone	10 g
Beef extract	10 g
K_2HPO_4	1 g
$MgSO_4 \cdot 7H_2O$	0.5 g
Agar	20 g

Potato slices, lightly rubbed with chalk, provide a good medium for many pigmented bacteria.

Corpe (1951) suggested the following method for detecting pigmented soil bacteria: 5 g of soil or sediment are placed in a sterile petri dish and spread in an even layer with a sterile spatula. Sterile (boiled beforehand) grains of rice are scattered on the surface of the soil or sediment. The dishes are incubated for 5 days. Pigmented bacteria grows on the rice grains. This method may be used for detecting pigmented bacteria in sediments only. It has proven to be of little use for detecting pigmented bacteria in water.

Rhodopseudomonas Medium

Distilled water	1000 ml
Peptone	2 g
Yeast extract	2 g
Calcium lactate	5 g
Lecithin	2 g

pH 7.6

Rhodospirillum rubrum Medium

Distilled water	1000 ml
Sodium lactate	5 g
Yeast extract	5 g
K_2HPO_4	5 g
Potassium acetate	0.4 g
$CaCl_2 \cdot 6H_2O$	0.02 g

The medium is neutralized to pH 7.2 with KOH.

The red pigments which contain bacteriopurpurin and bacterioviridine, characteristic of the two genera *Rhodopseudomonas* and *Rhodospirillum*, are formed better under anaerobic conditions in the presence of organic substances and lactate.

The above media are only a few of many that have been published. A compilation of media and methods has been prepared by Skerman.

CULTURING LUMINOUS BACTERIA

Luminous bacteria are, in general, isolated from salt-water media. Some grow in simple media, others in more complex media (Omelianski, 1940). Therefore, a number of media have been proposed for culturing luminous bacteria.

1. Distilled water	1000 ml
Fish muscle, namely, fresh perch	500 g
NaCl	30 g
Peptone	10 g

Asparagine	5 g
K_2HPO_4	1 g
$MgSO_4 \cdot 7H_2O$	0.5 g
Agar	20 g

2. | Tap water | 1000 ml |
NaCl	30 g
Peptone	10 g
Glycerin	5 g
Gelatin	100 g

3. | Tap water | 1000 ml |
Peptone	10 g
Glycerin	20 g
NaCl	30 g

4. | Fish broth | 1000 ml |
NaCl	30 g
Glycerin	10 g
Peptone	10 g
Asparagine	5 g
Gelatin	5 g

5. | Tap water | 900 ml |
Yeast water	100 ml
Dextrin	5 g
Peptone	10 g
NaCl	30 g
Agar	18 g

6. | Seawater | 1000 ml |
Peptone (Difco)	5 g
Yeast extract (Difco	0.5 g
Glycerol	3 ml
Agar (Difco)	11 g

A minimal medium for culturing luminous bacteria is given by Farghaly (1950).

Farghaly Medium (modified by Nealson, Platt and Hastings)

NaCl	30 g
$Na_2HPO \cdot 7H_2O$	7 g

KH_2PO_4	1 g
$(NH_4)_2PO_4$	0.5 g
$MgSO_4$	0.1 g
Glycerol	3 ml
Distilled water	1000 ml

METHODS OF CULTURING STREPTOMYCETES

The streptomycetes group of microorganisms is widely distributed in water masses. They may comprise 15% to 40% of all colonies which develop in nutrient media such as beef-peptone agar. They are found both in gently flowing masses of water (e.g. lakes and reservoirs) and in flowing waters (rivers and streams). The water mass contains relatively small amounts of streptomycetes; the sediments, however, are often rich in them. Mass development of streptomycetes in sediments (more than 1 million per gram of dry sediment) has been a reason for the formation of earthy odors in the water column.

Streptomycetes have a well developed, branching mycelium with hyphae which are not separated into cells (i.e. without transverse partitions). The hyphae of streptomycetes are usually thin (0.6–0.8 μ); only a few species have wider filaments (up to 1.2 μ in diameter). The length of the filaments varies: mycelia with both long and short hyphae may be distinguished (Fig. 146, A).

On solid nutrient media, streptomycetes form skin-like or cartilaginous colonies which grow into the agar; the colonies are so dense that they are removed with a needle only in their entirety. An aerial mycelium forms above the surface of such a colony. The colonies are easily distinguished from colonies of other microorganisms by the characteristic mycelium; the surface of a streptomycete colony is floury, fluffy, or velvety. Sporangiophores, filaments which carry spores, form on the branches of the aerial mycelium. Sporangiophores for different species of *Streptomyces* are helical or straight, and single or gathered into verticils (Fig. 146, B). Spores—always in short chains—form at the ends of the sporangiophores.

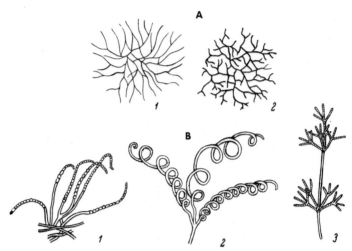

Fig. 146. Mycelium of streptomycetes (*A*) and sporangio-phores (*B*). *A: 1,* with long hyphae; *2,* with short hyphae. *B: 1,* straight; *2,* spiral; *3,* vorticular.

Streptomycetes may be detected in water and sediments by the plate count method of inoculating a nutrient medium.

Streptomycetes vary greatly in their nutrient require-ments. Some grow in complex organic substances; others grow well on nutrient media composed of simple chemical compounds. The vast majority of streptomycetes grow not only under favor-able conditions, but also under some rather unfavorable condi-tions. Streptomycetes break down proteins, urea, amino acids, and simpler nitrogenous substances. Proteins, peptones, and amino acids are the best sources of nitrogen for streptomycetes. However, they also use nitrates and ammonium salts, and some species use nitrites as well. Ammonium sulfate is a more suitable source of nitrogen for streptomycetes than is ammonium chloride.

Streptomycetes can use various carbon compounds, both nitrogenous and non-nitrogenous, as a carbon source. Carbo-hydrates (glucose, maltose), certain organic acids (e.g. acetic, malic, citric), polyhydric alcohols (mannitol, glycerin), and starch are the best sources of carbon for streptomycetes. They also use other alcohols, hydrocarbons of the aromatic series, hemicellulose, and lignin; some species use cellulose and rubber.

Streptomycetes which break down petroleum products (kerosine, gasoline, naphthalene, paraffin) and plant waxes have been described.

The ability of streptomycetes to break down different organic compounds determines their significance in nature, i.e. in water masses, sediments, composts, and so forth. In water masses streptomycetes effect the decomposition of organic plant and animal remains and the liberation of ammonia from complex proteins. Several investigators have noted the favorable influence of the streptomycetes on plant growth. Streptomycetes form different pigments: violet, red, rose, black, and brown. Many color the medium from dark brown to black. There are data which indicate that streptomycetes may cause gray sediments to turn black—this process is accompanied by the transfer of calcium and iron to upper layers of the sediment. They can contribute to the precipitation of calcium as $CaCO_3$.

A weakly alkaline medium is most favorable for this group. However, some species of *Streptomyces* grow at low pH.

Some streptomycetes form exchange products which are capable of inhibiting the growth of bacteria, including other species of *Streptomyces*. Many of them are capable of lysing bacterial cells, especially spore-formers. Some species of *Streptomyces* suppress the growth of *Azotobacter* species, others enhance it. The mass development of streptomycetes in sediments has great significance for the growth of other bacteria.

The following criteria may be used to identify streptomycetes: form of the colonies in solid nutrient media; formation of aerial mycelium which rises above the colony; growth in liquid nutrient media (streptomycetes in liquid media never form turbidity, but grow as separate flocs or colonies which then flow together into a film on the surface of the medium); and characteristics of the sporangiophore on the aerial mycelium (Fig. 146, *B*). Filaments stain gram-positive in young cultures.

Many media are suggested for culturing streptomycetes.

1. Czapek Medium

Sucrose (glucose or glycerin)	30 g
$NaNO_3$	2 g
K_2HPO_4	1 g

MgSO$_4$ · 7H$_2$O	0.5 g
KCl	0.5 g
FeSO$_4$ · 7H$_2$O	0.01 g
Agar	15 g
Chalk	3 g
Distilled water	1000 ml
pH 7.0–7.3	

2. Glucose-Asparagine Agar

Glucose	10 g
Asparagine	0.5 g
K$_2$HPO$_4$	0.5 g
Agar	15 g
Distilled water	1000 ml
pH 6.8	

Beef extract (2 g per liter) may be added.

3. Glycerin-Asparagine Agar

Glycerin	35 ml
Sodium aspartate	3.5 g
Ammonium lactate	6.5 g
K$_2$HPO$_4$	2.5 g
CaCl$_2$ · 6H$_2$O	0.1 g
NaCl	1 g
MgSO$_4$ · 7H$_2$O	0.3 g
Agar	20 g
Distilled water	1000 ml
pH 7.0	

4. Tyrosine Agar

Glucose	10 g
Tyrosine	1 g
(NH$_4$)$_2$SO$_4$	0.5 g
K$_2$HPO$_4$	0.5 g
Agar	15 g
Distilled water	1000 ml
pH 7.0	

5. Beef-Peptone Agar with 10 g of Glucose or 15 g of Glycerin Added

6. Beef-Peptone Gelatin

7. Glucose-Peptone Agar

Peptone	10 g
Beef extract	5 g
Glucose	10 g
KH_2PO_4	1 g
$MgSO_4 \cdot 7H_2O$	0.5 g
KCl	0.5 g
Agar	20 g
Distilled water	1000 ml

To reduce formation of aerial mycelium, the amounts of organic components in the medium may be decreased from one-half to one-tenth the given amounts.

8. Glucose-Peptone Gelatin

9. Aqueous Gelatin

Gelatin	100–200 g
Tap water	1000 ml
pH 7.4	

10. Starch Agar A

Soluble potato starch	10 g
K_2HPO_4	0.3 g
$MgCO_3$	1 g
NaCl	0.5 g
$NaNO_3$	1 g
Agar	15 g
Distilled water	1000 ml

11. Starch Agar B

Soluble potato starch	2 g
K_2HPO_4	0.5 g

$MgSO_4 \cdot 7H_2O$	0.2 g
$CaCl_2 \cdot 6H_2O$	0.05 g
$NaNO_3$	0.05 g
Asparagine	0.05 g
$Fe_2(SO_4)_3 \cdot 9H_2O$	trace
Washed agar	20 g
Distilled water	1000 ml
pH 7.4	

12. Potato Agar (mainly of historical interest)

Cleaned potato	500 g
Peptone	10 g
Beef extract	10 g
Agar	15 g
Tap water	1000 ml
pH 7.0	

Potato is cleaned and weighed and cut into small pieces; 350 ml of water are added, and the medium is boiled for 45 minutes. The extract is passed through a thin cotton cloth.

13. Potato-Glucose Agar (of historical interest)

Cleaned potato	300 g
Glucose	5 g
Agar	20 g
Tap water	1000 ml
pH 6.8–7.2	

14. Agar with Yeast Extract

Yeast extract	10 g
Glucose	10 g
NaCl	1 g
$MgSO_4 \cdot 7H_2O$	0.25 g
$FeSO_4 \cdot 7H_2O$	0.01 g
Agar	15 g
Distilled water	1000 ml

15. Soybean Agar

Soybean meal	10 g
Glucose	10 g
NaCl	1–5 g
$CaCO_3$	1 g
Agar	15 g
Distilled water	1000 ml

16. Soil Extract Agar

Beef extract	3 g
Peptone	5 g
Soil extract	1000 ml
Agar	15 g
pH 7.0	

The soil extract is prepared by heating 1 kg of garden soil with 2.5 liters of tap water for 1 hour in an autoclave at 120 C. It is filtered while hot.

17. Krasil'nikov Medium 1

River or tap water	1000 ml
K_2HPO_4	3 g
$MgCO_3$	0.3 g
NaCl	0.2 g
KNO_3	1 g
$FeSO_4 \cdot 7H_2O$	0.001 g
$CaCO_3$	0.5 g
Sugar	20 g

18. Krasil'nikov Medium 2

Distilled water	1000 ml
Glucose	10 g
KH_2PO_4	2.4 g
K_2HPO_4	5.7 g
$(NH_4)_2SO_4$	2.6 g
$MgCl_2 \cdot 6H_2O$	1.2 g
$ZnSO_4 \cdot 7H_2O$	0.01 g

$FeSO_4 \cdot 7H_2O$ 0.01 g
$CuSO_4 \cdot 5H_2O$ 0.006 g
$MnCl_2 \cdot 6H_2O$ 0.008 g
Sodium lactate   11 g

19. Krasil'nikov Medium 3

Distilled water   1000 ml
Glucose   10 g
$(NH_4)_2HPO_4$ 2 g
NaCl   2 g
K_2HPO_4 1 g
$MgSO_4 \cdot 7H_2O$ 0.5 g
$CaCl_2 \cdot 6H_2O$ 0.2 g
$FeSO_4 \cdot 7H_2O$ 0.002 g
$ZnSO_4 \cdot 7H_2O$ 0.01 g

20. Krasil'nikov Medium 4

Tap water   1000 ml
Saccharose   20 g
KNO_3 0.1 g
$MgSO_4 \cdot 7H_2O$ 0.05 g
KCl   0.07 g
KH_2PO_4 0.1 g
$Ca(NO_3)_2$ 0.25 g

Chapter 12

Submerged Slide Culture

Submerged slide culture exposes microbial floras and makes it possible both to determine the morphological composition of the bacterial flora of the water mass and sediments, and to probe the microbial biocenoses and interrelationships among separate species of these biocenoses. By this method surface layers of sediments reveal "imprints" of microorganisms in the sediments. In water samples, this method allows one to obtain not only a representation of the activity of the aquatic bacteria and of their ecological interrelations, but also the reaction of microbes to changes in the external medium.

OVERGROWTH (SUBMERGED) SLIDES

Those forms of bacteria which do not appear in culture but are present in water masses can be exposed by the submerged slide method. The usual nutrient media, used in routine investigations do not meet the nutritional requirements of many microorganisms; consequently, they do not grow. The use of submerged slides allows detection in water masses of a great number of diverse forms (Fig. 147), including species of *Caulobacter*, *Hyphomicrobium vulgare*, *Hyphomonas polymorpha*, *Rhodomicrobium vannielii*, *Pedomicrobium*, *Blastocaulis*, and *Blastobacter*.

Submerged slide culture requires use of thoroughly degreased, clean slides which are fastened at set distances along a cable. The cable carrying the slides is lowered into the water, with a weight at the bottom end and a buoy at the top (Fig. 148). Two slides, on each end of which numbers have been

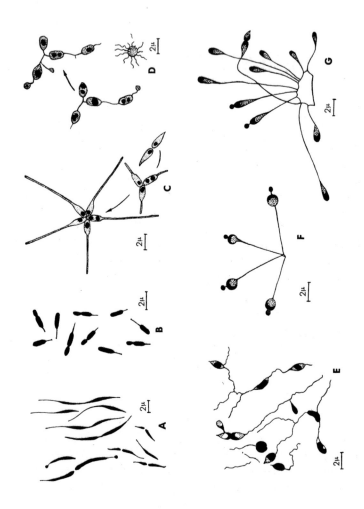

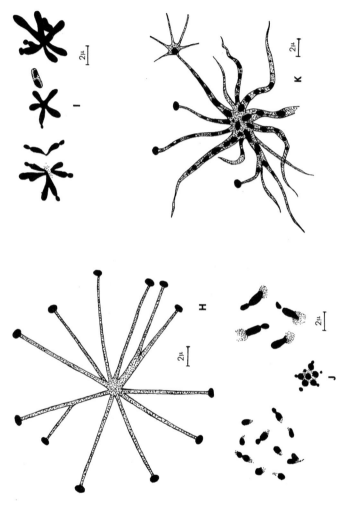

Fig. 147. Caulobacteria (Zavarzin, 1961). A and B, *Caulobacter* sp., C, *Hypomicrobium vulgare*; D, *Rhodomicrobium vannielii*; E, *Pedomicrobium* sp.; F. *Blastocaulis sphaerica*; G, *Blastocaulis* sp.; H, *Bl. kljasmiensis*; I and J, *Blastobacter* spp.; K. *Metallogenium personatum*. Arrows indicate sequence of development phases.

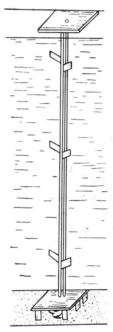

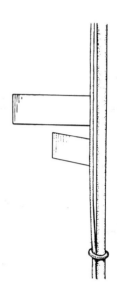

Fig. 148. Diagram of
arrangement of slides
at various depths.

Fig. 149. Arrange-
ment of slides in
water (Salimovskaya
and Rodina, 1936).

marked with a diamond pen, carborundum, or Pobedit (a tool
alloy composed of sintered tungsten, cobalt, and titanium car-
bides), are fastened at each designated level. The numbers and
corresponding levels are recorded.

There are various ways of setting the slides in the water.
According to one method (Salimovskaya and Rodina, 1936), two
rubber tubes of a different diameter are used simultaneously
(Fig. 149). The one with the wider opening serves as a sheath
for the cable, which is passed through it. The other, with the
smallest possible inner opening, serves as fastener for the slides.
For this purpose, two slits are cut at specific intervals (1 m or
more) along the smaller tube and the slides are inserted in the
slits as the rubber is stretched. The second tube is fastened to
the first with rubber rings at several intervals. With such an
apparatus, the slides remain intact, even when exposed in the
water mass to strong turbulence.

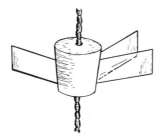

Fig. 150. Arrangement of slides on corks in water.

The slides may be attached in another way, e.g. in slits made in corks which are slipped on the cable at specific intervals (Fig. 150).

ZoBell and Allen (ZoBell, 1946) suggest using special devices made from rubber, plastic, or another material for submerging slides. Each such device carries six slides. Selection of material for construction of the apparatus is very important. It is best to use plastic; wood is undesirable because certain compounds are leached from wood into the water and this directly influences the composition of the microflora developing in its immediate vicinity.

A device suggested by Karzinkin can be used to arrange slides in water and sediment. It consists of a slab 20×25 cm wide and 0.5 cm thick. Wooden borders with grooves for holding 24 slides are attached along the long edges, from the top and bottom sides (Fig. 151). Twelve slides are placed in the lower grooves and are submerged in the sediment. The other twelve are placed in the upper grooves and remain in the bottom layer of water when the instrument is placed in the water mass. A weight is fixed to the upper part of the slab. The instrument is lowered on a cable, which may also be used to carry slides into the water by one of the described methods.

To avoid the effects of the wood, all wooden parts may be covered with paraffin.

If it is impossible to place the slides directly in the sediment, they can be submerged in a sediment sample immediately after it is taken, but the sample must be kept under a layer of water during incubation of the slides in the sediment. It is necessary to consider, of course, that conditions in this case are different from the natural situation.

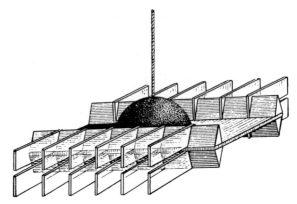

Fig. 151. Arrangement of slides in sediment (Karzinkin, 1934).

The apparatus with the slides is left in the water for a specific time, depending on the purpose of the investigation. The rate of overgrowth of the slides by microbes depends on the temperature of the water, nutrient content (most importantly, organic), currents, and current strength. Overgrowths develop rapidly in the summer in warm waters with a high content of organic nutrient and slowly in cold, oligotrophic water masses. The average submerging period for the slides is 3–4 days.

Slides which remain in water for a period of time sufficient for overgrowth are removed; precautions being taken against washing off weakly attached forms; and the slides are immediately placed on a vertical position in small flasks. In the laboratory the slides are dried, and then fixed with either absolute alcohol or formalin vapors. The latter is preferable, because it does not disturb the weakly attached forms.

The slides are stained in order to observe better the bacterial biocenoses. For this purpose a 5% solution of erythrosin in 5% carbolic acid water is generally used. Staining is performed by submerging the slides in staining jars. The duration of staining is 30–40 minutes. After staining the slides are washed by carefully passing them through a series of jars containing distilled water. The slides are then dried and examined with a microscope under oil immersion. Before staining with erythrosin, one of the two slides set at each water level is treated with a 2–5% solution of potassium ferrocyanide for 2–3 minutes, and then with 5% hydrochloric acid for 1–2 minutes. This is done to distinguish more easily the iron bacteria which develop on the slides. When this treatment is applied, iron bacteria stain red and the iron sheaths and deposits are blue.

The forms which attach themselves to the submerged slides can be isolated for further study and identification. In this case, the slides are taken out of the water and immediately placed in sterile moist chambers; they are then transported to the laboratory, where a very thin layer of a solid nutrient medium is poured over them. Because this procedure requires experience it is useful to practice first. The slides, covered with agar, are placed inside moist chambers which are then incubated. When several slides are available from a given water level, different media can be used for each of the slides. The selectivity of the media ensures growth of certain bacterial groups. Thus, for example, if the slides are covered with Fedorov agar, colonies of *Azotobacter* grow. Growth of the colonies indicates accumulation of nitrogen-fixing-bacteria-like cells, which may then be identified on the submerged slides as *Azotobacter* species. Thus, these results testify to the presence and active metabolic state of this microorganism in the given water mass or sediment. It is possible to obtain cultures of these microorganisms by pouring an agar medium for isolating yeast over the slides.

If the slides must be lowered through a surface film which is especially rich in bacteria and through a layer of water —near the boat—which is contaminated with waste or oil, a special instrument, constructed by Sorokin (1963), may be used. This apparatus (Fig. 152) permits lowering the slide to the necessary depth in a closed vessel. Usually, however, such precautions are not necessary.

The submerged slide method may also be used in a modified form. These modifications are used in studies of bacterial periphytes in the undisturbed living state. Voskresenskii (1947) suggested replacing slides with cover glasses, i.e. they are placed in the groove of a glass frame and lowered into the water (Fig. 153). To prepare the frame, a strip of a compound for pasting cover glasses is placed on a strongly heated rectangle of thin glass. A heated strip of cover glass about 5 mm wide

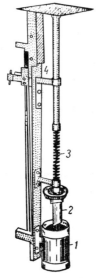

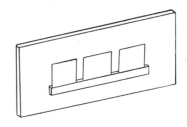

◀ *Fig. 152.* Sorokin apparatus for submerging slides. *1,* Vessel in which the slide is lowered; *2,* slide; *3,* spring; *4,* catch.

Fig. 153. Arrangement of cover glasses (Voskresenskii, 1947).

is then placed on the strip of paste. The cover glasses should lie tightly against the glass frame so that overgrowth may develop from one side.

To preserve the bacteria of overgrowth in the viable state, Voskresenskii recommends placing cover glasses in a small evaporating jar with water. The bottom of the evaporator except for the center portion is covered with a layer of paraffin for holding small glass tubes which will support the cover glasses. The bacteria are observed under water immersion. The side of the glass which is overgrown is submerged face down in the water.

CAPILLARY METHOD FOR STUDYING OVERGROWTHS
(PERFILYEV METHOD)

Perfilyev has devised a method for studying aquatic periphytes (Perfilyev and Gabe). Plane-parallel capillaries are used. This provides protection for young bacterial overgrowths from mechanical injury and desiccation. However, it must be noted that special conditions arise within the capillaries which are favorable for some species, but unfavorable for the development of a number of other species of bacteria.

Perfilyev termed his apparatus a capillary periphytometer (Fig. 154, *A*). The basic portion of the instrument is a
glass catch plate consisting of a slide 25.0 × 75.0 × 1.0–1.4 mm
in size. Two slotted glass mounting strips are fastened (by
flaming or with the aid of polymerized glue) facing towards
each other on the surface of the glass plate. The distance between the mounting planks is accurately determined and corresponds to the length (64 mm) of the removable part of
the glass trap bands of a rectangular cross section (2.0 × 0.5
mm) (Fig. 154, *B*). A freely sliding piece of rectangular glass
capillary 30 mm long is placed on each catch strip. This capil-

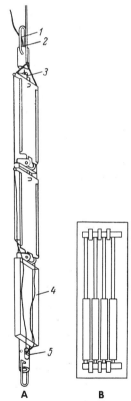

A   B

Fig. 154. Periphytometer (Perfilyev and Gabe, 1961). *A*, General view: *1*, end of cord which closes the cartridges; *2*, end of
cord which opens the cartridges; *3*, tube on which the cartridges are fastened; *4*, protective flaps; *5*, loop for suspending
weight. *B*, Catch plate of the periphytometer.

lary makes it possible to use high magnification, including immersion, when the primary stages of overgrowth are viewed under a microscope. Four catch strips provided with capillaries are placed on each catch plate of the periphytometer. Dividers are used so that the strips do not touch each other. The divider is a short piece (3 mm in length) of irregularly shaped capillary which has a C-shaped form in profile. The catch plates are placed in aluminum casettes consisting of two flaps—one hinged flap has grooves into which the glass catch plate is placed; the second flap has the form of a flat aluminum box which protects the capillary part of the catch plate from breaking when the instrument is lowered into the water and removed. The hinged flap serves to open and close the casette. The periphytometer is provided with a weight. The instrument is lowered closed; when the cable is raised, it opens. After the investigation is completed, the instrument is closed and hauled out of the water.

Perfilyev has constructed a peloscope [1] (Fig. 155) for studying the microflora of the area of sediment deposits produced by microorganisms. The instrument consists basically of a collection of 4 to 5 capillaries of the same length (1–2 cm). Such a packet of capillary units is fastened with a rubber ring and submerged vertically into the sediment so that the capillary units intersect the level in which the main microzones of transformation of the sediment are formed. To insure this, the packet of units is fastened to a glass support 65–70 mm long. The instrument is kept submerged for varying periods of time.

A variation of the instrument is the slotted peloscope. It consists of two rectangular glass plates which are joined by rubber rings with a gap between them arranged by placing two thin glass strips at the upper and lower ends.

If the purpose of the investigation is not only to study the microbial flora but also to obtain cultures by transfer into a nutrient medium, then one of the glass plates must be thin (no more than 0.16–0.17 mm thick), and the length and width must correspond to the dimensions of the circulating compartments or the culture vessels.

---

[1] Peloscope is from the Russian *peloskop,* the name of this Russian instrument.

When working with solid sediments, Perfilyev recommends that the lower end of the supporting plate be sharpened like a wedge to avoid breaking it when it is inserted into the sediment.

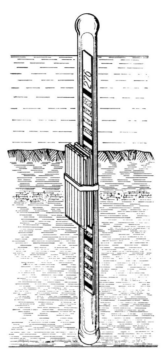

Fig. 155. Perfilyev and Gabe peloscope (Perfilyev and Gabe, 1961).

DETERMINING BACTERIAL CENOSES IN BODIES OF WATER

Determination of the composition of bacterial cenoses in water masses is absolutely essential, since unique communities of bacteria capable of existing under the prevailing conditions develop in each biotope of any water mass. To determine bacterial cenoses, direct microscopic methods are used most successfully: these include microscopic examination with an ordinary light microscope of submerged slides and of membrane filters used for direct count. Microscopic examination is also

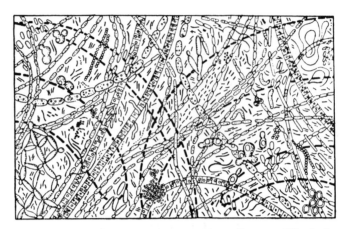

Fig. 156. Bacterial cenosis of contaminated water (Slavianka
River, below Pavlovsk).

made under ultraviolet light of living microorganisms stained
with fluorochromes immediately after samples are taken.

Two methods may be used when observing membrane
filters under a microscope: (1) a sketch of all microorganisms
in the field of vision, with examination of no less than 30 fields
of vision; (2) microphotograph of no less than 30 fields of vision
followed by a count of the microorganisms found in each field
of vision. In this way the predominant forms of microorganisms
are observed, as well as those forms which do not develop in
the media routinely used in microbiology. Counting the number
of the various forms makes it possible to establish the relationship of individual groups of microbes in various cenoses.

In addition to the data obtained in a microscopic examination of submerged slides, the predominance of various
species is verified and the character of their development and
their interrelationships in a cenosis are revealed. The peculiarity
of the communities of water and sediment bacteria, as established by the above-mentioned methods, is shown in Figs. 156
and 157.

By means of fluorescent microscopy the natural bacterial communities may be observed in their natural state. It
also enables determination of the natural bacterial groupings
in the microbial cenoses of a biotope. In sediments it is possible
to determine natural cenoses, as well as distribution of microorganisms between liquid and solid phases; the latter is im-

Fig. 157. Bacterial cenosis of contaminated soil (mouth of the Moscow River).

possible to show by any other method. To reveal microbial cenoses in sediments, it is sufficient to observe directly a small amount of the sediment in a fluorescent microscope (in incident light) when a fluorochrome and quencher are added. To reveal microbial cenoses in water, the samples first must be concentrated by filtration. This method allows observation and accurate visualization of the size and shape of cells under natural environmental conditions.

Examination of Live and Dead Cells in Bacterial Cenoses

In a number of cases in the study of bacterial cenoses, it becomes necessary to explain the state of the microorganisms in a particular community. Methods of differential staining and fluorescent microscopy can be used to observe live and dead cells.

Several staining methods have been suggested.

Peshkov Method

Necessary reagents include:

1. Fixative:	ethyl alcohol (96%)	60 ml
	glacial acetic acid	30 ml
	chloroform	10 ml

2. Giemsa stain: azure blue II, 1:1000
eosin, 1:1000
3. 0.25% Aqueous solution of light green dye

The staining procedure is as follows: (1) a smear or submerged slide is treated with the fixative for 20 minutes; (2) it is air-dried for 1 hour and then (3) submerged in Giemsa stain for 4 hours at 37 C or 24 hours at room temperature; (4) the smear is rinsed with neutral (pH 7.1) distilled water, (5) dyed for 20 seconds with the 0.25% aqueous solution of light green, and (6) rinsed with distilled water and dried. Results: microorganisms which were alive before fixation stain violet; those which were dead before fixation stain green.

Drobot'ko Method

Necessary reagents include:

1. 3% Aqueous solution of Congo red   100 ml
   Saturated solution of mercuric chloride   4 ml
2. Ethyl alcohol (96%)   100 ml
   Hydrochloric acid   3 ml
   Acidic fuchsin   to saturation

Staining procedure: (1) a smear is prepared and air-dried; (2) Congo red solution is added and left on for several minutes; (3) the dye is washed off and the specimen is rinsed for a long period of time; (4) the fuchsin solution is placed on the smear and left there for several minutes; (5) the smear is rinsed and dried. Results: the microorganisms which were alive before fixation stain red; those which were dead before fixation stain black.

Fluorescent Method

To reveal simultaneously both live and dead cells of microorganisms in the field of vision of a fluorescent microscope, treatment of the specimen with fluorochromes is recommended, using a mixture of primulin (1:20–1:40,000) and strongly diluted acridine orange (1:100,000). The live cells fluoresce green, and the dead cells yellow.

According to data of several authors, the best fluoro-chromes for revealing dead bacteria are primulin, thiazole yellow, and brilliant-dianile green (all diluted 1:100,000). When applying these fluorochromes (in the given dilutions), live bacteria barely fluoresce whereas dead bacteria fluoresce clearly. The following procedures are recommended by Meisel and his coauthors for counting live and dead bacterial cells in one and the same field of vision: the field of vision of the specimen with the fluorochrome is examined first under fluorescent illumination and the fluorescing dead cells are recorded. Then the same field of vision is examined with phase contrast, and all bacteria are noted, both live and dead. To do counts, it is most convenient to use a microscope provided with an opaque illuminator (OI-17).

BIOCONTROL WITH PROTECTIVE COATINGS
(ANTI-FOULING COMPOUNDS)

Any structure—regardless of the material from which it is constructed—located in water and vessels floating in the oceans, seas, or continental waters, may become fouled with bacteria, plants, and animals. Water pipes may become so clogged with these organisms that their internal opening may become completely closed. Ships which become fouled are slowed down, and overgrown buoys may sink.

The process by which ships and harbor structures become fouled begins with the attachment of microorganisms to the surface of the structure, followed by abundant growth of the attached forms. Algae, and later animals, settle on the bacteria, forming a continuous layer. Fouled vessels can be successfully dealt with at present with a passage of electric current. However, in a number of cases this method cannot be applied, and special paints which contain various toxic substances are used. The technology of preparation of paints which do not become fouled varies, as does their composition. Some can release a great amount of toxic material, immediately after which the output gradually declines; others yield toxic material gradually at a uniform rate. When analyzing the degree of toxicity of these coatings, they are usually placed on metal plates which are then submerged in water to the desired depth for a specific period of time. The presence of microorganisms can be determined by ordinary microbiological analysis: (1) by inoculation of the layer taken from a specific area of the surface of the plate with a sterile pad; (2) by fluorescent microscopy (treating a small piece of the deposit with a fluorochrome and observing it under a fluorescent microscope in incident light).

A laboratory test usually precedes a test under natural conditions. It is very convenient to use bacteria to determine the toxicity of paints which do not become overgrown or fouled. A method worked out by Rodina and colleagues in 1947 has been successfully applied by several laboratories. It is performed as follows: agar (usually beef-peptone or fish-peptone) which is heavily inoculated with bacteria is placed in a petri dish. When it solidifies, 5 or 6 small glass rings—about 1 cm high and stained beforehand on the surface with the paint or coating being tested—are placed in the petri dish.

The dish is placed in an incubator. As a result of the toxicity of the coating and the diffusion of ions into the agar, the bacteria around the ring cannot grow, and where no growth is present a zone can be measured. An equal thickness of the coating is an essential condition for obtaining comparable results. The most accurate method is to use a mixed combination of overgrowing microorganisms which develop under the natural conditions of the water. A combination of aquatic bacteria is significantly more resistant to toxic paint and coatings than are pure cultures of ordinary culture collection strains. However, pure cultures of aquatic bacteria may be used to obtain an indication of the relative strength of action of a particular compound on microorganisms. Preliminary research must be done on the isolation and characteristics of the more resistant strains responsible for fouling.

Chapter 13

Methods of Bacteriological Analysis of Water in Public Health

Two indicators are generally used in sanitary bacteriological determinations of water quality: (1) the number of heterotrophic microorganisms which grow within 24 hours at 37 C on agar prepared by standard methods; (2) the presence of bacterial indicators of fecal contamination of the water (American Public Health Association, 1960).

To determine the number of heterotrophic bacteria, inoculations are made in the usual manner. The sample is shaken thoroughly before inoculation and the amounts of water to be inoculated are placed in the petri dishes with a sterile pipette. Dilutions are used, or not used, depending on the expected results, i.e. the expected number of bacteria per milliliter. It is recommended that for each dilution the count plates be run in duplicate. The inoculation plates are incubated at 37 C.

Counts are made after 24 hours. The number of colonies which develop after 24 hours at 37 C is always less than the number of colonies which develop at 22 C after 48 hours.

In a survey of accepted European methods for determining the degree of contamination of water, Bick (1963) refers to the Bucksteeg and Thiele method (1959) as a very useful and easy enumeration of bacteria in water. Analyses are based on the reduction of 2,3,5-triphenyltetrazolium chloride (TTC) to red formazan by the reducing enzymes of living bacteria. A solution of TTC is added to the contaminated water. Through action of living and metabolically active bacteria, the reagent is reduced to red formazan. The red coloring is extracted with alcohol and the intensity of color is measured in a spectrophotometer. The results are then compared with a standard bacterial suspension containing a known number of cells. Bick indicates that these analyses make it possible to determine the number of bacteria more easily and quickly than by the pour plate method. At the same time, the accuracy of this method is greater than that of a count performed on plates of nutrient agar. In addition, the TTC method enables the investigator to obtain a more realistic picture of the biological activity of the contaminated water or sediment.

Fecal contamination of water is detected by the presence of *Escherichia coli* or *Enterobacter aerogenes*.

Organisms which are indicators of fecal contamination are usually characterized by the following: aerobic (facultatively anaerobic), oxidase negative, weakly motile, gram-negative, non-spore-forming rods. They do not liquefy gelatin, but do form indole, coagulate milk, and ferment—along with the formation of gases (hydrogen and CO_2)—glucose, lactose, maltose, fructose, arabinose, and mannitol. They do not ferment saccharose. On Endo agar these bacteria form characteristic red colonies with a golden metallic sheen, dark red colonies, and rose colonies with a dark center. Also, on Eosin-Methylene Blue (EMB) agar, *E. coli* colonies produce a characteristic green metallic sheen.

Taking a pure culture of *Escherichia coli* through a series of media making up the so-called complete series of tests is very awkward and sometimes wastes time. Consequently, at present a somewhat condensed analysis is frequently used.

The membrane filter method is technically the simplest and also provides the quickest results. It yields more accurate results because a count of colonies which grow on the filter is more accurate than an estimation made from data obtained by the dilution method. However, fermentation patterns following the Most Probable Number (MPN) method are also acceptable.

The results of analysis, by both methods, are expressed as the number of cells of the indicator organisms per liter of water (*coli* index) or as the smallest volume of water which contains 1 cell of *Escherichia coli* (*coli* titer).

Analysis of water by the membrane filter method (Razumov, 1953) consists of two stages: (1) culturing bacteria concentrated on a membrane filter in the Endo fuchsin-sulfite medium for 24 hours at 37 C; (2) growing bacteria, from colonies on the filters, at 43 C for 24 hours in tubes of a peptone-glucose or in lactose broth medium, and recording gas formation under these conditions. A diagram of the analysis is shown in Fig. 158.

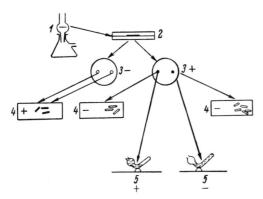

Fig. 158. Diagram of water analysis for indicators of fecal contamination. *1,* Water is filtered through a membrane filter; *2,* bacteria are grown on Endo medium; *3*⁺, presence of red colonies; *3*–, growth of colorless colonies; *4,* Gram staining: *4*+, positive; *4*–, negative; *5,* growth in glucose-peptone medium: *5*+, presence of fermentation; *5*–, absence of fermentation.

The first stage is as follows. The filter apparatus is sterilized. A sterile membrane filter (No. 1 or No. 2 or 0.45 μ Millipore membrane filter) is placed—with the mark or inscription up—on the filter holder using flamed forceps. The instrument is then assembled. A sterile pipette or graduated cylinder is used to place the designated volume of water into the funnel of the filtering apparatus. The collection flask is evacuated and the water is filtered.

The membrane filter with the settled bacteria is carefully removed with forceps and placed in a petri dish on the surface of Endo or EMB agar, which was poured and cooled beforehand. The filter must lie smoothly on the surface of the agar; for this purpose, it is gently placed on the agar and gradually pressed onto the agar (there should be no air bubbles between the filter and the medium). The numbered side of the filter containing the collected microorganisms must face up.

Several filters may be placed in one petri dish. The date, number of the sample, filter number, and amount of filtered water are recorded.

Endo agar is prepared as follows: beef-peptone agar (pH 7.4–7.5) is poured out in 50- or 100-ml quantities and sterilized in an autoclave. The following are prepared: (1) fuchsin solution (basic fuchsin, 10 g; 90% ethyl alcohol, 100 ml; the solution is filtered before use); (2) 10% solution of anhydrous sodium sulfite (Na_2SO_3) or a 20% solution of the same salt in crystal form ($Na_2SO_3 \cdot 7H_2O$); (3) 25% solution of lactose which is poured out in small portions and sterilized with flowing steam.

The fuchsin-sulfite medium is prepared as needed (on the current day or 1 to 2 days before the experiment), because it does not keep very long. When the medium is to be used, beef-peptone agar is melted, and 0.5 ml of the fuchsin solution is mixed with the sodium sulfite solution in a sterile test tube until a bright rose coloring is obtained. The lactose solution (2–3 ml) and the fuchsin-sodium sulfite mixture are introduced into the agar. Everything is mixed well and then poured into petri dishes. The prepared agar should have a cream color with a weak reddish tint. Endo and EMB agar are available in dehydrated, pre-mixed form from a variety of commercial bacteriological supply houses in the U.S. It usually is sold in prepared form as a powder. The commercial agar mixture is prepared by dissolving a specific amount in boiling sterile water. It is then poured into sterile petri dishes and dried overnight at 37 C before using. Directions for preparation and sterilization should be carefully followed.

Colonies which appear on a filter in Endo or EMB medium are counted after 24 hours. Only those colonies with the characteristic coloring are counted (Fig. 159). If only colorless colonies appear on the filter, then smears are made of two or three of them and Gram-stained. The absence of gram-negative non-spore-forming rods in the smears yields a final negative answer (–) to the presence of *Escherichia coli* in the given volume of water. If gram-negative rods are present, the second step is followed. If colored colonies are present, smears are made from several and they are Gram-stained. The second step of the analysis is then performed.

Fig. 159. Colonies of bacteria of the coliform group on a membrane filter. The colonies appear a characteristic color on the filter, depending on the medium used. Courtesy of the Millipore Filter Corporation.

When growth is very dense, colonies fuse and a mixed culture results. Transfers onto Endo or EMB agar are then made in order to obtain isolated colonies. Examination of smears under the microscope is repeated and the second step is followed.

In the second step of analysis, inoculations are made from colonies on a filter (and from those colonies from which smears were made) into a peptone-glucose or lactose broth medium prepared in test tubes with vials or into Dunbar tubes (Fig. 91). Dunbar tubes are no longer routinely used in diagnostic bacteriology so this method of analysis is chiefly of historical interest. The inoculations are placed in an incubator at 40 C for 24 hours. Gas formation in the inoculated tubes is considered positive (+) and confirms the presence of *Escherichia coli*. Absence of gas formation is interpreted as a negative result (−).

In the United States, the test for presence of members of the coliform group is done by the multiple-tube fermentation technique followed by the "presumptive," "confirmed" and "completed" test procedures as described in the A.P.H.A. A.W.W.A.-W.P.C.A. manual, *Standard Methods for the Exam-*

ination of Water and Wastewater, published by the American
Public Health Association, 1790 Broadway, New York, N.Y.

Peptone-Glucose Medium

Tap water	1000 ml
Peptone	10 g
Sodium chloride	5 g
Glucose	5 g

pH 7.4–7.6

Lactose Broth

Beef Extract	3 g
Peptone	5 g
Lactose	0.5 g
Distilled water	1000 ml

pH 6.9

Examination of water for the presence of *Escherichia
coli* may be facilitated by use of the fluorescent method. This
method was developed by Meisel and Strakhova (1955). The
first stage of analysis begins with the filtration of water, which
is performed in the usual manner. The membrane filters should
be stamped with a special rubber stamp before they are prepared for filtration so that each may be divided with lines
into eight equal sectors. This facilitates the quantitative count
of microcolonies appearing on the filters.

The filters, through which the water sample was filtered,
are placed on selective media in petri dishes in the usual manner. The dishes are placed in an incubator at 40–41 C for 5.5
hours. Special media designed for accelerated growth should
be used. After the colonies develop on the membrane filters,
the latter are removed from the agar and treated with a fluorochrome. A simple fluorochrome solution in a concentration of
1:1000 is prepared. The primary solution is diluted 2 or 3 times
with a physiological solution before use. Treatment with the
fluorochrome is as follows. A small amount of fluorochrome is
poured dropwise on the bottom of a sterile petri dish; on the
surface of this liquid is placed the filter with the back side down
(filters should float, as it were, on the surface of the fluoro-

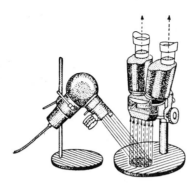

Fig. 160. Diagram of arrangement and path of rays in fluorescent examination of membrane filters (Meisel and Strakhova, 1955). *Solid lines,* blue light rays; *broken lines,* fluorescent illumination. Yellow-glass light filters are in the oculars of the magnifier.

chrome). Staining lasts for 5–10 minutes, after which the filters are removed from the fluorochrome solution and placed in a dry petri dish. Microcolonies on the filter are counted under a magnifying lens by illuminating the membrane filter from above with the bright blue light of an illuminator. This light induces fluorescence. An intense pencil of light from the light source should fall on the filter (Fig. 160). A blue filter, which retains thermal rays, or a cuvette containing a solution of diammonium sulfate, is placed in front of the light source. Yellow light filters are placed on the oculars of the magnifier. The stereoscopic magnifier MBS-1 with 25× power can be used.

Microcolonies of *Escherichia coli* give off fluorescence of a varying color depending on the fluorochrome used. With acridine orange, a greenish yellow fluorescence is given off, therefore a damping filter is used; with orange fluorescence, a dry filter is used.

Treating membrane filters with fluorochromes does not essentially injure the colonies. The majority of bacteria remain viable during this process; therefore, the second stage of the analysis can be performed. A capillary pipette (Pasteur pipette with a finely drawn out end) is used to remove the microcolonies from the filter, working under a magnifier or dissection microscope. A transfer is made from a microcolony to a slide for preparation of a smear and subsequent Gram staining.

Another portion is transferred to a micro-test tube containing a peptone-glucose medium with a cotton wad floating in the medium. The micro-test tubes are placed in an incubator at 43 C for 2–4 hours. Colonies of the typical strains of *Escherichia coli* yield a negative Gram stain and ferment the medium. Fermentation is detected by the great quantity of fine gas bubbles which float under the cotton wad and often even permeate it.

To accelerate growth of *Escherichia coli*, a nutrient medium is used which contains a reduced amount of agar. The medium is prepared from beef extract, yeast lysate, and an extract of barley sprouts. This medium was devised at the Rublevskaia water station (RWS). Parafuchsin is added to the medium to suppress auto-fluorescence of the membrane filters.

RWS Medium

Tap water	up to 1000 ml
Barley sprouts	20 g
Beef lysate	20 g
Yeast autolysate from brewer's yeast	45 ml
Agar	6–7 g
Sodium sulfite (anhydrous)	600 mg
Lactose	15 g
Vitamin B_1	40 mg
Saturated alcohol solution of parafuchsin	4 ml

If brewer's yeast is not available, the yeast autolysate may be replaced with yeast extract in the amount of 20 g.

Chapter 14

Radioactive Tracer Methods[1]

The biological use of radioactive isotopes as tracer atoms is based on the assumption that living organisms do not differentiate between isotopes of one and the same chemical element. As a result radioactive elements participate in all chemical reactions as do the stable isotopes.[2] This has been confirmed by a large volume of published data for many elements, although isolated observations have been made of non-identical behavior of radioactive and non-radioactive isotopes—the so-called isotope effect (the isotope effect is explained by the difference of atomic masses, i.e. their mass number). For example, data are available that show the unequal relative speed of assimilation of $C^{12}O_2$ and $C^{13}O_2$ (the rate is faster for the lighter isotope) and the difference in degree of accumulation of C^{12} and C^{14} when CO_2 is assimilated (3–6% more C^{12} accumulates).

Radioactive isotopes are characterized by the emission of ionizing radiation, which is recorded and measured with appropriate instruments. The nuclei of radioactive isotopes are unstable, and radioactive decay occurs within them. In most cases, radioactive isotopes are obtained artificially by bombarding the stable isotopes with powerful streams of neutrons, protons, and so forth, in special apparatus. Each element has both stable and radioactive isotopes, as shown in the following ex-

---

[1] The procedures provided in this chapter are rather simple, but direct. A good reference, among the many available, is the book *Principles of Radioisotope Methodology* by G. D. Chase and J. L. Rabinowitz (Burgess Publishing Company, Minneapolis, 1968).

[2] Stable isotopes are those whose atomic nuclei do not decay with the release of energy.

amples (Sokolov and Serdobol'skii, 1954):

Stable—H^1, H^2, C^{12}, C^{13}, N^{14}, N^{15}, O^{16}, O^{18}
Radioactive—H^3, C^{10}, C^{14}, N^{13}, N^{16}, O^{14}, O^{15}, O^{19}

The following types of radiation are observed when radioactive elements decay: α, β, γ. α-Rays are a stream of positively charged particles with a mass of 4, equivalent to helium nuclei. The penetrating power of α-rays is very small; an α-particle passes through a layer of air no more than 11 cm thick, or a layer of water no more than 0.15 mm thick. β-Rays are a stream of electrons (β^-) arising from the nucleus. Their penetrating power is significantly higher than that of α-rays. β-Rays are capable of penetrating a layer of air several meters thick and a layer of aluminum about 5 mm thick. γ-Rays are similar in nature to X-rays: γ-rays penetrate significantly more deeply than β^- particles, and their effect is not localized near the source of emission. Some elements may decay with the emission of a positron, a particle equivalent to the β^- except that it has a positive charge, β^+. That is, some radioactive elements emit α-, β^--, and γ-rays (e.g. radium and its products); others emit β-rays alone (e.g. P^{32}); a third group emits α-rays alone (lithium, berillium). During radioactive decay, some radioisotopes emit positrons (β^+), although electrons (β^-) are more often emitted. The latter is most convenient for counting tracer atoms.

The speed of radioactive decay varies for different elements. The decay of the nuclei of some atoms takes place over infinitesimal periods of time; for others, it takes very long. Theoretically, the total decay of all the nuclei of a radioactive isotope with any emission (with a long decay time) is completed after an infinite amount of time. Consequently, it is extremely difficult to take this time into account. The half-life, the time during which 50% of the current number of atoms of the isotope decays, is easier to determine.

Radioactive isotopes with a very short half-life are not suitable for experimental work. Thus, for example, one of the most important elements in the life of organisms—nitrogen—has four radioactive isotopes: N^{12}, N^{13}, N^{16}, and N^{17}, but they have a very short half-life (a few seconds) and, therefore, can not be used for experimental work. The heavy, stable isotope N^{15} is used instead of the others as a tracer. The heavy stable isotopes of hydrogen and oxygen (H^2 and O^{18}) may also be used instead of the radioactive forms. It must be noted that working with stable isotopes is significantly more complicated than with radioactive isotopes except under certain circumstances, for example, H^2.

Both radioactive and stable isotopes are used as tracers for a number of elements important in metabolism. These elements include carbon, sulfur, and several others.

When deciding which isotope to use in a given experiment, various factors must be considered: the chemical nature of the labeled compound; dilution; availability of counting equipment; duration of the experiment; and the possible effect of the radioactivity itself.

A definite advantage of radioactive isotopes is the possibility of using large dilutions: for example, 10^{-6} and even 10^{-9}. The permissible limit of dilutions for stable isotopes is significantly lower: 10^{-4}. However, stable isotopes, by virtue of being stable, offer a positive attribute.

The concentration of stable isotopes is usually determined in a special apparatus, the mass spectrometer, by measuring change in the movement of particles in electromagnetic fields. (For measuring deuterium, one can use the falling drop method.) The extent of deflections for each particle is determined by the ratio of its mass to its charge.

Accuracy of calculation is, generally speaking, significantly higher when radioactive isotopes are used. Thus, with a mass spectrometer calibrated for masses in the range of 4–60, measurements can be made for N^{15} which are correct within 2–3%, 4–5% for C^{13}, and 4–5% for O^{18} (Verkhovskaya, 1955).

It must also be remembered that measurement of the concentration of stable isotopes requires isolation of very pure biological samples.

The curie is the unit most often used for quantitative comparison of radioactivity. This unit was employed in early studies to compare the emission of radioactive elements with the emission of radium, which served as a standard for many years. At present, the curie is defined as the radioactivity of that amount of a radioactive material in which the same quantity of disintegrations occurs in 1 second as in 1 g of radium,

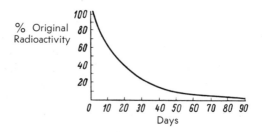

Fig. 161. Curve of radioactive decay of P³². *Abscissa,* days; *ordinate,* radioactivity as percentage of original radioactivity. The curve, although correct, can be better presented as a semilogarithmic plot.

i.e. 3.7×10^{10} disintegrations per second. A millicurie is 1000 times less radioactivity (it equals 3.7×10^{7} disintegrations per second). A microcurie is one-millionth of a curie (3.7×10^{4} disintegrations per second).

Another unit, the Rutherford (radioactivity of any radioactive substance which yields 10^{6} atomic disintegrations per second and equals 1/37 of a millicurie), was substituted for the curie. However, this unit of measurement is no longer used.

The radioactivity of the specimen does not remain unchanged; it decreases uninterruptedly from the continuing radioactive emission of the isotope. Therefore, the rate of radioactive decay of the specimen, i.e. the half-life, must be known and from this, a curve showing decrease of radioactivity of the specimen with time is plotted. This curve permits calculation of the degree of radioactive decay over a given period of time. When the experiment is to be run, the radioactivity of the reagent containing the isotope is usually measured at the beginning of the experiment, using the counts measured and the half-life as given in tables and curves in appropriate manuals. A decay curve (Fig. 161) is plotted, with the original level of radioactivity set at 100% (as provided on the label of the reagent).

DETECTION PROCEDURE

Microorganisms that have taken up radioactive isotopes are detected and the amount of uptake of radioactive com-

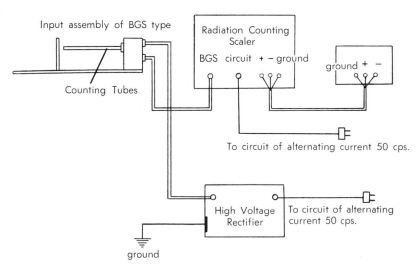

Fig. 162. Diagram of the B-2 counter for detecting radioactivity (Troshin, 1956).

pound is determined with a radiation counter. The autoradiography method is also used. A variety of radiation counters are used to measure the radioactivity of substances containing radioactive isotopes. Presently in use in Russia is the B-2, which consists of a counter, an input assembly (BGS) which amplifies pulses originating from the counter, a mechanical counter, a high-voltage rectifier (VSE) from which voltage is supplied to the counter, a counting scaler, and a timing device. A diagram of this counter is shown in Fig. 162. A variety of counting equipment is available in the United States. Major suppliers of such equipment at the present time include Packard Instruments Company, Nuclear-Chicago, and Tracerlab.

There are various Geiger-Müller counter tubes: aluminum, glass, and end window. When working with radioactive isotopes which emit β-particles of high radioactivity (P^{32}, Na^{22}, Na^{24}, K^{42}), aluminum counters are used. When working with isotopes which have low emission (C^{14}, Ca^{45}, S^{35}), end window counters are used; in these counters, the weak β-particles pass into the counter through a special small window closed with a thin (2 ± 1 mg per square centimeter) mica sheet.

The Geiger-Müller counter is a thin-walled tube, along

the axis of which is stretched a thin tungsten filament isolated from the body of the counter. The air in the tube is
replaced with a mixture of argon and alcohol vapors or other
desired counting gases with appropriate quenching gas. High
stabilized voltage of 700–1000 v is applied to the tungsten
filament and a large difference in potential is created between
the filament and wall of the counter. The wall becomes a
cathode, the filament an anode. A β-particle, passing through
the wall or window into the counter, ionizes the gas molecules.
Negatively charged ions move toward the anode; positively
charged particles move toward the cathode. The negatively
charged particles on the way to the filament increase in number at increasing rates, and, upon reaching the filament, they
cause the formation of an electric pulse. The pulses of the
current which arise in the counter do not possess sufficient
power to activate the mechanical counter. Therefore, they are
first amplified in an amplifier, and a mechanical counter located at the exit of the amplifier counts the pulses.

   Usually the radioactivity of samples from material undergoing analysis is determined by comparing the recorded
pulses of these samples per unit time with the recorded pulses
of a reference radioactive substance, with a known radioactivity
included in the experiment. The following conditions must be
observed when measurements are made.

   1. The radioactivity of experimental and reference samples must be within the limits of the ability of the counter to
register them (i.e. radioactivity should not be too low).

   2. Experimental and reference samples must be the same
shape, the same material, and of the same weight per unit area.
To achieve these specifications, samples are placed in small
dishes of uniform size (planchets, either flat or cup-shaped)
which are stamped out from small sheets of copper, brass, or
aluminum (approximately 1 mm thick, with a diameter half
the diameter of the cylindrical counter or a little smaller than
the window of the end window counter) and then dried. When
working with liquids, the sample is concentrated by careful
heating, but not boiling. The small dishes are wrapped in aluminum foil about 0.001 mm thick, or they are sealed with foil.

3. Radioactivity of bacteria is often measured directly on membrane filters, through which the liquids (such as water or media) containing bacteria are filtered; or the bacteria are concentrated by filtration to a specific volume, an aliquot of which is then measured out with a pipette, placed on a brass plate, and dried.

A sample must be filtered with a funnel which has a filtering area smaller than the area of the window of the end window counter used in measuring the radioactivity of the filters. After filtration, the filters are dried and, if radioactive carbonate is used (see above), the filters must be treated to remove mineral particles containing a radioactive isotope, which may be present on the filters. The particles are removed with hydrochloric acid after the specimen is filtered. The filters are treated with hydrochloric acid by one of three methods:

a) One milliliter of 2% solution of hydrochloric acid is passed through the filter.

b) The filters are transferred to filter paper and wetted with several drops of 2% hydrochloric acid.

c) The filter is placed with its bottom side on the surface of a 0.1 N solution of hydrochloric acid in a petri dish and allowed to float on the surface of the acid for 2–3 minutes.

The filters are then placed on discs of filter paper soaked in distilled water to remove the residual hydrochloric acid. The next step is to air-dry the filters. They are placed on filter paper in a petri dish and dried in a dessicator above calcium chloride or soda lime. Radioactivity is measured with an end window counter. Filters often curl up during drying; it is therefore recommended that special holders be used to avoid this. Preservation of the horizontal surface is essential for an accurate count.

4. To obtain comparable results for all samples, the radioactivity of the samples is measured under the same geometric conditions. The center of the specimen must be under the center of the window of the end window counter and the

distance from the surface of the sample to the wall of the counter must always be the same.

5. All experimental and reference samples must contain the radioactive isotope in the same chemical form and be of equal weight per unit area. To meet this requirement a minimum amount of sample is used and the number of counts per unit of weight or volume is calculated. For accurate results, a correction is introduced for specimen absorption of β-particles (self-absorption).

6. The so-called background count must be calculated, i.e. the number of counts per unit time recorded by the counter when the sample is not in the counter. This value must be subtracted from the counts obtained with the sample. The background count is continually measured because of its inconstancy. If experiments are to be done with weakly radioactive samples, to decrease the natural background count, the counter is placed in a lead chamber, i.e. a thick-walled (5–7 cm) lead box which protects the counter from outside (cosmic) radiations.

The length of time allotted for counting must be sufficient to obtain reliable data (after background subtraction).

The probable error of measurements is obtained by the formula

$$E_n = 0.67 \ \sqrt{N},$$

where N = number of counts.

The probable error is calculated in per cent by the formula

$$E_\% = \frac{67 \ \sqrt{N}}{N} \ .$$

This formula applies when dealing with high activity and low background. One should consult the statistics for determining the error of the counts when the background plays a significant role in a total count. Appropriate statistics are available in texts on radioisotope tracer methodology.

Probable Error of Measurements (Hevesy, 1950)

Error %	Number of Counts
1	4445
2	1114
5	177
10	45
20	11

Another method of measuring radioactivity is by autoradiography. This method is based on the effect of the emissions of radioactive substances on a photosensitive emulsion, this effect being analogous to that of light rays. With autoradiographs, it is possible to observe the distribution of radioactive elements in an organism and in a cell.

To perform autoradiography, the radioactive sample is placed on a photosensitive plate and is tightly pressed to it; the sample and the plate are then wrapped well in black paper. The material is left in a dark, dry place, usually for several days, after which the plate is developed. The time of exposure is determined empirically. It depends on the photosensitivity of the plate, the amount of radioactive isotope present, and the nature of the isotope emission. Highly sensitive X-ray film is usually used to prepare autoradiographs.

Sample material for autoradiography must be carefully prepared. The sample must be dry and located in one plane. The more radioactive the bacterial cell, the sharper the track which the cell leaves on the emulsion of the highly sensitive film.

Autoradiography permits the measurement of radioactivity in separate cells and locations within sections of the cell where the tracer accumulates.

The inclusion of radioactive elements into cells of microorganisms takes place in general as a result of metabolic uptake in nutrition and growth (adsorption and diffusion processes may have some significance). After entering the cell, tracer compounds undergo the same changes as non-tracer compounds. The metabolism of tracer compounds in cells is determined by the above method. When resolving complex questions, e.g. determining how uptake and transport of isotope occurs in the cell, the isotope method is combined with chemical and physical methods of separation and determination of cell structures (chromatography, ion-exchange resins, biologically active substances).

The very good accuracy of the tracer method permits resolution of various questions about the biology and ecology of microorganisms and the part they play in the transformation of matter in a water mass. Meisel (1955) mentions the following questions in microbiology which can be resolved by the isotope method:

1. explanation of the methods and rates of exchange and cycling of individual elements and their compounds in microbial cells

2. revelation of the way different elements penetrate and are included and metabolized within the cell

3. discovery of the pathways and mechanisms of biosynthesis of various substances by bacteria (e.g. proteins, amino acids, carbohydrates, vitamins, antibiotics, lipids)

4. establishment of the methods of catabolism and anabolism of various substances as a result of bacterial enzymatic action

5. explanation of the mechanisms of action on microorganisms of biotic and antibiotic substances, as well as of various physical agents

6. study of the character of the interrelationships among separate microbial species, and between microorganisms and plants and animals

In aquatic microbiology, the tracer method is used to determine the coefficients of concentration (by microorganisms and other components of the water masses) of various isotopes that enter, or may enter, bodies of water. This method is also used to elucidate the means by which radioactive isotopes penetrate the bodies of aquatic animals, to explain the nutritional significance of bacteria and dissolved organic matter, and to resolve the question of feeding selectivity of aquatic animals. The tracer method is also helpful in studies of isotope distribution in a water mass, the magnitude of assimilation of carbon dioxide by bacteria, and the role of microorganisms in elimination of radioactive elements from water. Thus a variety of wide-ranging, yet highly significant, problems become amenable to attack with the availability of radioisotope methodology. The technique, of course, can be much more sophisticated in application than presented in this chapter.

METHODS FOR DETECTING CARBON DIOXIDE FIXATION BY BACTERIA

A special method is employed to measure the magnitude of CO_2 fixation by bacteria, i.e. the Steemann-Nielsen method (1952) for determining the amount of photosynthesis in a water mass. The results of determinations were evaluated by several authors who had performed such experiments (Kuznetsov, 1955; Sorokin, 1955, 1957, 1958, 1960a; Kuznetsov and Romanenko, 1963). Work done recently on verifying the suitability of this method for determining chemosynthesis in water masses indicates that, by this method, the total (or, more precisely, to a greater extent the heterotrophic) assimilation of carbon dioxide by bacteria can be determined (Zharova, 1963; Romanenko, 1963). In fact, the very arrangement of the experiments for the Steemann-Nielsen method excludes the possibility of all autotrophic aerobes developing in the vessels. Therefore, the method may only be used to determine the total assimilation of carbon dioxide by bacteria. This applies to analyses both in water and in sediment deposits of water masses.

The assimilation of carbon dioxide by bacteria is usually measured in water samples taken from various depths. The samples are collected in a glass container and the water is filtered through a membrane filter of the preliminary, or prefilter, pore size (see Chapter 3) to remove phyto- and zooplankton. The filtered water is then poured into bottles (two bottles for each depth level) of about 100-ml capacity. The bottles used in the experiment must be chemically clean. The carbonate present in the sample must be chemically determined before the uptake experiments are done. The bottles are placed in small black bags. Only at this point should there be added to each bottle, by pipette, 1 ml of radioactive sodium carbonate or bicarbonate, i.e. carbon is the trace element ($Na_2C^{14}O_3$ or $NaHC^{14}O_3$), with a specific radioactivity of about 5 μc or 10^6 counts per minute per milliliter when measured in a counter.

The isotope solution used is prepared from a commercially available compound. A bottle with a ground glass stopper is carefully cleaned with an acid-dichromate solution. Into the bottle is poured a specific amount of distilled water, freed of CO_2 by boiling, and a 1% solution of KOH to a concentration of 2 ml of alkali per liter of water. The alkali is required to prevent volatilization of the $C^{14}O_2$. The contents of the ampoule containing the radioactive isotope are emptied into the bottle; the stopper is then inserted and the bottle is shaken. The prepared solution is stored in the dark. Before it is introduced into the test bottles, the solution must be filtered through membrane filter No. 1 or No. 2.

After the isotope is added, the test bottles are closed with ground glass stoppers so that there are no air bubbles under the stoppers. The small bags containing the bottles are tightly closed and are lowered into the natural body of water under study (a bottle at each appropriate depth) and left for 24 hours. One to two control bottles, into which 1 ml of 40% formalin is added together with the radioisotope, are set for each series of test bottles.

When the bottles are removed, their contents are fixed with formalin and the bacteria in the water are filtered onto membrane filter No. 1 or No. 2. Usually a significant amount of water (40–60 ml) is filtered. The filters are dried, treated as above, and the radioactivity is determined.

The magnitude of the daily assimilation of CO_2 by bacteria is computed according to the formula of Steemann-Nielsen for determining degree of photosynthesis:

$$C_a = \frac{rC_{carb.}}{Rt},$$

where C_a is the rate at which carbon dioxide is assimilated (in milligrams of C per liter); $C_{carb.}$ is the amount of carbon as free carbon dioxide and carbonates in the water; R is the radioactivity of the water in the bottle after the isotope is added (counts per minute per liter); r is the radioactivity of the bacteria on the filter (in counts per minute per liter of filtered water); and t is time of exposure (in days). A correction must be allowed for the reduced rate of assimilation of C^{14} compared with C^{12} (5–6% from the assimilation value). The proportion r/R indicates which portion of the C^{14} added was taken up into the bacterial cells by the end of the time period in which the bottles were exposed.

The initial radioactivity of the water (R) is determined under the same conditions as those under which the radioactivity of the test filters is determined, i.e. the same filtration apparatus, the same counter, and so on are used. For this analysis, barium chloride is added to combine with the C^{14}. The precipitated barium carbonate is collected on a filter by filtration. The layer of precipitate on the filter must be thin. If it is thick, self-absorption of the particles takes place and the number of counts measured decreases. However, the op-

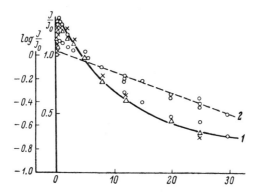

Fig. 163. Self-absorption curve for $BaC^{14}O_3$, obtained with an end window counter. *1*, Curve expressing relationship of measured radioactivity to thickness of the $BaC^{14}O_3$ layer; *2*, part of the same data in the form of a semilogarithmic graph, permitting determination of R. *Abscissa,* thickness of the $BaC^{14}O_3$, mg/cm² layer; *ordinate,* ratio of measured radioactivity at a given thickness of the specimen (J) to the radioactivity of zero thickness (J_0). (From *"Metodicheskoe posobie po opredeleniiu pervichnoi produktsii organicheskogo veshchestva v vodoemakh radiouglerodnym metodom,"* 1960 [Method Manual on determining primary production of organic matter in water masses by the radiocarbon method].)

posite process of self-scattering may take place, leading to an increase in the number of recorded counts. In these instances, corrections are introduced. Either the self-absorption curves already published (as for example, Fig. 163) or correction curves obtained for given experimental conditions are used.

Uptake Kinetics in Ecosystems.   Because organic compounds are present in natural waters in such low concentrations, the usual bioassay methods are limited to vitamins and other micronutrients. The organic compounds available as energy source are almost always below 10 mg/liter, the concentration necessary to produce measurable turbidity. Hobbie and Wright (*Limnol. Oceanogr., 10:* 471–474, 1965) have developed a bioassay for glucose that uses the kinetics of substrate uptake, rather than growth, as the response reaction. Radioactively labeled glucose uptake is used to measure glucose concentration in natural waters, but, more importantly, the curve relating uptake velocity to substrate concentration permits detection of two different types of mechanisms of uptake: specific transport systems effective at very low substrate concentrations, traced to the bacteria; and a diffusion mechanism, effective only at higher substrate concentrations, to the algae. A standard technique, employing glucose-C^{14} or other organic compounds, has been developed by Hobbie and Wright (*Mitt. Internat. Verein. Limnol., 14:* 64–71, 1968) which permits measurement of nutrient turnover rates and approximation of substrate concentrations in an ecosystem. Thus velocity of uptake of a substrate in nature can be used as a measure of heterotrophic activity. This application of organic radioisotope methodology is finding wide acceptance by microbial ecologists, particularly those interested in dynamics of an ecosystem.

Precautionary Measures.   Work with radioactive isotopes demands strict adherence to the rules of safety for protection from radiation in a dosage higher than the maximum allowed dosage. Dangerous levels of several radioactive isotopes are given in Fig. 164. Naturally, the smallest possible amounts of a radioactive isotope should be used as a tracer. The work must be conducted in a separate, specially equipped laboratory with good exhaust ventilation. Special clothing may also be required. It is always essential to use special forceps and pi-

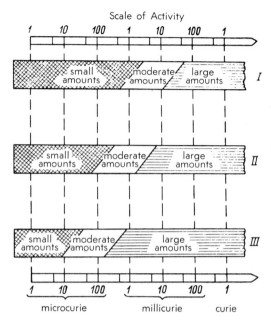

Fig. 164. Toxic levels of some radioactive isotopes.
I, Low-hazard: Na^{24}, K^{42}, Cu^{64}, Mn^{52}, As^{76}, As^{77}, Kr^{85}, Hg^{197}. *II,* Moderately dangerous: H^3, P^{32}, Na^{22}, S^{35}, Cl^{36}, Mn^{54}, Fe^{54}, Co^{60}, Sr^{89}, Nb^{95}, Ru^{103}, Ru^{106}, Te^{127}, Te^{129}, I^{131}, Cs^{137}, Pr^{143}, Nd^{147}, Au^{198}, Au^{199}, Hg^{203}, Hg^{205}, Ba^{140}, La^{140}, Ce^{141}. *III,* Extremely dangerous: C^{14}, Ca^{45}, Fe^{55}, Sr^{90}, Y^{91}, Zr^{95}, Ce^{144}, Bi^{210}.

pettes with syringes or bulbs fitted to them. When working with highly radioactive samples, a shield of thick Plexiglas must be placed between the samples and the person conducting the experiment. A periodic check of the isotope concentration in the air must be made routinely.

Radioactive wastes must be collected in special containers and the wastes and containers must be disposed of according to rules prescribed for public health safety (in the United States, Atomic Energy Commission publications).

Glassware must first be washed with a special cleaning mixture—either a solution of sodium carbonate or a solution of the following composition (Polianskii mixture): water, 10 liters; nitric acid (concentrated), 675 ml; saturated solution of oxalic acid, 1 liter; NaCl, 200 g. The glassware is then washed in the usual manner. When working with C^{14}, the glassware must be washed with the sodium carbonate solution.

Chapter 15

Determining pH
and Oxidation-Reduction Potential

DETERMINING pH OF MEDIA

Growth of microorganisms takes place within specific ranges of pH in a given medium. Therefore, it is essential to adjust the pH in nutrient media and to measure pH when taking water samples.

The pH of an aqueous solution of any compound is defined as the ratio between the concentration of hydrogen ions (H^+), which determine acidic properties, and the concentration of hydroxyl (alkaline) ions (OH^-). The pH of an aqueous solution is considered neutral when the concentration of hydrogen ions equals the concentration of hydroxyl ions.

To establish the pH of an aqueous solution of any compound, the concentration of hydrogen ions is measured, because the concentration of hydroxyl ions is related to the concentration of the former: when the concentration of one type of ions increases, a strictly corresponding decrease of the other ions occurs. The pH of the medium is expressed not as an absolute value of the concentration of hydrogen ions, but in the form of a hydrogen indicator, designated by the symbol pH.

pH is the negative logarithm of the concentration of hydrogen ions (C):

$$pH = -\log C_{H^+} = \log \frac{1}{C_{H^+}};$$

for example, in distilled water, the concentration of hydrogen ions is 10^{-7}:

$$pH = -\log 10^{-7} = \log \frac{1}{10^{-7}} = \log 1 - \log 10^{-7} = 0 - (-7) = 7.$$

The values of pH higher than 7 occur in alkaline solutions, below 7 in acidic solutions. A unit change in pH corresponds to a tenfold change in the hydrogen ion concentration since the hydrogen index is a power of 10 given with the opposite sign. Thus at pH 6 the concentration of hydrogen ions is ten times greater than at pH 7; at pH 8 it is ten times less than in a neutral medium.

pH is determined in two ways: colorimetrically and electrometrically.

Colorimetric Determination of pH

Colorimetric determination of pH is based on the property of certain organic substances (dyes) to change color, depending on the pH of the medium. These dyes are called indicators.

Indicators are either weak acids or weak bases, whose non-dissociated molecules have one color, and whose ions have another color. If, for example, an indicator is a weak acid, then it will be found as non-dissociated molecules in an acidic medium and as ions in an alkaline medium.

In practice, two sets of indicators are used: Clark and Lubs dichromatic indicators and Michaelis monochromatic indicators. The former are most suitable for this purpose (Table 13).

Aqueous alkaline indicator solutions are the most stable. Indicator (0.1 g) is weighed out on an analytical balance and made up as an aqueous or alcoholic solution, as required for solubility. One method for preparing an aqueous solution is as follows. The indicator dye is ground in a porcelain mortar with a 1/20 N solution of sodium hydroxide. The amount of NaOH varies according to the indicator to be dissolved and is given in Table 13. To obtain a 0.04% indicator solution, distilled water, to 250 ml, is added to the alkaline solution. For a 0.02% solution, distilled water to 500 ml is added. The solutions

Table 13. Dichromatic indicators
(Clark and Lubs)

Indicator	Color Change	pH Interval	1/20 N Solution NaOH (ml per 0.1 g of indicator)
Thymol blue	red-yellow	1.2–2.8	4.3
Bromophenol blue	yellow-blue	3.0–4.6	3.0
Methyl red	red-yellow	4.4–6.0	7.4
Bromocresol purple	yellow-purple	5.2–6.8	3.7
Bromothymol blue	yellow-blue	6.0–7.6	3.2
Phenol red	yellow-red	6.8–8.4	5.7
Cresol red	yellow-red	7.2–8.8	5.3
Thymol blue	yellow-blue	8.0–9.6	4.3

are kept in bottles with ground glass stoppers and are stored in a cool place.

Colorimetric determination of pH is based on the comparison of the color of the medium containing the indicator with a standard reference scale of transitional colors of the indicator in a series of standard solutions corresponding in pH to the range within which the solution to be determined falls.

Buffer mixtures are made up and used as standard solutions. A characteristic feature of buffer mixtures is their ability to resist pH shift. The hydrogen concentration in these mixtures does not change when diluted with water (some shift occurs when a dilution of 1000 times or greater occurs). When standard solutions are prepared to measure pH, their pH is first determined potentiometrically with a pH meter.

Preparing pH indicator solutions for field use, when a pH meter cannot be used, is critical and it is best that the final solutions be verified by potentiometric measurement. Scales for measuring pH in the USSR are prepared according to the specifications of the State Hydrologic Institute (Leningrad, V. O., 2 liniia, dom 23).

Analysis is done by comparing the color of the test solution with the color of the indicator in buffer solutions as follows: After rinsing a test tube (matched glass tubes) three times with the test solution, the solution is poured to the mark, i.e. the same volume is added as is added to the reference test tubes (usually 10 ml). The indicator is immediately poured into the test tube. The test tube is stoppered; the contents are mixed; and the color compared with the color of the standard solution of the same pH as read from the indicator in the solution being measured. If the color in the test tube extends beyond the effective limits of the indicator, the procedure is repeated until the colors coincide with a reference pH indicator tube (the pH of the water sample is read as equal to the pH of the standard test tube of the same color). If the water contains colored material, pH can be determined colorimetrically, since it is possible to compensate for discoloration.

When working with water or sediment samples, a "salt" or salinity correction is included in the final result of the analysis. When measurements are made at a temperature higher than 18 C, or the temperature at which the pH meter is usually set, a temperature correction must also be made.

The pH of media (especially turbid and highly colored agar media) can be adjusted by using indicator papers such as litmus paper. These permit a rapid, though not very accurate, measurement. Reagent (indicator) papers produced in Russia, *Phan* papers produced in Czechoslovakia, or pH indicator papers available commercially in the United States may be used.

Electrometric Determination of pH

Potentiometers (e.g. LP-5 or LP-58, Fig. 165) are used for electrometric determination of pH. In the U.S.A., a wide variety of pH meters are available commercially from such firms as Beckman Instruments and Photovolt Corporation. Battery-operated pH meters are available for field work. The description and instructions for use of a pH meter are usually provided with the instrument itself. The simplest and most accurate method is to measure pH with a glass electrode.

A glass electrode (Fig. 166) is a thin sheath (glass 0.03–0.05 mm thick) made from special electrode glass (Kriukov, 1955; Alekin, 1959). The use of a glass electrode is based on the formation of a potential difference at the boundary of the phases between the glass and the solution.

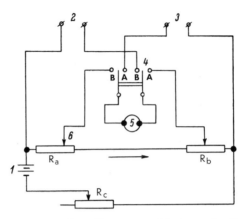

Fig. 165. Diagram of a pH meter. *1*, Elements feeding chains of resistors: R_a, measuring resistor; R_b, adjusting resistor; R_c, regulating resistor. *2*, Measured voltage. *3*, Normal element. *4*, Switch in various positions: *A*, when working current is set; *B*, when measuring. *5*, Galvanometer. *6*, Moveable contact.

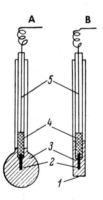

Fig. 166. Glass electrodes (Alekin, 1959). *A*, Electrode with bulb; *B*, electrode with sealed end. *1*, Extremely thin glass layer, soldered to the end of the tube; *2*, silver chloride electrode; *3*, 0.1 N solution of HCl; *4*, mercury column for contact; *5*, copper wire.

The measuring procedure when a calibrated glass electrode is used is simple. The water sample is poured into a small beaker, and into it is placed a glass electrode which has been rinsed twice with this same water and another electrode containing a saturated solution of KCl. The electromotive force is measured with a potentiometer.[1] The concentrations of a large number of cations (Ca^{++}, Cu^{++}, Li^+, Na^+, Ag^+, K^+, and NH_4^+) and anions (Cl^-, ClO_4^-, NO_3^-, F^-, Br^-, $S^=$, and I^-), as well as dissolved gases (CO_2 and O_2), can be determined with specific electrodes (Beckman Instruments; Orion) and a conventional pH meter.

DETERMINING OXIDATION-REDUCTION POTENTIAL

The growth of microorganisms in the ocean is closely related to oxidation-reduction conditions within the water or sediment. More specifically, the nature of the microbes found in a given body of water depends on the Eh of the environment. These conditions, which depend on the rate of oxidation-reduction reactions, cannot be characterized only by oxygen content. According to contemporary views, oxidation-reduction reactions involve loss and gain of electrons. Matter which loses a negative charge is oxidized and that which gains a charge is reduced. Thus there exist both oxidizing and reducing processes which proceed without the participation of either oxygen or hydrogen, although in nature the oxidizing agent is usually oxygen, and the reducing agent is usually hydrogen. The presence of an oxidizing agent and a reducing agent in a medium creates the oxidation-reduction (O/R) potential. There may be several or even many such O/R systems in natural environments. For example, oxidized and reduced iron ions, Fe^{+++} and Fe^{++}, may occur in a water sample, along with oxidized and reduced manganese ions, Mn^{++++} and Mn^{++}, and so forth. Also important to consider are $S^=$ and S°. In an O/R system, the transfer of electrons will continue until an equal

---

[1] There are many makes and variations of pH meters available in the United States, both for fieldwork and for the laboratory. The measurement of pH is, generally speaking, a rather uncomplicated, but important, procedure in aquatic microbiology. Many of the modern instruments are equipped with automatic temperature and salinity compensators.

concentration of oxidized and reduced forms is established, or, in other words, until the oxidation-reduction potential (O/R) is balanced.

The oxidation-reduction potential characterizes the oxidation-reduction state of the medium: the higher the concentrations of the oxidizing agent, the greater the potential. The O/R indicates in which direction a given oxidation reaction of a given solution will go if mixed with another solution. Thus, when a solution with a higher potential is mixed with a solution having a lower potential, the former will oxidize the latter. This process will continue until a state of equilibrium is needed. When determining the O/R in a complex mixture, it is sufficient to determine the potential of any simple system. Because there must be H^+ and OH^- ions in any aqueous solution (whereas other ions may not be present), it is usually convenient to determine the O/R by using the $H^+ \rightleftharpoons H_2$ system.

The relationship between the magnitude of the potential and the concentration of the oxidizing agent and reducing agent is summarized in the Nernst equation:

$$E = E_0 + \frac{RT}{nF} \ln \frac{\text{oxidizing agent}}{\text{reducing agent}} \text{ or } E = E_0 + \frac{RT}{nF} \ln \left(\frac{Ox}{Red} \right),$$

where E_0 is the potential when equal concentrations of oxidant and reducer are present; R is the gas constant; T is the absolute temperature; F is the charge produced by 1 gram-equivalent of matter (equals 96,500 coulombs); n is the number of electrons transferable in accordance with the formula of the reaction; Ox is the concentration of the oxidized form of the reacting substance; and Red is the concentration of its reduced form.

The oxidation-reduction potential with respect to hydrogen is indicated by Eh. When measuring Eh the difference of potentials between two electrodes is determined: one is the reference electrode and the other is the "unknown." The Eh is further calculated with respect to the standard hydrogen electrode, the potential of which is taken as zero. This concept derives from the fact that, when an electrode is submerged in a 1 N solution of hydrogen ions while the pressure of the gaseous hydrogen equals 1 atm, the concentration of both the oxidized and the reduced form of hydrogen is 1.

The Nernst equation is then expressed as:

$$Eh = 0.029 \log \left(\frac{(H^+)}{(H_2)} \right),$$

because at 20 C and when the natural logarithms are converted to base 10, $2.303RT/2F = 0.029$.

This value changes with respect to temperature:

15 C	0.0285	21 C	0.0291
16 C	0.0286	22 C	0.0292
17 C	0.0287	23 C	0.0293
18 C	0.0288	24 C	0.0294
19 C	0.0289	25 C	0.0295
20 C	0.0290		

Because hydrogen participates in many O/R reactions, the Eh value in the system depends not only on the ratio of oxidized to reduced component but also on pH. Therefore, when measuring Eh it is always necessary to determine the pH and to indicate at which pH the Eh was measured.

To obtain comparable data on the oxidation-reduction conditions in solutions with different pH values, another index, rH_2, was suggested, but is no longer used routinely.

The discussion above has been concerned with oxidation-reduction processes that are very simple and uncomplicated. These are distinguished by the fact that (1) they are reversible, and (2) they proceed rapidly.

The biologist usually encounters irreversible systems when a substance which influences the Eh of the solution is oxidized or reduced and cannot be returned to its original state. Such substances include sugars and organic acids. In these cases, the above formula

$$Eh = E_o + \frac{RT}{nF} \ln \frac{Ox}{Red}$$

does not fully correspond to the actual state of the O/R systems. However, it has been demonstrated by a number of authors that the O/R of an irreversible system may be characterized by a potential analogous to the Eh of a reversible system. This potential is the Eh of the weakest reducer which can reduce the given irreversible system. Just as the Eh of the reversible system, the Eh of an irreversible system changes with respect to pH.

There are two methods of determining the oxidation-reduction potential: colorimetrically and electrometrically.

Colorimetric Determination of O/R

Clark in his earlier work in bacteriology first suggested a set of indicators which change color when the oxidation-reduction conditions change, and which cover the greater part of the Eh scale. These indicators are, for the most part, colorless in the reduced state and colored in the oxidized state. However, each indicator encompasses a very restricted range of Eh. Thus, use is made of a large number of such indicators, each of which has a limited range of application. Because of the peculiarities of indicators and their impracticability in the presence of air (the solutions are oxidized and the O/R changes in air), the indicators are used only to determine Eh under certain specific circumstances. However, these indicators can be useful in determining Eh of anaerobic cultures. In this case, agar media, poured into tall, narrow test tubes, should be used, since oxygen will only penetrate to a depth of a few millimeters below the surface of the agar. It is sufficient to add 0.2% agar to promote growth of anaerobes, which do not grow within several millimeters of the agar surface. Sterile indicator solutions are added to the nutrient medium directly before inoculation. To establish the extent to which the test culture reduces the Eh, several indicators have to be used in a series of inoculations. A list of such indicators is given by Rabotnova (1957).

Electrometric Determination of O/R

This method is basically simple and is not technically different from measurement of pH. For these measurements, it is best to use the same pH meter (LP-5, LP-8, or a commercially built American model)used for determining pH.

When determining O/R, the difference in potentials between two electrodes is measured. The platinum electrode is placed in the test solution and the difference of potentials (or, as it is also called, the electromotive force) between it and the reference electrode—which is also placed in the liquid— is measured. Usually calomel electrodes are used as standard or reference electrodes.

After the difference in potential between the platinum and calomel electrodes is determined, the difference in potential between the platinum electrode of the water being tested and the standard or reference hydrogen electrode is measured. The difference is the O/R or Eh of the solution.

The electromotive force (emf) or the difference of potentials between the platinum and calomel electrodes is measured with a potentiometer by the usual compensation method.

The measure of the electromotive force is read at the moment of compensation; the reading is taken in millivolts. Information on the construction and use of the instrument is usually provided by the manufacturer. After measurement is made, Eh is calculated according to the following formula:

$$Eh = Ex + Ec,$$

where Ex is the magnitude of the measured potential, i.e. the potential of the electrode being tested in comparison with the calomel electrode. However, because Eh is the potential of the electrode compared with the potential of a standard hydrogen electrode at the specific temperature at which the measurement was made, then it is necessary to introduce a correction for the difference between the calomel and hydrogen electrodes (Ec). Because the calomel electrode differs from the hydrogen electrode at 20 C by +250 mv, it is necessary to add 250 mv to the reading obtained during measurement (regardless of the sign of the potential).

In aquatic microbiology, measurements of the O/R are made in water and sediments and in laboratory cultures of microorganisms.

Determining Eh in Water

Determining the Eh directly in a water sample is extremely complicated because of a number of difficulties such as having to make electrometric analyses in situ and the time required to calibrate the potential between the electrode and the test medium. Thus, it is understandable that data on O/R of water and sediment is sparse.

When determining O/R of a water sample, the water sample should not be exposed to the air. It is best to use an instrument that allows direct determination of the Eh in the water mass and at various depths, as has been done by a number of investigators (Nekhotenova, 1938, and others). From results of tests conducted directly in bodies of water, Nekhotenova demonstrated that the potential is established after 15–30 minutes. It takes a little longer to determine the potential when the first measurement is made, regardless of whether the electrode is placed on the surface or under it. This is understandable since time is needed to set up the system. When the second measurement is made, the potential is established more rapidly. Therefore, to measure the potential at different depths, the instrument does not have to be taken out of the water after each measurement, but is merely lowered or raised to the depth at which the potential is to be measured (Nekhotenova, 1938).

To determine the Eh of water in the laboratory, samples must be collected with an instrument such as a bathometer, which prevents the sample from coming into contact with air. Also, water samples taken are transferred as quickly as possible to the laboratory, where the water to be analyzed is taken

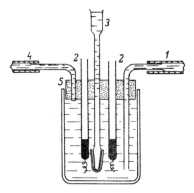

Fig. 167. Container for determining Eh in water (Kuznetsov and Romanenko, 1963). *1,* Delivery tube; *2,* platinum electrodes; *3,* small siphon of the calomel electrode; *4,* outlet tube; *5,* stopper with fixed electrodes.

directly from the bathometer and placed in a special container (Fig. 167) without introducing air bubbles.

Determining Eh in Sediments

The O/R in benthic deposits of water masses indicates the nature of their biochemical and physicochemical processes. Sediment samples should be collected with a sounding corer, stratometer, or similar coring device. For a layer-by-layer determination of Eh, the sediment core samples are transferred to a special Plexiglas container of the same diameter. This permits a careful examination of the structure of the sediment, and subsequently analyses are made with material taken from the central portion of the core sample. Electrodes are inserted through special apertures in the lid of the container into the core sample in such a way that each of them lies completely in the desired layer or occupies several microzones. Usually several platinum electrodes, in addition to a calomel electrode, are placed in the core sample at specific distances from one another.

To determine Eh in sediments, the standard calomel electrode is replaced by a rod-shaped calomel electrode, as suggested by Kriukov (1955), since the pores of the standard ground glass joint of the calomel electrode may clog with soil particles. In the electrode suggested by Kriukov, contact with the sample is attained through a thin fiber of asbestos at the end of the electrode. This keeps the electrode free from contamination and facilitates conductivity.

All electrodes must be placed in sediment samples no less than 15 minutes before measurements are made. After the specified time has lapsed, the electrodes are connected to the potentiometer and readings are taken. This is then repeated with the second, third, and subsequent platinum electrodes, each of which is located at a specific depth in the core sample. Measurements are repeated after 1.5 hours and 2 hours. The potential attains a constant value usually after 2 hours from the moment the electrode was inserted into the core sample. The potential is considered to have reached a stable value if it does not fluctuate more than 2 mv in 5 minutes. The temperature of the sediment must also be measured to cal-

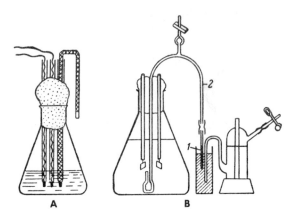

Fig. 168. Instruments for culturing bacteria and measuring Eh in a nutrient medium. *A*, Rabotnova (1957); *B*, Nekrasov (1933): *1*, locking tube with gypsum; *2*, level of saturated KCl solution.

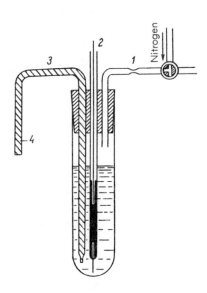

Fig. 169. Vessel for measuring Eh in a culture liquid in an atmosphere of nitrogen (Raynaud and Viscontini, 1945). *1*, Point where tube is sealed after vessel fills with nitrogen; *2*, platinum electrode; *3*, bridge; *4*, agar with saturated solution of KCl.

culate the Eh. The pH is measured with a glass electrode placed directly in the sediment.

The magnitude of the potentials of two electrodes placed in a sediment at a small distance from each other may be somewhat different, even in sediments which are homogeneous in composition. Small differences in moisture and aeration, the presence or absence of decomposing organic substances, and varying levels of microbial growth will affect these measurements. These and other conditions may create localized changes in the O/R. In homogeneous sediments, localized changes in the O/R usually do not exceed 20–30 mv.

Determining O/R in Cultures of Microorganisms

The electrometric method allows continuous scanning of the changes in the oxidation-reduction conditions during growth of a culture. This method makes it possible to obtain a curve of the change of Eh with time without interfering with the biochemical processes. Various instruments have been devised for this purpose. Rabotnova (1957) indicates that taking samples from a culture inevitably leads to agitation of the medium and the microorganisms, which, in turn, will produce a shift in Eh, Rabotnova recommends measuring Eh in cultures where growth has already ceased because of substrate depletion. Alternatively, the cultures may be killed beforehand by adding thymol, which itself has no effect on the Eh or pH, 18–20 hours before analysis. However, immersion of the electrode in a culture is the most accurate method of determining the Eh during growth. At least two, preferably three electrodes, should be used, so that an average of the readings may be taken.

Special pieces of apparatus (Fig. 168) have been constructed for measuring Eh directly in cultures of microorganisms. Some authors recommend removing any air in solution, to avoid influence of oxygen from the air, when Eh analyses are made. According to Rabotnova, the best method of "deaeration" is to blow nitrogen that has been completely purified of all traces of oxygen through the test medium. Raynaud and Viscontini (1945) constructed a special instrument for such analyses (Fig. 169). Measurements made in a vacuum also yield good results (Rabotnova, 1957).

Bibliography

Alekin, O. A. 1959. Metody issledovaniia fizicheskikh svoistv i khimicheskogo sostava vody [Methods of investigating the physical properties and chemical composition of water]. *Zhizn' presnykh vod*, t. IV, ch. 2. Izd. AN SSSR, M.-L.: 213–300 [*Zhizn' presnykh vod*, vol. IV, part 2. Publishing House of the USSR Academy of Sciences, Moscow and Leningrad. Pp. 213–300].

Allen, O. N. 1957. *Experiments in Soil Bacteriology*. Burgess Publishing Co., Minneapolis. Pp. 1–117.

Anderson, J. I. W., and W. P. Heffernan. 1965. Isolation and characterization of filtrable marine bacteria. *J. Bact.* 90: 1713–1718.

Augier, J. 1956. Recherches sur la mesure de l'activite hemicellulolytique des sols. *Ann. Inst. Pasteur* 90(2): 161–170.

Baranov, I. V. 1956. Kombinirovannyi batometr VNIORKh dlia otbora prob vody i grunta [Combined bathometer of the All-Union Scientific Research Institute of Lake and River Fisheries for sampling water and sediment]. *Nauchno-tekh. buill. VNIORKh*, No. 1–2: 67–69 [*Scientific Technical Bulletin of the All-Union Scientific Research Institute of Lake and River Fisheries*, no. 1–2, pp. 67–69].

Bartholomew, J. W. 1962. Variables influencing results, and the precise definition of steps in Gram staining as a means of standardizing the results obtained. *Stain Techn.* 37(3): 139–155.

Belozerskii, L. N., and N. I. Proskuriakov. 1951. *Prakticheskoe rukovodstvo po biokhimii rastenii* [*Practical manual on the Biochemstry of Plants*]. Izd. "Sovetskaia nauka" M.: 1–388, ["Soviet Science" Publishing House, Moscow. Pp. 1–388].

Bick, H. 1963. A review of Central European methods for the biological estimation of water pollution levels. *Bull. Org. Mond. Sante* 29: 401–413.

435

Brough, F. K. 1950. A rapid microtechnique for the determination of nitrate reduction by microorganisms. *J. Bact.* 60(3): 365–366.

Bucksteeg, W., and H. Thiele. 1959. Die Beurteilung von Abwasser und Sohlamm mittes TTC (2-3-5-Triphenyltetrazolium-chlorid). *Gas u Wasserfach* 100(36): 916–920.

Bukatsch, F. 1958. Versuche zur Sichtbarmachung von Gei elaggregaten der Bakterien im Fluorescenzmicroskop. *Arch. Mikrobiol.* 31(1): 11–15.

Callao, V., and E. Hernandez. 1961. Differenciacion de colorantes en sus colonias. *Microbiol. Esp.* 14(4): 247–252.

Callao, V., and E. Montoya. 1960a. The use of dyes to distinguish species of *Azotobacter*. *J. Gen. Microbiol.* 22(3): 657–661.

Callao, V., and E. Montoya. 1960b. Effectos producidos por el telurito potasico sobre el desarrollo de los *Azotobacter* en medios de cultivo solidos. *Microbiol. Esp.* 13(4): 351–360.

Carter, C. T., and A. L. Smith. 1957. *Principles of Microbiology*, 3rd ed. C. V. Mosby Co., St. Louis. Pp. 1–665.

Cataldi, M. S. 1940. Aisamiento de *Beggiatoa alba* en cultivo puro. *Rev. Inst. Bacteriol. Dept. Nacl. Hig.* (Buenos Aires) 9 (4): 393–423.

Chase, G. D., and J. L. Rabinowitz. 1968. *Principles of Radioisotope Methodology*, 3rd. ed. Burgess Publishing Co., Minneapolis.

Cholodny, N. G. 1953. (also spelled Kholodnyi) *Zhelezobakterii* [*Iron Bacteria*]. Izd. AN SSSR, M.: 1–222 [Publishing House of USSR Academy of Sciences, Moscow. Pp. 1–222].

Clifton, C. E. 1950. *Introduction to the Bacteria.* McGraw-Hill Book Co., New York. Pp. 1–528.

Cohen, J. S., and R. H. Burris. 1955. A method for the culture of hydrogen bacteria. *J. Bact.* 69(3): 316–319.

Colwell, R. R., and R. D'Amico. *A Practical Manual for the Numerical Taxonomy of Bacteria.* W. H. Freeman & Co., San Francisco.

Colwell, R. R., and W. J. Wiebe. 1970. Core characteristics for use in classifying aerobic, heterotrophic bacteria by numerical taxonomy. *Bull. Georgia Acad. Sci.* 28: 165–185.

Corpe, W. A. 1951. A study of the widespread distribution of *Chromobacterium* species in soil by a simple technique. *J. Bact.* 62(4): 515–517.

Depoux, R. 1953. Sur l'utilisation des antibiotiques pour l''isolement en culture pure des *Azotobacter* à partir de la terre. *Ann. Inst. Pasteur* 84(3): 645–647.

Derx, H. G. 1951. *Azotobacter insigne* sp. n. fixateur d'azote a flagellation polar. *Proc. Kon. Nederl. Akad. Wet.* 54(4): 342–350.

Drachev, S. M., A. S. Razumov, S. V. Bruevich, B. A. Skopintsev, and M. T. Golubeva. 1953. *Metody khimicheskogo i bakteriologicheskogo analiza vody* [*Methods of Chemical and Bacteriological Analysis of Water*]. Medgiz, M.: 1–279 [State Publishing House of Medical Literature (Medgiz), Moscow. Pp. 1–279].

Dugdale, R., V. Dugdale, J. Neess, and J. Goering. 1959. Nitrogen fixation in lakes. *Science* 130(3379): 859.

Farghaly, A. H. 1950. Factors influencing the growth and light production of luminous bacteria. *J. Cellular Comp. Physiol.* 36: 165–183.

Faust, L., and R. S. Wolfe. 1961. Enrichment and cultivation of *Beggiatoa alba. J. Bact.* 81(1): 99–106.

Fedorov, M. V. 1951. *Rukovodstvo k prakticheskim zaniatiiam po mikrobiologii* [*Manual for Practical Studies in Microbiology*]. Izd. s.-kh. lit., M.: 231 [Publishing House of Agricultural Literature, Moscow. P. 231].

Guseva, K. A. 1956. Metody ekologo-fiziologicheskogo issledovaniia vodoroslei [Methods of Ecological-Physiological Investigation of Algae]. *Zhizn' presnykh vod*, t. IV, ch. 1. Izd. AN SSSR, M.-L.: 122–159 [*Zhizn' presnykh vod*, vol. IV, part 1. Publishing House of USSR Academy of Sciences, Moscow and Leningrad. Pp. 122–159].

Hevesy, G. 1950. *Radioaktivnye izluchateli* [*Radioactive Indicators*]. Izd. inostr. lit., M.: 1–539 [State Publishing House of Foreign Literature, Moscow. Pp. 1–539].

Identification Methods for Microbiologists. 1966, 1968, 1969. Society of Applied Bacteriology. Academic Press, New York.

Imshenetskii, A. A. 1949. Optimal'nye pitatel'nye sredy dlia desul'furiruiushchikh bakterii [Optimal nutrient media for sulfate-reducing bacteria]. *Mikrobiologiia* 18(4): 324–331.

Imshenetskii, A. A. 1953. *Mikrobiologiia tselliulozy* [*Microbiology of Cellulose*]. Izd. AN SSSR, M.: 1–439 [Publishing House of the USSR Academy of Sciences, Moscow. Pp. 1–439].

Imshenetskii, A. A., and E. L. Ruban. 1953. Poluchenie chistykh kul'tur Nitrosomonas [Obtaining pure cultures of *Nitrosomonas*]. *Mikrobiologiia* 22(4): 376–384.

Ingellman, B., and H. Laurell. 1947. The preparation of silicic acid gels for the cultivation of microorganisms. *J. Bact.* 53(3): 364–365.

van Iterson, W. 1958. *Gallionella ferrunginea* Ehrenberg in a different light. *Verh. Kon. Nederl. Akad. Wet.* AFD. Natuurkunde, Tweede Reeks 52(2): 1–185.

Ivanov, M. V. 1955. Metod opredeleniia produktsii bakterial'noi biomassy v vodoeme [Method of determining production of bacterial biomass in a water mass]. *Mikrobiologiia* 24(1): 79–89.

Jensen, H. L. 1955. *Azotobacter macrocytogenes* n.sp., a nitrogen-fixing bacterium resistant to acid reaction. *Acta Agricult. Scand.* 5(2): 280–294.

Jensen, V. 1961. Rhamnose for detection and isolation of *Azotobacter vinelandii* Lipman. *Nature* 190(4778): 832–833.

Jensen, V. and E. J. Petersen. 1955. *Taxonomic Studies on* Azotobacter chroococcum *Beijerinck and* A. beijerinckii. *Lipman.* Roy. Vet. and Agr. Coll., Copenhagen. Yearbook 1955: 107–126.

Kalina, G. P. 1962. Ul'trafil'tratsiia v mikrobiologii. Metody i apparatura. *Obshchaia mikrobiologiia,* t. II. Mnogotomnoe rukovodstvo po mikrobiologii, klinike, i epidemiologii infektsionnykh boleznei [Ultrafiltration in Microbiology. Methods and Apparatus. *General Microbiology,* vol. II of multivolume manual on the microbiology, clinical aspects, and epidemiology of infectious diseases]. Medgiz, M.: 400–416 [State Publishing House of Medical Literature (Medgiz), Moscow. Pp. 400–416].

Karyakin, Yu.V. 1936. *Chistye khimicheskie reaktivy* [*Pure Chemical Reagents*]. Khimteoretizdat, L.: 1–617 [Khimteoretizdat, Leningrad. Pp. 1–617].

Karzinkin, G. S. 1934. K izucheniiu bakterial'nogo perifitona [On the Study of Bacterial Periphytes]. *Trudy Limnologich. st. v Kosine,* v. 17: 21–45 [*Transactions of the Limnological Station in Kosino,* no. 17, pp. 21–45].

Kholodnyi. See Cholodny

Knaysi, G. 1951. *Elements of Bacterial Cytology,* 2nd ed. Comstock Publishing Associates, Ithaca, New York. Pp. 1–375.

Koleshko, O. I. 1960. Stroenie iunykh kolonii *azotobaktera* [Structure of young colonies of *Azotobacter*]. *Mikrobiologiia* 29 (2): 293–295.

Komarova, L. I. 1949. Pipetka dlia vydeleniia kul'tur iz odnoi kletki [Pipette for isolating cultures from one cell]. *Mikrobiologiia* 18(4): 370.

Kondratieva, E. N. 1963. *Fotosinteziruiushchie bakterii* [*Photosynthetic Bacteria*]. Izd. AN SSSR, M.: 1–314 [Publishing House of the USSR Academy of Sciences, Moscow. Pp. 1–314].

Korde, N. V., and N. I. Pyavchenko. 1950. Pribory dlia vziatiia prob ozernykh otlozhenii [Instruments for Sampling Lake Sediments]. *Trudy Sapropel. Labor.* AN SSSR, v. IV: 115–119 [*Transactions of the Sapropelic Laboratory* of the USSR Academy of Sciences, no. IV, pp. 115–119].

Kotelev, V. V. 1958. K metodike vydeleniia iz pochvy mikroorganizmov, razlagaiushchikh organofosfaty [On technique of isolating from soil those bacteria which decompose organophosphates]. *Dokl. Vses. akad. s.-kh. nauk im. Lenina,* v.

9: 17–18 [*Reports of the Lenin All-Union Academy of Agricultural Sciences*, no. 9, pp. 17–18].

Krasil'nikov, N. A. 1949. *Opredelitel' bakterii i aktinomitsetov* [*Guide to the Identification of Bacteria and Actinomycetes*]. Izd. AN SSSR, M.-L.: 1–829 [Publishing House of USSR Academy of Sciences, Moscow and Leningrad. Pp. 1–829].

Krasil'nikov, N. A., and V. I. Kotelev. 1957. Kachestvennoe opredelenie fosfataznoi aktivnosti nekotorykh grupp pochvennykh mikroorganizmov [Qualitative determination of the phosphatase activity of certain groups of soil microorganisms]. DAN SSSR 117(5): 894–895 [*Reports (Doklady)* of the USSR Academy of Sciences 117(5): 894–895].

Kriss, A. E., and E. A. Rukina. 1952. Biomassa mikroorganizmov i skorost' ikh razmnozheniia v okeanicheskikh glubinakh [Biomass of microorganisms and rate of their reproduction in ocean depths]. *Zh. Obshch. Biol.*, t. 13, No. 6–7: 346–362 [*J. Gen. Biol.* 13(6–7): 346–362].

Kriukov, P. A. 1955. Izmerenie velichiny pH so stekliannym elektrodom. Sb. *Sovremennye metody khimicheskogo analiza prirodnoi vody* [Measuring pH with a Glass Electrode. Coll.: *Contemporary methods of chemical analysis of natural water*]. Idz. AN SSSR, M.: 7–13 [Publishing House of the USSR Academy of Sciences, Moscow. Pp. 7–13].

Kudryavtsev, V. I. 1954. *Sistematika drozhzhei* [*Taxonomy of Yeast*]. Izd. AN SSSR, M.: 1–426 [Publishing House of USSR Academy of Sciences, Moscow. Pp. 1–426].

Kushnir, Yu.M. 1958. *Sovetskie elektronnye mikroskopy. Pribory i tekhnika eksperimenta* [*Soviet Electron Microscopes. Instruments and Experiment Technique*]. Izd. "Znanie," M.: 1–31 ["Knowledge" Publishing House, Moscow. Pp. 1–31].

Kuznetsov, S. I. 1955. Ispol'zovanie radioaktivnoi uglekisloty C^{14} dlia opredeleniia sravnitel'noi velichiny fotosinteza i khemosinteza v riade ozer razlichnykh tipov. Sb. *Izotopy v mikrobiologii* [The Use of Radioactive Carbon Dioxide C^{14} for Determining the Relative Magnitude of Photosynthesis and Chemosynthesis in a Series of Diverse Lakes. Coll.: *Isotopes in microbiology*]. Izd. AN SSSR, M.: 126–135 [Publishing House of the USSR Academy of Sciences, Moscow. Pp. 126–135].

Kuznetsov, S. I. and V. I. Romanenko. 1963. *Mikrobiologicheskoe izuchenie vnutrennikh vodoemov. Laboratornoe rukovodstvo* [*Microbiological Study of Inland Waters. Laboratory Manual*]. Izd. AN SSSR, M.-L.: 1–129 [Publishing House of USSR Academy of Sciences, Moscow and Leningrad. Pp. 1–129].

Lamanna, P. D., and M. F. Mallette. 1953. *Basic Bacteriology.* Williams & Wilkins Co., Baltimore. Pp. 1–677.

Lambin, S., and A. German. 1961. Precis de microbiologie, t. 1. Masson, Paris. Pp. 1–458.

Leifson, E. 1951. Staining, shape and arrangement of bacterial flagella. *J. Bact.* 62(4): 377–389.

Levshin, V. L. 1956. *Liuminestsentsiia i ee tekhnicheskoe primenenie [Fluorescence and Its Technical Application].* Izd. AN SSSR, M.: 1–47 [Publishing House of USSR Academy of Sciences, Moscow. Pp. 1–47].

Lewis, N. M., O. D. McNail and R. C. Summerfelt. 1963. A device for taking water samples in sterile bottles at various depths. *Ecology* 44(1): 171–173.

Lockhart, W. R. and J. Liston, Eds. *Methods for Numerical Taxonomy.* 1970. American Society for Microbiology, Bethesda, Maryland.

Lodder, J., and N. J. W. Kreger-Van-Rij. 1952. The Yeasts. A Taxonomic Study. North-Holland Publishing Co., Amsterdam. Pp. 1–173.

Lovelace, T. E., and R. R. Colwell. 1968. A multipoint inoculator for petri dishes. *Appl. Microbiol.* 16: 944–945.

Malek, I. 1956. Protochnyi metod razmnozheniia mikrobov [Continuous culture method]. *Mikrobiologiia* 25(6): 659–667.

Manual of Microbiological Methods. 1957. Society of American Bacteriologists. McGraw-Hill, New York and London. Pp. 1–315.

Matthews, A. D. 1961. A simple method for the cultivation of anaerobes. *Canad. J. Med. Tech.* 23(4): 145–151.

Meisel, M. N. 1955. Primenenie izotopov v mikrobiologii. Sb. *Izotopy v mikrobiologii* [Application of Isotopes in Microbiology. Coll.: *Isotopes in Microbiology*]. Izd. AN SSSR, M.: 5–18 [Publishing House of USSR Academy of Sciences, Moscow. Pp. 5–18].

Meisel, M. N., and V. A. Strakhova. 1955. *Uskorennyi liuminestsentnyi metod obnaruzheniia bakterii (obnaruzhenie i kolichestvennyi uchet* kishechnoi palochki) *[Accelerated Fluorescent Method of Detecting Bacteria (Detection and Quantitative Count of* Escherichia coli)]. Izd. AN SSSR, M.: 1–48 [Publishing House of USSR Academy of Sciences, Moscow. Pp. 1–48].

Metodicheskoe posobie po opredeleniiu pervichnoi produktsii organicheskogo veshchestva v vodoemakh radiouglerodnym metodom [Method Manual on Determining the Primary Production of Organic Substance in Water Masses by the Radiocarbon Method] 1960. Izd. Belorussk. gos. univ., Minsk: 1–26 [Publishing House of Byelorussian State University, Minsk. Pp. 1–26].

Meyers, G. E. 1958. Staining iron bacteria. *Stain Techn.* 33(6): 283–285.

Mikhailov, I. F., and S. I. Dyakov. 1961. *Liuminestsentnaia mikroskopiia [Fluorescent Microscopy]*. Medgiz, M.: 1–221 [State Publishing House of Medical Literature (Medgiz), Moscow. Pp. 1–221].

Mishustin, E. N. 1948. O roli sporonosnykh bakterii v pochvennykh protsessakh [The role of spore-forming bacteria in soil processes]. *Mikrobiologiia* 17(3): 201–207.

Mogilievskii, G. A. 1953. *Mikrobiologicheskii metod poiskov gazov i neftianykh zalezhei [Microbiological Method of Prospecting for Gases and Petroleum Deposits]*. Gostoptekhizdat, M.-L.: 1–56 [State Publishing House for Literature on Fuels (Gostoptekhizdat), Moscow and Leningrad. Pp. 1–56].

Naumova, A. N. 1961. Mineralizatsiia fosfororganicheskikh soedinenii rizosfernymi i pochvennymi bakteriiami [Mineralization of organophosphorus compounds by rhizosphere and soil bacteria]. *Transactions Inst. Mikrobiol.* 11: 222–232.

Nechaeva, N. B. 1953. Obrazovanie metana mikroorganizmami [Methane formation by microorganisms]. *Mikrobiologiia* 22 (4): 456–471.

Neess, J. C., R. C. Dugdale, V. Dugdale, and J. J. Goering. 1962. Nitrogen metabolism in lakes. I. Measurement of nitrogen fixation with N^{15}. *Limnology and Oceanography* 7 (2): 163–169.

Nekhotenova, N. I. 1938. Elektrometricheskoe opredelenie okislitel'no-vosstanovitel'nogo potentsiala v vodoemakh [Electrometric determination of oxidation-reduction potential in water masses]. *Mikrobiologiia* 7(2): 186–197.

Nekrasov, N. I. 1933. Okislitel'no-vosstanovitel'nyi potentsial v biologii [Oxidation-reduction potential in biology]. *Usp. Biol. Khimii* v. 10: 72–118 [*Prog. of Biol. Chem.* 10: 72–118].

Niderl', D., and V. Niderl'. 1949. *Mikrometody kolichestvennogo organicheskogo analiza [Micromethods of Quantitative Organic Analysis]*. Goskhimizdat, M.-L.: 1–276 [State Publishing House for Literature on Chemistry (Goskhimizdat), Moscow and Leningrad. Pp. 1–276].

van Niel C. B. 1944. The culture, general physiology, morphology and classification of non-sulfur-purple and brown bacteria. *Bact. Rev.* 8: 1–118.

Niskin, S. 1962. Deep Sea Res. 9: 501–503.

Odoevskaia, N. S. 1961. Novye sredy dlia *Bac. megaterium var. phosphaticum* v pochve [*New media for Bac. megaterium var. phosphaticum in Soil*]. *Biull. Nauchno-Tekhn. Informatsii po S.-kh. Mikrobiologii* v. 10: 30–32 [*Bulletin of Sci-*

entific-Technical Information on Agricultural Microbiology, no. 10, pp. 30–32].

Omelianski, V. L. 1940. *Prakticheskoe rukovodstvo po mikrobiologii* [*Practical Manual on Microbiology*]. Izd. AN SSSR, M.-L.: 1–427 [Publishing House of USSR Academy of Sciences, Moscow and Leningrad. Pp. 1–427].

Ortel, S. 1957. Bienengift als nähbröden zur Trennung grampositives und gramnegatives Bakterien. *Zbl. Bakt.* [Orig.] 168: 313–317.

Pekhov, A. P. 1962a. Elektronnaia mikroskopiia mikroorganizmov. *Obshchaia mikrobiologiia*, t. II. Mnogotomnoe rukovodstvo po mikrobiologii, klinike i epidemiol. infekts. boleznei [Electron-Microscopy of Microorganisms. *General Microbiology*, vol. II of multivolume manual on the microbiology, clinical aspects, and epidemiology of infectious diseases]. Medgiz, M.: 73–108 [State Publishing House of Medical Literature (Medgiz), Moscow. Pp. 73–108].

Pekhov, A. P. 1962b. *Elektronnomikroskopicheskoe issledovanie bakterii i fagov* [*Electron Microscope Investigation of Bacteria and Phages*]. Medgiz, M.: 1–224 [State Publishing House of Medical Literature (Medgiz), Moscow. Pp. 1–224].

Perfilyev, B. V., and D. R. Gabe. 1961. *Kapilliarnye metody izucheniia mikroorganizmov* [*Capillary Methods of Studying Microorganisms*]. Izd. AN SSSR, M.-L.: 1–534 [Publ. House of USSR Academy of Sciences, Moscow and Leningrad. Pp. 1–534].

Peshkov, M. A. 1962. Svetovoi mikroskop i ego raznovidnosti. *Obshchaia mikrobiologiia*, t. II. Mnogotomnoe rukovodstvo po mikrobiolog., klinike i epidemiol. infekts. boleznei [The Light Microscope and its Varieties. *General Microbiology*, vol. II of multivolume manual on microbiology, clinical aspects, and epidemiology of infectious diseases]. Medgiz, M.: 11–72 [State Publishing House of Medical Literature (Medgiz), Moscow. Pp. 11–72].

Pochon, J., and P. Tardieux. 1962. Techniques d'analyse en microbiologie du sol. *Ed. de la Tourelle*, S. Mande:1–111.

Postgate, J. R. 1959. Differential media for sulphur bacteria. *J. Sci. Food Agric.* 10(12):669–674.

Rabotnova, I. L. 1957. *Rol' fiziko-khimicheskikh uslovii* (pH *i* rH$_2$) *v zhiznedeiatel'nosti mikroorganizmov* [*The role of Physico-Chemical Conditions* (pH *and* rH$_2$) *in the Vital Activity of Microorganisms*]. Izd. AN SSSR, M.: 1–274 [Publishing House of USSR Academy of Sciences, Moscow. Pp. 1–274].

Raynaud, M. 1949. Les bacteries anaerobies pectinolytiques. *Ann. Inst. Pasteur* 77(4):434–470.

Raynaud, M., and M. Viscontini. 1945. Le potentiel d'oxydoreduction au cours de la régénéretion des milieux employes pour la culture des anaérobies. *Ann. Inst. Pasteur* 71(1–2): 172–187.

Razumov, A. S. 1932. Priamoi metod ucheta bakterii v vode [Direct count method for bacteria in water]. *Mikrobiologiia* 1(2): 131–146.

Razumov, A. S. 1953. Metody sanitarno-bakteriologicheskikh issledovanii. v kn.: S. M. Drachev, A. S. Razumov, S. V. Bruevich, B. A. Skopintsev, i M. T. Golubeva: *Metody khimicheskogo i bakteriologicheskogo analiza vody* [Methods of sanitary-bacteriological investigations. *In* S. M. Drachev, A. S. Razumov, S. V. Bruevich, B. A. Skopintsev, and M. T. Golubeva, *Methods of Chemical and Bacteriological Analysis of Water*]. Medgiz, M.: 226–271 [State Publishing House of Medical Literature (Medgiz), Moscow. Pp. 226–271].

Repetigny, J., and S. Sonea. 1958. Detection microscopique de la fluorescence primaire chez les bacteries. *Canad. J. Microbiol.* 4(1): 17–23.

Repetigny, J., S. Sonea, and A. Frappier. 1961. Etudes microfluorometrique des microorganismes. La fluorescence primaire. *Ann. Inst. Pasteur* 101(3): 353–366.

Rittenberg, D., A. S. Keston, F. Roseburg, and R. Schoenheimer. 1939. The determination of nitrogen isotopes in organic compounds. *J. Biol. Chem.* 127: 291–299.

Rittenberg, B. T., and S. C. Rittenberg. 1962. The growth of *Spirillum volutans* Ehrenberg in mixed and pure cultures. *Arch. Mikrobiol.* 42(2): 138–153.

Rodina, A G. 1956. Metody mikrobiologicheskogo issledovaniia vodoemov [Methods of microbiological investigation of water masses]. *Zhizn' presnykh vod*, t. IV, ch. 1. Izd. AN SSSR, M.-L.: 7–121 [*Zhizn' presnykh vod*, vol. IV, part 1. Publishing House of USSR Academy of Sciences, Moscow and Leningrad. Pp. 7–121].

Rogosa, M., M. I. Krichevsky, and R. R. Colwell. 1971. Method for coding data on microbial strains for computers. (Edition AB). *Int. J. System. Bacteriol.* 21 (suppl): 1A–184A.

Romanenko, V. I. 1963. Potential'naia sposobnost' mikroflory vody k geterotrofnoi assimiliatsii uglekisloty i k khemosintezy [Potential capacity of aquatic microflora for heterotrophic assimilation of carbon dioxide and for chemosynthesis]. *Mikrobiologiia* 32(4): 668–674.

Ruban, E. L. 1961. *Fiziologiia i biokhimiia nitrifitsiruiushchikh mikroorganizmov* [*Physiology and Biochemistry of Nitrifying Microorganisms*]. Izd. AN SSSR, M.: 1–173 [Publishing House of USSR Academy of Sciences, Moscow. Pp. 1–173].

Saissac, R., L. M. Brugiere and M. Raynaud. 1952. Activite pectinolytique de quelques bacteries anaérobies. *Ann. Inst. Pasteur* 82(3): 356–361.

Salimovskaja-Rodina, A. G. 1936. Opyt primeneniia metoda plastinok obrastaniia k izucheniiu bakterial'noi flory vody [Application of submerged slides method to the study of aquatic bacterial flora]. *Mikrobiologiia* 5(4): 487–493.

Schlegel, H. G., and N. Pfennig. 1961. Die Anreicherungskultur einiger Schwefelpurpurbakterien. *Arch. Mikrobiol.* 38(1): 1–39.

Seliber, G. L. (ed). 1962. *Bol'shoi praktikum po mikrobiologii,* pod red. Seliber [*Manual on Microbiology,* ed. Seliber]. Izd. "Vysshaia shkola" M.: 1–490 ["Higher School" Publishing House, Moscow. Pp. 1–490].

Sieburth J. McN. 1963. A simple form of the Zobell bacteriological sampler for shallow water. *Limnology and Oceanography* 8(4): 489–492.

Sierra, G. 1957. A simple method for the detection of lipolytic activity of microorganisms and some observations on the influence of the contact between cells and fatty substrates. *Antonie Leeuwenhoek* 23: 15–23.

Silverman, M. P., and D. G. Lundgren. 1959. Studies on the chemoautotrophic iron bacterium *Ferrobacillus ferrooxidans.* An improved medium and a harvesting procedure for securing high cell yields. *J. Bact.* 77(5): 642–647.

Skerman, V. B. D. 1969. *Abstracts of Microbiological Methods.* Wiley-Interscience, New York.

Skerman, V. B. D. 1967. *A Guide to the Identification of the Genera of Bacteria,* 2nd ed. Williams & Wilkins Co., Baltimore.

Skerman, V. B. D. 1968. A new type of micromanipulator and microforge. *J. Gen. Microbiol.* 54: 287–297.

Slavnina, G. P. 1947. Razrabotka liuminestsentnogo metoda obnaruzheniia bakterii, okisliaiushchikh uglevodorody [Concerning the fluorescent method of detecting bacteria which oxidize hydrocarbons]. *Izv. Gl. Upr. Geol. Fondov.* v. 3: 80–81 [*Bulletin of Main Directorate of Geological Funds,* no. 3, pp. 80–81].

Sokolov, A. V., and N. P. Serdobol'skii. 1954. Metodika primeneniia "mechenykh atomov" pri agrokhimicheskikh issledovaniiakh. Sb. *Agrokhimicheskie metody issledovaniia pochy* [Application of Tracers in Agricultural-Chemical Investigations. Coll.: *Agricultural-chemical methods of investigating soils*]. Izd. AN SSSR, M.: 318–340 [Publishing House of Sciences, Moscow. Pp. 318–340].

Sorokin, Yu.I. 1955. O bakterial'nom khemosinteze v ilovykh otlozheniiakh [Bacterial chemosynthesis in sediments]. *Mikrobiologiia* 24(4): 393–399.

Sorokin, Yu.I. 1957. Rol' khemosinteza v produktsii organicheskogo veshchestva v vodoemakh. Podlednyi khemosintez v vodnoi tolshche Rybinskogo vodokhranilishcha [The role of chemosynthesis in the production of organic substances in water reservoirs. Chemosynthesis within the water layer of the Rybinsk water reservoir in winter]. *Mikrobiologiia* 26(6): 736–744.

Sorokin, Yu.I. 1958. Rol' khemosinteza v produktsii organicheskogo veshchestva v vodoemakh. Izuchenie khemosinteza v ilovykh otlozheniiakh s pomoshch'iu C^{14} [The role of chemosynthesis in the production of organic substance in reservoirs. Stuyding chemosynthesis in silt depositions with the aid of C^{14}]. *Mikrobiologiia* 27(2): 206–213.

Sorokin, Yu.I. 1960a. Opredelenie izotopnogo effekta pri usvoenii mechenoi uglekisloty v protsesse fotosinteza i khemosinteza [Determination of the isotopic effect during assimilation of labeled carbon dioxide in the course of photosynthesis and chemosynthesis]. *Mikrobiologiia* 29(2): 204–208.

Sorokin, Yu.I. 1960b. Batometr dlia otbora prob vody na bakteriologicheskii analiz [Bathometer for sampling water for bacteriological analysis]. *Biull. Inst. Biol. Vodokhr.*, No. 5: 48–50 [*Bulletin of Institute of Biology of Water Reservoirs*, no. 5, pp. 48–50].

Sorokin, Yu.I. 1963. Ob istinnoi prirode novogo klassa mikroorganizmov *Krasil'nikovia Kriss* [On the validity of the new class of microorganisms Krasil'nikovia Kriss]. *Mikrobiologiia* 32(3): 425–433.

Spivak, G. B. 1954. Nekotorye obshchie svoistva elektronnomikroskopicheskikh ustroistv. v kn.: *Elektronnaia mikroskopiia* [Some general characteristics of electron microscope arrangements. In *Electron Microscopy*]. Gos. izd. tekhniko-teoret. lit., M.: 11–24 [State Publishing House of Technical-Theoretical Literature, Moscow. Pp. 11–24].

Stadtman, T. C., and H. A. Barker. 1951. Studies on the methane fermentation. *J. Bact.* 61(1): 67–71.

Standard Methods for the Examination of Water and Waste Water, ed. 11. 1960. American Public Health Association.

Stanier, R. Y., Doudoroff, M., and Adelberg, E. 1970. *The Microbial World*, 3rd edition. Prentice-Hall, Englewood Cliffs, N.J.

Stapp, C. 1940. Azotomonas insolita ein neuer aerober Stickstoffbindender Mikroorganismus. *Centr. Bakt.* II Abt., Bd. 102, H. 1/3: 1–19.

Steemann-Nielsen, E. 1952. The use of radioactive carbon (C^{14}) for measuring organic production in the sea. *J. Coseil* 18: 117–140.

Stolbunov, A. K., and F. P. Ryabov. 1965. Novaia model' bakteri-

ologicheskogo batometra [New model of bacteriological bath-
ometer]. *Voprosy gidrobiol.*, M.: 400 [*Problems in Aquatic Biology,* Moscow. P. 400].

Sushkin, N. G. 1949. *Elektronn mikroskopyi [Electron Microscope].* Gos. izd. tekhniko-teoret. lit., M.: 1–276 [State Publishing House of Technical-Theoretical Literature, Moscow. Pp. 1–276].

Tarusov, B. N. 1954. *Osnovy biologicheskogo deistviia radio-aktivnykh izluchenii [Fundamentals of the Biologic Effectiveness of Radiation].* Medgiz, M.: 1–140 [State Publishing House of Medical Literature (Medgiz), Moscow. Pp. 1–140].

Tauson, V. O. 1925. K. voprosu ob usvoenii parafina mikro-organizmami [On the subject of the assimilation of paraffin by microorganisms]. *Zh. Russk. Bot. Obshch.* 9: 161–176 [*J. Russ. Bot. Soc.* 9: 161–176].

Tauson, V. O. 1928. O bakterial'nom okislenii neftei [Concerning the bacterial oxidation of petroleums]. *Neft. Khoz.* XIV(2): 220–230 [*Petroleum Economy* 14(2): 220–230].

Thirst, M. 1957. Gelatin liquefication. *J. Gen. Microbiol.* 7(2): 396–400.

Tidwell, L., C. D. Heather, and C. Merkle. 1955. An autoclave-sterilized medium for the detection of urease activity. *J. Bact.* 69(6): 701–702.

Timakov, V. D., and D. M. Goldfarb. 1958. *Osnovy eksperimen-tal'noi meditsinskoi bakteriologii [Fundamentals of Experimental Medical Bacteriology].* Medgiz, M.: 1–347. [State Publishing House of Medical Literature (Medgiz), Moscow. Pp. 1–347].

Troshin, A. S. 1956. Metod radioaktivnykh indikatorov i ego pri-meneie v gidrobiologii [Radioactive tracer method and its application in aquatic biology]. *Zhizn' presnykh vod,* t. IV, ch. 1. Izd. AN SSSR, M.-L.: 414–437 [*Zhizn' presnykh vod,* vol. IV, part 1. Publishing House of USSR Academy of Sciences, Moscow and Leningrad. Pp. 414–437].

Verkhovskaya, I. N. 1955. O nekotorykh osobennostiakh metoda izotopnykh indikatorov. Sb. *Izotopy v mikrobiologii* [Concerning Certain Special Features of the Tracer Technique. Coll.: *Isotopes in microbiology*]. Izd. AN SSSR, M.: 227–233 [Publishing House of USSR Academy of Sciences, Moscow. Pp. 227–233].

Volarovich, M. P., and V. P. Tropin. 1963. Izuchenie mikroflory torfov metodom elektronnoi mikroskopii [An electron-micro-scope study of peat microflora]. *Mikrobiologiia* 32(2): 281–287.

Voskresenskii, K. A. 1947. Novye priemy izucheniia mikrotsenozov [New methods of studying microcoenoses]. *Biull. Mosk.*

Obshch. Ispytat. Prirody 52: 35–43 [*Bull. Moscow Soc. Natur.* 52: 35–43].

Wiken, T. 1940. Untersuchunger über Methangarung unddie dabei wirksamen Bakterien. *Arch. Microbiol.* 11(3): 312–315.

Williams, O. B. 1955. *Laboratory Manual for General Bacteriology,* 3rd ed. Hemphill's Book Stores, Austin, Texas. Pp. 1–114.

Wilska, A. 1954. Observations with the anoptral microscope. *Mikroskopie* 9: 1–80.

Winogradsky, S. N. 1952. *Mikrobiologiia pochvy. Problemy i metody* [*Soil microbiology. Problems and Methods*]. Izd. AN SSSR, M.: 1–792 [Publishing House of USSR Academy of Sciences, Moscow. Pp. 1–792.]

Zavarzin, G. A. 1958. O vozbuditele vtoroi fazy nitrifikatsii. Morfologiia vozbudtelia vtoroi fazy nitrifikatsii [The inducer of the second phase of nitrification. Morphology of the inducer of the second phase of nitrification]. *Mikrobiologiia* 27(6): 679–686.

Zavarzin, G. A. 1961. Pochkuiushchiesia bakterii [Budding bacteria]. *Mikrobiologiia* 30(5): 952–975.

Zhadin, V. I. 1956. Metodika izucheniia donnoi fauny vodoemov i ekologii donnykh bespozvonochnykh [Methods of studying benthic fauna of water masses and ecology of benthic invertebrates]. *Zhizn' presnykh vod* t. IV, ch. 1. Izd. AN SSSR, M.-L.: 279–382. [*Zhizn' presnykh vod,* vol. IV, part 1. Publishing House of USSR Academy of Sciences, Moscow and Leningrad. Pp. 279–382].

Zharova, T. V. 1963. Assimiliatsiia uglekisloty geterotrofnymi bakteriiami i ee znachenie pri opredelenii khemosinteza v vodoemakh [Assimilation of carbon dioxide by heterotrophic bacteria and its significance in determining chemosynthesis in water masses]. *Mikrobiologiia* 32(5): 843–849.

Zhdanov, D. A. 1955. *Sovremennye metody i tekhnika morfologicheskikh issledovanii. Pod. red.* Zhdanova [*Contemporary Methods and Technique of Morphological Investigations.* Ed. by Zhdanov]. Medgiz, L.: 1–259 [State Publishing House of Med. Lit. (Medgiz), Leningrad. Pp. 1–269].

ZoBell, C. E. 1946. Marine microbiology of Massachusetts. *Chronica botanica,* New York: 1–240.

Author Index

449

Subject Index